FULL BODY BURDEN

Kristen Iversen received a Ph.D. in English from the University of Denver. She is Director of the MFA Program in Creative Writing at the University of Memphis and also Editor-in-Chief of *The Pinch*, an award-winning literary journal. Iversen has two sons and currently lives in Memphis.

www.kristeniversen.com

PRAISE FOR *FULL BODY BURDEN*:

'As [Iversen] and her primary school classmates were being taught to duck and cover pathetically when the Soviets finally struck, they were all along coming under silent attack from their own government. Against a bracingly realised backdrop of the rural American mountain west, Iversen recounts a superficially untroubled childhood of horse-riding, jumping in lakes and kissing boys. Yet it is turned chilling by our steadily mounting knowledge, and her happy ignorance at the time, of what was going on just a few miles upwind without the consent of those being polluted' John Swain,

Literary Review

'Intimate . . . Powerful . . . A potent examination of the dangers of secrecy . . . A serious and alarming book with its share of charming moments'

New York Times

'A striking tale of innocence in a time and a place of great danger'

Atlantic

KRISTEN IVERSEN

Full Body Burden

Growing Up in the Shadow of a Secret Nuclear Facility

VINTAGE BOOKS
London

Published by Vintage 2013

2 4 6 8 10 9 7 5 3 1

First published in Great Britain in 2012 by Harvill Secker

'Plutonian Ode' from Collected Poems 1947-1997 by Allen Ginsberg. Copyright © 2006, the Allen Ginsberg Trust, used by permission of The Wylie Agency (UK) Limited.

Vintage
Random House, 20 Vauxhall Bridge Road,
London SW1V 2SA

www.vintage-books.co.uk

Addresses for companies within The Random House Group Limited can be found at: www.randomhouse.co.uk/offices.htm

The Random House Group Limited Reg. No. 954009

A CIP catalogue record for this book
is available from the British Library

ISBN 9780099571858

The Random House Group Limited supports the Forest Stewardship Council® (FSC®), the leading international forest-certification organisation. Our books carrying the FSC label are printed on FSC®-certified paper. FSC is the only forest-certification scheme supported by the leading environmental organisations, including Greenpeace. Our paper procurement policy can be found at:
www.randomhouse.co.uk/environment

Printed and bound in Great Britain by Clays Ltd, St Ives PLC

For my family: my siblings, Karin, Karma, and Kurt;
my father; and in loving memory of my mother.
Most of all, this book is for Sean and Nathan,
who have lived with it from the beginning.

I suppose my thinking began to be affected soon after atomic science was firmly established. Some of the thoughts that came were so unattractive to me that I rejected them completely, for the old ideas die hard, especially when they are emotionally as well as intellectually dear to one. It was pleasant to believe, for example, that much of Nature was forever beyond the tampering reach of man—he might level the forests and dam the streams, but the clouds and the rain and the wind were God's.

—Rachel Carson

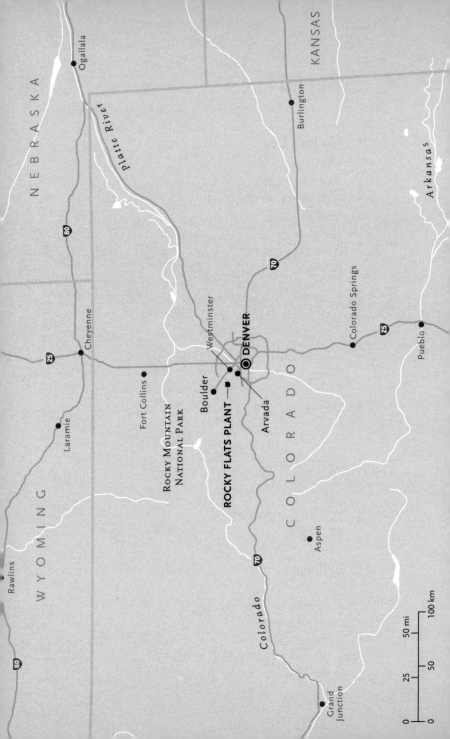

CONTENTS

BODY BURDEN: THE AMOUNT
OF RADIOACTIVE MATERIAL
PRESENT IN A HUMAN BODY,
WHICH ACTS AS AN INTERNAL
AND ONGOING SOURCE OF
RADIATION

Plutonium button at Rocky Flats

FULL BODY BURDEN

MOTHER'S DAY

1963

It's 1963 and I'm five. I lie across the backseat of the family car, sleeping with my cheek pressed against the vinyl. My mother sits in the front with baby Karin and my father drives, carefully holding his cigarette just at the window's edge. This is how I remember my mother and father: smoking in a cool, elegant way that makes me want to grow up quick so I can smoke, too. It's evening and I'm tired and cranky. The spring day has been spent on a long drive through the Colorado mountains, a Sunday ritual.

We turn the corner to our home on Johnson Court, the square little house my parents bought when my father left his job as an attorney for an insurance company and set up his own law practice. The neighborhood is made up of winding rows of houses that all look like ours: a front door and a picture window facing the street, two windows on each side, and a sliding door in the back that opens to a postage-stamp backyard. We have a view of the mountains and one tree.

"Uh-oh," my mother says.

"Jesus." My dad stops the car. I scramble to my knees to look.

Our house is smoldering. One side is gone. A fire truck and a police car with streaking red lights stand in the driveway.

My dad jumps out and my mom reaches over and pulls up the parking brake. "Dick," she says, "I'm taking Kris to the neighbor's." My mother is always good in a crisis.

Mrs. Hauschild is waiting at her door. She takes a pair of pajamas from her daughter's room—we're almost the same age—and she beds me down in the basement in a sleeping bag. "She'll be fine here," Mrs. Hauschild says. "She doesn't need to see all that commotion." She suggests they both have a drink and a cigarette. My mother nods.

"Someone must have left the lamp on in Kris's bedroom," my mother says as they walk up the stairs. "The drapes caught on fire."

I repeat these words in my head until I come to believe I set the fire myself. I can still picture my bedside lamp, the brass switch, the round orange globe always warm to the touch.

Years later—decades, in fact—my father laughs when I tell him this story. "You didn't cause that fire, Kris," he says. "Your mother and I did. We had been sitting and talking in the living room, having a drink together, and we left a burning cigarette in the ashtray. Neither of us noticed. The drapes in the living room caught fire first." The flames never reached my room.

This is how I want to remember my parents: still talking to each other, even when the world was tumbling down around their ears.

WE RENT a basement apartment for a month and then move back to our rebuilt house. Nothing is ever said about the fire. Nothing is ever said about dark or sad or upsetting events, and anything that involves liquor is definitely not discussed. My parents are elegant drinkers. My mother can make a Manhattan with just the right splash of whiskey and vermouth. My father takes his bourbon straight on ice. After dinner, once my mother has tucked us into bed, my parents make cocktails and play cribbage to determine who has to do the dishes. From my bedroom I can hear my mother's soft laugh. Sometimes there's a stack of unwashed plates in the sink when we leave for school in the morning.

Soon another baby is born: my sister Karma. This is not a hippie name, despite the fact that we live close to Boulder. My mother insists on naming her daughters after her Norwegian heritage: Kristen, Karin, Karma.

At the top of the hill behind our house stands the Arvada cemetery. The year 1863 is etched in a stone marker at the entrance. The cemetery works like a magnet. As soon as our mother puts us out into the yard for the afternoon—just like the kids and grandkids on the family farm back in Iowa, who were expected to fend for themselves for the day—Karin and I scramble over the fence and head for the hill. We are our own secret club, and Karma joins us as soon as she is old enough to toddle along. Sometimes the other neighbor girls—Paula, Susie, and Kathy—are allowed into the club as temporary members. We trek across the field behind the row of backyards and through the old apple orchard and get up to the creek, where we balance a flat plank across the shallow, sluggish water and tiptoe across. Water spiders dance across the surface and tiny minnows scatter when we push our toes into the muddy bottom.

At the crest of the hill stand row after row of headstones. Some are tall, others flat against the ground. Some have the names of children or images of their faces etched in the stone, and we stay away from those. We run up and down the rows, shrieking and gathering up the plastic flowers. We pile all our flowers in the middle and sit in a circle around them. We look down the hill to our house and imagine our mother, big and round, lying on her bed and waiting for the next baby, a boy at last, she's sure of it. A little farther, we can see the Arvada Villa Pizza Parlor and the Arvada Beauty Academy. Between our neighborhood and the long dark line of mountains stands a single white water tower, all by itself. The Rocky Flats water tower. There is a hidden factory there.

That hidden factory is the Rocky Flats Nuclear Weapons Plant, a foundry that smelts plutonium, purifies it, and shapes it into plutonium "triggers" for nuclear bombs. The plant also recycles fissionable material from outmoded bombs. A largely blue-collar link in the U.S. government's nuclear bomb network, Rocky Flats is the only plant in the country that produces these triggers—small, spherical explosives that provide an atomic bomb's chain reaction. The triggers form the heart of every nuclear weapon made in America. From 1952 to 1989, Rocky Flats manufactures more than seventy thousand plutonium triggers, at a cost of nearly $4 million apiece. Each one contains enough breathable particles of plutonium to kill every person on earth.

Rocky Flats' largest output, however, is radioactive and toxic waste. In all the decades of nuclear weapons production, the nuclear weapons industry produces waste with too little thought to the future or the environment. The creation of each gram of plutonium produces radioactive waste, virtually all of which remains with us to the present day.

But no one in our community knows what goes on at Rocky Flats. This is a secret operation, not subject to any laws of the state.

The wind blows, as it always does. I imagine the bones of pioneers and cowboys beneath our feet. The chill of evening begins to creep up the hill; the air turns cold when the sun dips.

"Let's go!" Karin yells, and we jump to our feet and roll and tumble down the hill. We bounce across the plank and race across the field, full speed, before the sun sets and the ghosts come out.

IN THE beginning, Rocky Flats is called Project Apple. In 1951, years before I'm born, a group of men from the Atomic Energy Commission (AEC) meet in an old hotel off the beaten track in Denver. No press, no publicity. Their job is to find a site to build a secret bomb factory that will carry out the work that first began with the Manhattan Project, the covert military endeavor that developed the first atomic bomb during World War II.

Until now, all nuclear bombs in the United States have been custombuilt at the weapons research and design laboratory at Los Alamos, New Mexico, with materials supplied from the plutonium production facility at the Hanford site in eastern Washington State and the uranium enrichment facility at Oak Ridge, Tennessee. But with the heightening Cold War—a high state of military tension and political conflict with the Soviet Union and its allies that will continue for decades—the United States wants to mass-produce nuclear weapons. They need a roll-up-your-sleeves, get-down-to-business, high-production bomb factory. An assembly line.

AEC officials choose a site on a high, windy plateau not far from the growing cities of Arvada, Boulder, and Denver—cities that can provide workers and housing. Landowners are forced to sell their land to the government, and construction on Project Apple begins immediately.

A few months later, the *Denver Post* breaks the news of the new plant with the headline THERE IS GOOD NEWS TODAY: AEC TO BUILD $45 MILLION A-PLANT NEAR DENVER. Announcement of the plant catches everyone by surprise, including state and city officials, and the news breaks like a thunderbolt over the community. Though owned by the AEC, the plant will be operated by Dow Chemical, a private contractor that will be indemnified against any accident or mishap. The Rocky Flats Nuclear Weapons Plant will become the workhorse of an AEC complex of weapons facilities that eventually includes thirteen sites from Nevada to Kansas to South Carolina. Each AEC facility will be involved in its own particular aspect of the design, manufacture, testing, and maintenance of weapons for the U.S. nuclear arsenal.

Components and processes will be divided up around the country, but Rocky Flats will be one of two sites designed to produce the fissionable plutonium "pits" at the core of nuclear bombs. (After 1965 it will be the only site.) The whole system depends upon Rocky Flats.

Construction of the plant is rushed.

Few people know the deal is in the works. Not even the governor has an inkling. Colorado's top elected officials are not informed that the plant will be built until after the decision is made and there's no going back. But Denver welcomes the windfall. No one knows what the factory will produce. No one cares. It means jobs. It means housing. Contractors, the local power plant, and local businesses all look forward to the "juicy plum" to be known from now on as Rocky Flats.

It's the Cold War. The bombs dropped on Hiroshima and Nagasaki in 1945 may have ended one war, but they started another. The perceived Soviet threat is an ever-present shadow in American life. The Atomic Energy Act of 1946 creates an impenetrable wall of secrecy around the

U.S. nuclear establishment. All government decisions and activities related to the production of nuclear weapons will be completely hidden. Information about nuclear bombs, toxic and radioactive waste, environmental contamination, and known and unknown health risks to workers and local residents is all strictly classified.

And no one asks questions.

An editorial in the *Denver Post* predicts that Rocky Flats will be "a source of satisfaction to all residents who have an abiding faith in Colorado's destiny and future greatness." The newspaper reports that workers on the project will be safer than "downtown office workers who have to cross busy streets on their way to lunch."

The announcement is made simultaneously in Denver, Los Alamos, and Washington, D.C. The plant site in Jefferson County has been chosen for "operational values," including the fact that the land is nothing but an old rocky cow pasture, "virtual waste land." Officials from the AEC emphasize that no atom bombs or weapons will be built at Rocky Flats, only some unspecified component parts. The plant will not give off "dangerous wastes" or use large quantities of water, gas, and electricity. When questioned further by reporters, AEC spokesman Dick Elliott states adamantly, "Atomic bombs will not be built at this plant."

One small but devastating error escapes notice. The site criteria specifically state that the wind passing over the plant should not blow toward a major population center. But there is a mistake in the engineering report. Engineers base their analysis on wind patterns at Stapleton Airport, on the other side of Denver, where winds come from the south. Rocky Flats is well known for extreme weather conditions—rain, sleet, snow, and especially the prevailing winds, including chinooks that travel down the eastern slope of the Rockies from the west and northwest, directly over Rocky Flats and straight toward Arvada, Westminster, Broomfield, and Denver. Called "snow eaters," chinook winds occur when the jet stream dips down and hits the fourteeners—the 14,000-foot mountains west of Denver—where they lose their moisture. The winds warm as they race down the lee side of the mountain range, and by the time they reach flat land, they're hot and often exceed 100 miles per hour. Snow melts overnight. Sometimes chinooks snap telephone poles, blow out windshields, and overturn vehicles in the area around Rocky Flats.

One employee who notices the error is Jim Stone. An engineer hired to help design Rocky Flats before it opens, Stone is a careful and thorough man. Born during the Depression, he was sent to a Catholic orphanage when his parents couldn't afford to raise him. His path to becoming an engineer has been hard won, and he brings years of experience to his job at Rocky Flats. He warns against the location of the plant "because Denver is downwind a few miles away." He is ignored.

The name Rocky Flats is taken from the dry, rolling land dotted with sage and pine trees, a name chosen by early homesteaders who raised cattle and hay. Now it will no longer be ranchland. The money is in housing. Jefferson County and the entire Denver area are booming. Just over half a million in 1950, by 1969 the population of the Denver metro area has more than doubled. Jefferson and Boulder counties are two of the fastest-growing counties in the entire country. Thomas Mills, the mayor of Arvada, worries about housing. Rocky Flats plans to hire at least a thousand permanent workers immediately, and unlike in other nuclear towns, such as Los Alamos, workers will not be housed on-site. "The housing situation is rough here. We'll receive the brunt of all that traffic to the plant because we're on the only direct route to it," Mills says. "The city is comprised mostly of small homes. There really is only one large apartment house. . . . It's going to cause us lots of headaches." By the first week of March 1951, extensive new home construction has begun.

The plant is surrounded by two tiers of barbed-wire fence stretching ten miles around the circumference of the core area. The first tier, three feet high, is to keep cattle out. The second tier, nine feet high, is electrified and patrolled by guards with guns, high-powered binoculars, and, eventually, tanks. With the exception of a two-story administration building, the plant's buildings are built low to the ground, in ravines cut deep into the soil. The factory is almost completely invisible from the road. By early 1952, things are in full production. By 1957, nearly

1,600 people work at Rocky Flats. Radioactive and toxic waste have to be dealt with from the beginning. Effluence is run through a regular sewage disposal plant and empties into nearby Woman Creek. Solid and liquid waste is packed into fifty-five-gallon drums. Much of what remains is incinerated. What spews from the smokestacks of the production buildings is expected to disperse by the time it reaches the outer limits of the plant boundary.

The product that comes off the factory line at Rocky Flats is a well-kept secret.

By 1969, more than 3,500 people work at the plant. No other nuclear bomb factory has ever been located so close to a large and growing population.

WE BEGIN what we do best as a family: collecting pets. They come and go. Fluffy, a gray tabby who melts in my arms when I rock her on the backyard swing, lasts only a few weeks before a neighbor's dog gets her. Melody is a sweet-natured calico cat who disappears almost as quickly; when my sister Karma sees a photo of a similar-looking cat in a glossy magazine, she tells me that Melody has run off to become a famous cat model. We drive a dachshund to neurosis by chasing him around the house. Fritzi is then sent to the home of an elderly couple to recover. He never returns. My mother takes us to the Arvada Pet Store and buys me a green parakeet I name Mr. Tweedybopper. Karin gets a tiny turtle, Tom, in a plastic moat, and Karma gets a pair of hamsters. When they succumb to the various hazards of our household—Mr. Tweedybopper catches a draft, Tom Turtle dehydrates, and the hamsters successfully plot an escape—we visit the pet store again.

My father endures our ever-expanding household with little comment. He spends Saturdays—the only day we see him—mowing the backyard in Bermuda shorts, black socks, and worn penny loafers. My sisters and I dance along behind him in the clipped path, the scent of the grass thick, sweet, and heady. Saturday is also trash day. We help Dad pack up all the household trash and take it out to our incinerator, a cement-block monument in the backyard, blackened from use. We take turns pushing trash in through the trapdoor at the front. Everything goes—cans, paper, plastic, food, coffee grounds. Dad lights a match and we watch the pieces catch and burn and the oily smoke curl up into the sky.

WITH THE birth of my brother, Kurt, the house reaches its limit. My father says he doesn't have room to think, and my mother claims she's losing her mind. Our Sunday drives take us out by Rocky Flats, through empty landscapes of planned housing developments, dirt roads drawn in chalk, and squares of land separated by wooden spikes with fluttering orange ribbons. Bulldozers push piles of earth and dig rows of deep foundations like a vast potter's field. My parents sit up late at night at the kitchen table, looking at blueprints and adding up numbers.

"Guess what, kids," my mom says. "We're moving to a new house."

Our house begins with a deep rectangular pit. My mother drives us out in the station wagon, a long green lizard of a car with no seat belts, so we can watch. No one back then has seat belts; if they do, they don't use them. My father takes pride in not buckling up.

Carpenters arrive in weatherbeaten pickups. The soil is rocky and the workers cuss. We aren't supposed to hear, even if it is in Spanish. There is a lot of pounding. I remember the bones: two-by-fours reaching to the sky, anchored in concrete.

Our skeletal house stands on nearly two acres at the end of a road that dips down to a small hill, where our driveway begins. Not a long driveway, but long enough to set us apart from everyone else. There is no grass or trees, only mud. We look out from the freshly poured concrete of our front porch and see lines of spindly houses: streets laid out for pavement and front yards of raw earth waiting for sod, doors and windows, mortar and bricks. All the pieces ready to be put together. Some families have already moved in with their dogs and tricycles and motorcycles and an occasional horse stabled in the backyard.

The developer calls it Bridledale. My mother calls it heaven. Bridledale represents the golden dream of suburban life and all its postwar promises.

The bills begin to mount and our new house is still not finished. My father spends more time at the office. Some evenings, if he's home from work, we go to the McDonald's near the old bowling alley, where the dry cleaners used to be. Now two shiny arches loom yellow in the sky. "What does this represent?" my dad asks. He never waits for a response. "This represents change," he says. The sign out front shows how many hamburgers have been sold. Millions. *Who eats all those hamburgers?* we wonder. "Out of the car," Dad orders. He's in a hurry. He's always late and he's always in a hurry. The world gallops two steps ahead of him and he never catches up.

We stand at the shiny counter while he orders. Six cheeseburgers. Six Cokes. Six orders of fries. The room is clean and efficient and people stand politely in line. The clerk crisply folds the top of each white bag, and my dad carries them to the car and stacks them together on the front seat, where no one is allowed to sit.

"Can we have just a bite?" Karma asks.

"No."

"A fry?" Kurt, now a toddler, is sandwiched between his sisters. His hair is shaved close across the top of his head, a bright blond fuzz.

"No." Dad smiles. He's pulled off his tie, and the crisp shirt he put on this morning is crumpled and damp. "Sit tight."

My mother forbids us to eat any of it until we get home, lest only empty white sacks arrive. It's ten minutes there and ten minutes back and temptation is strong. My dad has a game on the radio turned up loud, and the four of us sit cheek by jowl in the backseat, fighting over property lines. Occasionally the game is interrupted by the irksome buzz of the Emergency Broadcast System. Dad mutters along with the game, but eventually his hand wanders up over the back of the seat, fingers pacing like spider legs. "Who wants a pinch?" We squeal. The hand descends, waving, searching for an elbow or knee. "Who needs a tickle?"

On the way home we stop at Triangle Liquor, where an amiable man stands at the counter, a black-and-white television flickering behind him. He looks out to the parking lot, counts heads, and adds the right number of cherry suckers to the bag while my dad digs for his wallet. Time is short. We grab as many french fries as we can before he strides out, slides back into the seat, hands out suckers, and tucks the brown paper bag with the big square bottle beneath his seat.

When my mother asks me later if we stopped at the liquor store, I say no. I know the rules. I know what not to say, what subjects are taboo, and what secrets must remain secrets.

PEOPLE COME to see my father with all sorts of problems, and his law practice grows. Divorces. Speeding tickets. Drug charges. DUIs. I think he must be very wise. He works in a small brick office with few windows and comes home only to sleep. His waiting room is filled with overflowing ashtrays and people whose faces are rough and tired. Within walking distance of his office is a Dolly Madison ice cream parlor and a smoke-filled bar. On Saturdays we go with Dad to work so our mother can get some time to herself. After we spend a couple of hours banging the keys and spinning the ball on his secretary's worn Selectric typewriter, Dad gives us money and the four of us file down the street for chocolate sundaes while he heads to the local bar. Sometimes he just sits at his desk and drinks straight from the bottle in his desk drawer. We finish our ice cream and patiently wait until he tells us to get into the car.

My mother doesn't like my dad to bring clients to the house, but soon some of their possessions begin to appear. A clock, a set of dishes, a car that sputters and burns oil and has to be hauled away. If people can't pay their bills, he takes whatever they can give. Sometimes all they can give is a promise, and that's okay, too.

One day a client drives up in an old truck pulling a shaky, single-stall horse trailer and unloads a tall, ancient sorrel horse named Buster. "Now you kids can learn to ride," my dad declares. Both he and my mother spent their childhood summers on family farms in Iowa. Every family needs a horse, they say. Even in the suburbs. For twenty dollars we can keep Buster in a nearby field until our new house is ready. One of the best things about Bridledale is that we can have horses.

Buster turns out to be a dubious gift, his back so bony and sharp no one can endure sitting on him bareback. We think we're saving him from the glue factory, but he's so far gone that he spends only a couple of weeks in our care before he's loaded back into the shaky trailer and taken away.

But the damage is done. I want a horse now, badly. A real horse. My grandmother in Arizona sends me a collection of tiny white porcelain horses and they prance across the ledge of my windowsill in full equestrian *joie de vivre*. I don't care for dolls or dresses or Easy-Bake Ovens. I dream of pintos and palominos, Morgans and Thoroughbreds and Tennessee Walkers.

I hear whispered conversation in the kitchen regarding plans for my birthday party. "She still remembers the rocking horse she lost in the fire," my mother says.

There is a long pause.

"I know a man with a horse," my dad says. "A good horse. And he owes me something."

THE BEST way to watch the stars is lying flat on my back, in the backyard on our big trampoline cool with dew. Our house is far enough out from the city that the night sky is as black as soot and the stars shimmer in tiny pinpricks, with the veil of the Milky Way spiderwebbing across the sky. Sometimes the moon is nothing more than a thin curl of ribbon, and other nights it's round and full and portentous, a pregnant beacon. And yet I know all its brilliance is borrowed. The moon has no light of its own; it pirates its light from an invisible sun.

The other beacon in that night is Rocky Flats. The lights from Rocky Flats shine and twinkle on the dark silhouette of land almost as beautifully as the stars above, but it's a strange and peculiar light, a discomforting light, the lights of a city where no true city exists. It, too, is portentous, even sinister—if only one could have the ability to see beyond the white glimmer, to see what is really there.

In the daylight, we can see the Rocky Flats water tower from our back porch. "What is Rocky Flats?" I ask my mother.

"I don't know," she says. "It's run by Dow Chemical. I think they make cleaning supplies. Scrubbing Bubbles or something."

Neither of us likes housework very much, so we leave it at that.

✹

THE DAY Tonka arrives, the field behind our house smells of melted snow even though spring flowers poke through the mud. Tonka comes in a two-horse trailer pulled by a white pickup and he is everything Buster was not. Young. Frisky. And he's never had a bit in his mouth.

"He's not quite broke yet," Glen explains. Glen is a cowboy, the real thing, and we know he's in some kind of deep, secret trouble if he's working off a debt for my dad. His girlfriend comes along. She's short and pretty and sits on the tailgate of his truck. My mother wonders aloud if Glen's wife is at home.

Tonka is the most gorgeous creature I've ever seen. Brown and white patches splash across his coat. He has a long cream stripe across his left shoulder and a narrow white blaze down his nose. His legs are so white it looks like he's wearing silk stockings.

"Hey, Krissy," Glen calls. I hate that name. I jump off the fence I'm straddling with my sisters, and he hands me a piece of horse candy. With his feathery lips Tonka nibbles at my neck and arms and then plucks the candy from the palm of my hand. "He likes you," Glen says.

"Let's get her on!" my dad says. My mother waves from the back patio where she's getting the birthday cake ready.

"Well," Glen says, "I guess she can ride bareback." He swings me up across Tonka's smooth brown back and hands me the reins. "Just hang on tight, honey. Grip with your knees." The bridle is nothing more than two strips of leather and a rawhide cord across Tonka's nose. "Just give him a little neck rein to make him turn. You know how to do that?"

I shake my head.

"Just press the reins across this side if you want to go left and this other side if you want to go right. Pull straight back and he'll stop."

"Okay."

Tonka flattens his ears back toward me as if he doesn't like what he hears.

"Just don't let him know you're nervous. Remember, you're in control."

I nod.

"Off you go." Glen makes a clucking sound, like a chicken.

Tonka doesn't move.

"Give him a little nudge with your heel, honey," Glen says.

Tonka doesn't seem to know what to do.

"Here then," my dad offers, and reaches out and swats him hard across the rump.

Tonka leaps straight up like a grasshopper and suddenly we're lurching across the field. His back is as smooth and slick as a watermelon. I clap my legs against his sides and Tonka understands this as a command for a dead run. Fenceposts fly past. I lunge forward onto his neck and try to find something to grab.

"Hang on!" my sisters yell.

"Grab one rein and turn his head in!" Glen calls. "Make him turn and stop! Pull a rein!"

I've lost the reins. I catch a blurred glimpse of my mother, who is shouting something about not running through barbed wire.

I twist my hands into his mane. And then I see it coming. The one apple tree—the big one—where we pick green apples filled with wormholes to take home to our mother. The trunk is old and gnarled and there are two large limbs, one on each side.

Tonka is a veteran of tricks. Later it will occur to me that this was not his first performance. Just before we make contact, he drops his head and slides under the bough, smooth as a limbo dancer. The branch hits me straight across the chest, full force.

I fall flat on my back. Is this death? I can't breathe. By the time Glen's face is peering into mine, I manage a gasp.

"Don't worry," he consoles me. A wide grin spreads across his face. "Everyone gets the wind knocked out of them once or twice."

"You fall off, you have to get right back on," my dad adds, jogging up.

I lift my head and Tonka trots over, swinging his head from side to side in a kind of celebratory shake. He sniffs me over in a friendly way. Do I have more horse candy?

"She can wait until tomorrow to climb on again," my mother calls. "Birthday cake first."

I stand. The ground feels a little shaky beneath my boots. Glen hands me the reins. Tonka gives me a nudge and obediently follows me back to the house.

For the first time I'm in love.

A few weeks after Tonka arrives, my grandmother comes for a visit. Opal is tall and elegant and wears sweaters from her travels in Norway with my grandfather, a Lutheran minister who has recently passed away. Opal's shoes and handbag match perfectly and she speaks her mind, an unusual trait in our family. She doesn't approve of the glass of bourbon my father often has in his hand and she's not sure how she feels about her daughter's marriage, even after four kids. But she wants to see the new house we're building out by Rocky Flats, even if it's not much more than a hole in the ground.

My parents have a new Kodak and my mother wants a photo of the four of us kids with Tonka. A photo Opal can take back with her, a photo we can use for our Christmas card. I lure Tonka to the fence with horse candy and pull the bridle over his ears. We decide not to use a saddle so we can get more kids on his back. My mother aims the camera. My dad swings Kurt up first. "You sit in back," he says, his voice a little too loud, the edges of his words tumbling one over the next. "You're the little guy. It's easier on the horse's kidneys."

"Okay." Kurt grins a toothless grin.

"Karma next," Dad says. He swings her up and Karma's long legs hang down Tonka's sides. Tonka gets a little jumpy.

Karin goes up next. She grips the mane at the base of Tonka's neck. "Whoa," she says as Tonka sidesteps. It's crowded now. My plan is to stand in front and hold the reins, but suddenly I'm lifted from behind. "There you go!" Dad exclaims, and tosses me up, too high, too hard, too fast.

I have no time to protest. I land on Tonka's neck. He ducks his head, spooked, and I fall forward. Karin tumbles on top of me. Karma grabs Kurt and they slide off together, just before Tonka bolts.

My mother pulls Karin up off the ground. "This one's okay," she says. Karma and Kurt seem fine. I can taste dirt in my mouth. "Let me see that arm," Mom says. I hold up my right arm and it hangs at an angle. "I think you've broken it," she says.

"It looks like the bone is sticking out." Karin likes graphic details.

"I'll take her to the hospital," Opal says. My mother hands her the keys. We drive to the emergency room, Opal humming grimly behind the wheel. When we return hours later, I have a white plaster cast and a sling. Opal reports that not a single tear was shed. "She's pretty tough," she says.

"All my kids are tough," my mother says. We sit down to a dinner of hamburger casserole, canned peas, and Jell-O salad. No words are exchanged between my father and grandmother.

Later, as I lie in my bed with my arm in a cast—now covered with flower-power marker, thanks to my sisters and brother—I hear Opal arguing with my mother in the living room about my father, his drinking, how it's affecting the children. What is she going to do?

No one argues in front of the children. Nothing is said in front of the children. We know not to talk about our father's drinking even among ourselves.

MY ARM heals. Months pass. Our Sunday-morning drives continue as our new house nears completion. On this Sunday—May 11, 1969—the Colorado sun is clear and bright and it's Mother's Day. My sisters and I wear matching dresses and saddle shoes. Kurt has on a little sweater and tie, and my dad wears a clean shirt. At the restaurant, our favorite Italian place, my mother tells us to behave ourselves as we straggle from the car and gather around the fountain on the restaurant's patio. My dad digs into his pocket for pennies and we each make a wish before dropping one into the water. "This means that you'll always come back," my dad says. "Just like the fountain in Rome. It's like a curse."

"It means you'll always come back to a place that makes you happy," my mother corrects, and after an hour's wait—there are many families in the courtyard waiting to celebrate Mother's Day—we are seated at a table. My mother orders a Manhattan and gives me the liquor-soaked cherry. We eat big plates of spaghetti with fresh bread and butter and spumoni for dessert, so much that we have to sleep on the way home, the four of us slumped together in the backseat, property lines forgotten, our stomachs so full they ache.

TWO MILES away, in an underground plutonium processing building at Rocky Flats, a few scraps of plutonium spontaneously spark and ignite in a glove box.

A glove box is where plutonium triggers are made. The production line at Rocky Flats consists of a series of linked, sealed, stainless-steel glove boxes, up to sixty-four feet in length, in which plutonium is shaped by human hands. The glove boxes are designed to be kept at a slight vacuum so that any accidental leak will draw air into the box rather than allow plutonium particles to escape. Uniform-clad workers stand in front of the glove boxes and place their arms into heavy, lead-lined gloves and peer through an acrylic window to mold and hammer the plutonium "buttons" into shape. Running above the glove-box line is the chainveyor, an enclosed conveyor system that moves plutonium from task to task along the line. Tall, transparent plastic glove boxes move the plutonium up and down, between the glove-box line and the chainveyor, like dumbwaiters. Small stepladders are provided for workers, particularly women, who aren't tall enough to reach the arm portals. It's difficult and cumbersome work, with no small amount of risk, as plutonium is highly combustible.

There is no immediate alarm—the alarm has been disconnected to save space in the crowded production room. Production takes precedence over safety.

The spark goes unnoticed.

In sixteen years of operation, the plant has quietly doubled in size. More than three thousand employees work their daily shifts and then go home, where they can't talk about where they work or what they do. Few people have clearances to enter more than one building. No one knows exactly what happens at Rocky Flats. Workers in one area don't know what other workers do. The press doesn't know. It's all under the cloak of national security.

The half-buried 771 complex—several buildings designed to manufacture plutonium—is at the heart of the plant, surrounded by guard towers and barbed wire. A bluff hides it from the road. Hundreds of glove boxes snake across a floor area that encompasses two buildings: Building 776 and Building 771. The production floor is like a big, shiny kitchen stretching the length of two football fields.

The word *trigger* is almost euphemistic. In a nuclear warhead or hydrogen bomb, there are two steps: an initial fission explosion, called the "trigger," followed by a secondary fusion explosion. Each stage releases nuclear energy, and the two stages happen so quickly that they appear to be simultaneous. The principal isotope, or form, of plutonium in these bombs is plutonium-239. The trigger is cradled in conventional explosives, which compress the plutonium inward, creating a high enough temperature and strong enough pressure to initiate an atomic chain reaction. Roughly the size of a softball or grapefruit, this initial bomb, smaller than the Nagasaki plutonium implosion bomb, triggers thermonuclear fusion between tritium and deuterium, the two forms of heavy hydrogen, and is capable of leveling a small city by itself. But the detonation that creates this fission explosion then triggers the far more powerful fusion explosion of a hydrogen bomb—a mushroom cloud, as in the bomb that destroyed Nagasaki.

The plutonium triggers created at Rocky Flats form the explosive fissionable core essential to every nuclear weapon in the United States' arsenal. Yet each pit is an atomic bomb in its own right, of the same type as the Trinity and Nagasaki bombs.

Precise manufacture of the trigger is crucial, as any flaw or variation could cause a nuclear warhead to malfunction. A perfect, "diamond-stamped" trigger is the goal, again and again, whatever the risk.

On this day there are few workers due to the holiday. Only a skeleton crew is on hand. More than 7,640 pounds of plutonium—roughly enough for one thousand thermonuclear bombs—is held in the maze of glove boxes, pipes, tanks, and containers.

Small fires are common. When a plutonium chip sparks, the worker douses it with sand or drops it into machining oil to snuff it out. There is no automatic sprinkler system or floor drainage. Water is used on plutonium only as a last resort because water can cause plutonium to go "critical"—that is, it can create a spontaneous nuclear chain reaction that can be lethal to anyone within close proximity.

But no one sees this spark. The spark in the glove box grows into a flame.

Four security guards, Stan, Bill, Joe, and Al, are driving to work. They don't mind working on Sunday, even though it's Mother's Day. The shift is quiet and the pay is good. Like many Rocky Flats employees, they like to carpool. They know each other well. One of the best things about working at Rocky Flats is that it feels like family. Stan likes driving; he's behind the wheel of his new Chevy Corvair. Joe, who tops three hundred pounds, rides shotgun. Bill and Al are both tall and have folded themselves into the backseat as best they can.

Given the constant ravages of wind, rain, and snow, the road out to Rocky Flats can be rough. Old-timers tell stories of flat tires and overheated radiators in the summer, black ice and whiteouts in the winter. Back in the fifties, when the plant was being built, the weather sometimes made it impossible to work. The wind alone could push a man off his feet, and cattle knocked down outhouses while men were still inside. That's all changed now. The road is paved and the old guard shack is gone, replaced by a compound of more than ninety buildings, all hidden from the road by a bluff. Only the entrance gate is visible.

May can be as cold as February, but this afternoon is tentatively calm. Cottony clouds rest their bellies flat against a blue sky. Meadowlarks sing. The dry brown of winter has given way to foothills spotted with a few pine trees, hardy grass, and fragile wildflowers. Beyond the foothills a sharp-toothed ridge runs from north to south, with the dark, flat slabs of the Boulder flatirons in the distance. The road is nearly deserted. Families are at home or church or waiting for tables at restaurants.

Stan Skinger was twenty years old when he started working at Rocky Flats. He'd worked as a plumber during high school back in Illinois and traveled west for the wedding of a friend. His friend had a Colorado bride, but Stan fell in love with Colorado. He heard about the plant and applied for a job. He didn't know what he was getting into, and he didn't particularly mind. He just wanted to be in Colorado. Rock climbing, biking, skiing, he loved it all.

Like many employees, Stan started out as a janitor. The pay was good—no, great. And he didn't mind being a janitor. He got a kick out of the old coal miners who worked in the bowels of the plant where the plutonium work was done—the hot zone, they called it. And he liked meeting the new college kids who came in all cocky about climbing the corporate ladder. He kept an eye on the job postings, and when a position in Plant Protection opened up, he applied and became a guard. Just when his paycheck was getting really decent, he was drafted. Stan served two tours in Vietnam, the second time in Special Forces. He didn't like to talk about what he saw there. It changed him.

His wife and his job were waiting when he returned. But after two years in Vietnam, Stan wasn't sure he wanted to go back to Rocky Flats. He'd have to carry a gun and he felt a little jumpy. And he wasn't naïve. He knew what they did at Rocky Flats. It was a bomb factory. Most employees didn't want to think too much about that. No one used the word *bomb*. They had special words for the plutonium disks that rolled off the production line: triggers, pits, buttons. The bomb was called nothing more serious than a "device" or "gadget." The workers were making the parts, not pulling the trigger.

Stan wasn't the only one who felt uneasy. Like some of the old-timers who had been in the Navy and seen the nuclear weapons tests in the Pacific, or served in the Army and experienced some of the atomic bomb tests in Nevada, employee Jim Kelly—who started working at the plant in 1958, and eventually presided over the union—knew right from the start what they did at the plant. He knew the destructive power of the bomb. It was terrifying. He and other workers reconciled themselves with the notion that when the "device" left the plant, it couldn't explode. That was technically true, since the nuclear bombs from Rocky Flats were sent down to Amarillo, Texas, where they were packed into a nest of conventional explosives.

Jim admitted to himself that this argument was like somebody saying he worked in a dynamite factory, but he didn't make explosives because the blasting caps were made somewhere else. It was a way to deny what they were doing. It bothered him, but he kept it to himself. He didn't talk to his family about it, and like others he tried to repress the enormity of what was going on.

Other workers had similar sentiments. Dr. Robert Rothe, a nuclear physicist who performed approximately 1,700 nuclear experiments—many of them extremely dangerous—at a laboratory at Rocky Flats, felt "somewhat divorced from the actual nuclear weapon itself. In fact, I have hardly ever even seen any of the components for a nuclear weapon."

As far as Stan Skinger was concerned, the world had lost its innocence when the first atomic bomb was dropped. He had been three years old when that happened. But he remembered. He remembered his parents talking about it. You couldn't go back after something like that. It was a done deal. In a rational world, there would be no need for nuclear weapons. But human nature didn't allow people to be rational, he felt. At least not all at the same time.

He gave it some thought and decided to go back to Rocky Flats after all. There he met a kindred spirit, a guard named Bill Dennison, and they became fast friends.

Bill Dennison is a big, soft-spoken man, fifteen years older than Stan. He, too, keeps his war experiences to himself, although his are from a different war. After ninth grade he dropped out of school, left home, and spent his teens working on ranches in Colorado and Wyoming. At seventeen he joined the Army and was sent to Korea, where he served as a machine-gunner in an infantry company until, as he later described it, a mortar shell "blew him all over the field." He was surprised to find himself still breathing. Of the 120 men in his unit, he was one of only 36 who survived. He and his buddies were trapped for three days without water before they crawled far enough to find a stream. They drank and got sick. A few days later they reached a point upstream and realized the water was filled with rotting bodies.

Bill's health was never the same.

When Bill returned to the states in 1951, he needed a job. His older brother worked at Los Alamos, the laboratory in New Mexico that developed the first nuclear bomb. Los Alamos was a tight-knit, closed community—a company town, really—surrounded by a stunning landscape. Bill liked it. His brother told him to check out Rocky Flats. The pay was good and the work steady.

It turned out that Bill was old enough to fight for his country but too young to work for Rocky Flats. He had to wait a few months until he turned twenty-one and the government completed his background check. Finally, in August 1952, Bill became Rocky Flats employee number 972 and started work as a guard. It wasn't long before he was offered a promotion to chemical operator—a worker on the production line in the hot zone—and the raise that went along with it. He took the job.

Bill knew the basics of radiation: you couldn't feel it, you couldn't see it, you couldn't smell it, you couldn't taste it. You wouldn't know if you were exposed. But with enough exposure, you got sick. Too much exposure and you died. Like most employees, though, he wasn't too worried. There was a lot of talk about safety. Given what he'd been through already, it seemed a relatively small risk.

But Bill didn't last long as a chem op. He was surprised to discover that he didn't have the nerve to work the glove-box line, holding plutonium semi-spheres the size of small half-grapefruits in his lead-lined gloves. Lingering health problems made it hard for him to stand for long hours, and it was a very tense business. Sometimes things went wrong.

Bill asked to be reassigned to guard duty.

He understands better than most the problems Rocky Flats has had with off-site contamination. "I work out there," he tells people, "but I wouldn't live out there."

On this May day, the four men turn onto Indiana Street and reach the east entrance of the plant. Normally there's a line of cars at the gate at shift change, but because of the holiday, it's a short-shift day with minimum staffing. Bill glances up at the guard towers, where invisible figures watch over the six thousand acres of land bounded by strands of barbed-wire and No Trespassing signs. One might expect a top-secret nuclear weaponry facility to look like something out of a James Bond film—a fortress of gleaming metal and glass—but Rocky Flats is a cluster of shabby gray concrete buildings with a distinct government feel. Every building has a number. Every employee has a number.

The men pull up to the gate, ready to show their badges even though the guard usually recognizes their faces. But this time the guard waves them down. "Hold on, you guys," he says, his face tense.

Stan rolls down his window. "What's up?" he asks.

"There's a fire at the 771 complex."

Stan turns to Bill in surprise. That's the plutonium line.

"Better hurry," the guard says. "It's a bad one."

ON THIS particular afternoon, Willie Warling isn't thinking about work. He's headed down to the local bar for a beer. Maybe two. It's a beautiful day and he has the day off. He could use the relaxation. He's got a stressful job.

Willie works in the 771 complex—the Hell Hole, people call it. Chain link and razor wire surround the heavily guarded two-story building half-buried in a rocky gulch. It's the core of the plant, where plutonium is molded and shaped before it's sent to the Pantex facility in Texas to be put in bomb casings.

Willie didn't start out working as a radiation monitor. He began at Rocky Flats as a janitor, worked a couple of years as a shop clerk, and then moved into what they called health physics. A radiation monitor's job is to control contamination. Contamination on the surface of the skin can usually be scrubbed off, but if an alpha particle is inhaled or ingested, it lodges in the body and emits a high, localized dose of radiation. Internal alpha emitters like plutonium are more harmful per unit dose than gamma or X-ray radiation. The damage is permanent and ongoing. The lungs are especially vulnerable. Plutonium can ignite spontaneously when exposed to air, and as it burns, it turns into a very fine dust, similar to rust. This dust consists of intensely radioactive particles that remain in the air for long periods and are easily inhaled. Even a single particle of plutonium can lodge in the lungs and continuously expose the surrounding tissue. Cancer may result, although it can take years or even decades to manifest.

The weight of plutonium is measured in micrograms. A single microgram—that is, one millionth of a gram of plutonium—is considered by the DOE to be a potentially lethal dose. A needle in a haystack, a dot on the head of a pin, a flea in a cathedral. In 1945 the AEC (later called the DOE) defined the "tolerance level" for nuclear workers exposed to plutonium as one microgram. In other words, by the time you've reached your tolerance level, you've received a potentially fatal dose.

Willie's job is to make sure the plutonium stays put.

He suits up for work every day in a Halloween costume of sorts: full-face mask, cap, protective clothing, rubber gloves—sometimes two or three sets of rubber gloves—and often a tank of supplied air. He makes sure the other employees suit up correctly. He tests them before they go in to work on the glove boxes, and he tests them when they come out to go home.

The work gets hot in more ways than one. Willie sweats beneath his uniform, especially when there's been a spill and he has to stay on duty for hours without a break. Sometimes it takes two or three weeks to clean up just one spill. And it's a never-ending story of cleaning one thing up and something else going wrong. A valve leaks, a glove box leaks, a pipe breaks.

Willie works at Rocky Flats for almost a year before he begins to understand what is coming off the assembly line. When he does understand, he never speaks to his wife about it. Or his three kids. He never speaks to anyone about it. It's important work. He wants to protect the secrets of Rocky Flats, just like everyone else. National security is at stake.

But on this Mother's Day, Willie isn't thinking about work. He's thinking about having a beer.

THE SPARK in the glove box grows. The two utility operators on shift are busy attending to another area. The spark feeds on the steady supply of oxygen from the ventilation system and bursts into an intense flame. The

Plexiglas window on the glove box suddenly begins to burn, releasing hot, noxious gases. The lead-lined rubber gloves catch fire. The Benelex shielding—considered nonflammable—ignites. Fire fills the glove box and moves into the next in line. It snakes quickly, quietly through the linked glove-box lines of both Building 776 and Building 777.

At 2:27 p.m., a building heat detector finally triggers an alarm at the Rocky Flats fire station. Three firefighters are on duty: two men by the name of Skull and Sweet, and their captain, Wayne Jesser. A minute later, one of the utility operators returns to Building 776 and smells smoke. He is the first and only man to pull an alarm.

WILLIE WARLING is about to take a sip from his third beer of the afternoon when the bartender approaches. "Your wife is on the phone," he says.

"What?"

"Your wife." The bartender looks away.

Willie walks behind the register and picks up the receiver.

"Mom?" he asks. He can't remember exactly when he started calling her that. It must have been somewhere along the line between all the kids.

"Something's wrong," she says. "You need to come home."

"What are you talking about?"

"Your manager called. They're having a problem."

He drives home. It could be anything. Things happen all the time.

His wife meets him at the door. "He said to call back right away," she says. "Hurry."

Willie calls. His supervisor's voice is tense. "I need you to come in to work. We need you right now."

"Well." He pauses. This is a surprise. "I don't know whether I should come out there."

"Why is that?"

Willie clears his throat. "I should tell you I've had a couple of beers."

"We need you now, Willie."

He looks over to see if anyone is listening. "They always told us not to go to work if we've had any beer, any alcohol or drugs or anything," he says. "I don't know if I should come in."

"Listen, Willie," says the supervisor. "We need you. You come on out here right now."

"All right." Willie hangs up the phone. "I'm going into work, Mom," he says. "I should be home soon." Suddenly he feels sober.

THE THREE firemen jump into the fire truck and roar to the west side of the building. Sweet stays in the truck. Skull and Jesser climb into their protective suits, pull on their hard hats, strap on oxygen tanks, and head into the building. Dense black smoke fills the room. Flames shoot a foot and a half above the top of the glove-box line. The men can't see well enough to move forward. They look down at their feet and try to follow the emergency evacuation markings painted on the floor, ducking to avoid the hot, glowing beads of lead dripping from the radiation shielding above.

They've been trained not to use water on plutonium. Each man grabs a canister of liquid carbon dioxide and together they try to shoot down the flames.

It doesn't work. Jesser grabs two more CO_2 canisters off a second line and they try to shoot down the fire again. It has no effect.

The men hastily retreat to rethink their strategy. They burst through the doors into the fresh air and pull off their equipment. A couple more workers have arrived, and Skull and Jesser are checked for radiation.

They're blazing hot.

"You can't go back in there. You're contaminated," the radiation monitor shouts. Jesser can't tell who's behind the mask. *Is that Willie Warling?* he wonders.

"We've got to," Jesser yells. He jerks off his mask.

"You're hot, sir," the radiation monitor yells. "Screaming hot. You've got plutonium all over you. Put that mask back on and don't take it off again." The monitor is breathing hard. He can feel the heat coming from inside the building, right through his coveralls, and he's standing outside in the wind. He can't imagine what it's like inside. And there's smoke, black smoke, coming out of the stack of Building 776. Black smoke isn't supposed to come out of those filters.

BILL AND Stan arrive at the guard center at Building 21. The skeletal guard crew is in a state of panic. "A couple firefighters are down there," a guard says. "And they're hot. Already. They can't go back in."

"How big is this fire?" Stan asks.

"Can't tell. Big."

"Who's in there?"

"Not sure. The first responders are hot—they went in to check it out and weren't in full suits. The Health Physics guys won't let them back in."

"Who else is going in?" Bill asks.

"We're waiting for backup," the guard says. "Everyone's out. No one's here."

Stan demands, "So who else is going in?"

"No one. We're calling people in. We can't even get ahold of people. It's a holiday, remember?"

"You're calling off-duty guys?" Stan asks.

"Yeah. And we're trying to get more Survive-Air tanks from other districts. We need equipment."

"Are you joking?" another guard asks. "We don't have that kind of time."

Bill and Stan glance at each other. They have no special training other than the basic fire training that all guards receive—essentially, how to use a fire extinguisher.

"I guess that means us," Stan says. Bill nods. They suit up in full bibs, taping the bottoms of their coveralls to their booties with duct tape, and grab a couple of masks and air tanks.

Bill's mind is racing. He knows what can happen with this kind of fire. He saw it back on September 11, 1957, when the first big fire at Rocky Flats occurred. It's all still vivid in his mind.

☢

BACK THEN he was only twenty-five, and it was his fourth year as a guard at Rocky Flats. On that day, too, he arrived at the gate as usual, ready to walk his route in Building 91. "Bill, wait," the guard said. "Don't go to 91. They need you in Building 771 [then called Building 71]." The plutonium processing building.

"What's up?"

"Fire."

Bill suited up with one of the firefighters. His coveralls weren't the right size, and he used so much duct tape across the bottom of his pants that later it ripped all the hair off his legs. They got to the building. Everything looked fine from the outside.

"Where do we go?" Bill asked.

"Down that hatch." A supervisor pointed. "Just follow that fire hose down into the building. Two guys already went down, but we don't know what happened to them."

Bill and his partner climbed down into the passageway. They followed the hose to where the tunnel branched off in a Y. On instinct they followed the path to the right, opened a door, and walked into a wall of fire.

They closed the door.

"You scared?" his partner asked.

"No." Bill shook his head. He was good at keeping his emotions in check.

The men backtracked to the Y-point, where they met the other two men. "It's out!" the men cried. "We got it."

But they didn't get it. It was like a fire in a haystack, cool on the outside but a furnace within. This fire, too, had started in a glove box, in a plutonium skull, a thin casing left over from the mold for the molten metal. As with the later fire, there had been no alarm—heat-detecting sensory equipment was disabled when it slowed production.

The fire couldn't be stopped. Firefighters turned on the exhaust fans—an inadvertent mistake—which fanned the flames and carried hot gases into the main air exhaust system. The fire raged through the first bank of filters and then, suddenly, threatened all the filters that stretched across the roof, called the plenum. The roof and the entire complex were at risk.

The men knew not to use water on a plutonium fire. The risk of the blue flash of a criticality, or nuclear chain reaction, was too great. There would likely be no explosion—simply the blue flash signaling a surge of neutron radiation fatal to everyone in the immediate vicinity. But they were desperate. They began using water. For a moment it seemed to work. Then suddenly the air pressure dropped. There was silence, and then a deafening blast. Bill was rocked by the explosion. The force twisted the plenum's steel frame, destroying most of the filters, and blew the lead cap off the 152-foot smokestack. Flames shot more than two hundred feet above the rim. And the fire continued. For thirteen hours, unfiltered radioactive smoke poured out of the 771 smokestack—smoke filled with plutonium, americium, beryllium, acids, cleaning solvents, and other toxic contaminants. Bill was coated head to foot with plutonium—"crapped up," the workers called it—and one ear in particular required a vigorous scrubbing, even though he'd been wearing a mask and respirator. But he received the most contamination during the subsequent cleanup. Months and months of cleanup.

How much radioactive and toxic material escaped into the environment? No one knew, or will ever know, for sure. The 1957 fire was so hot it melted the top of the ten-story exhaust stack and destroyed the radiation sensors. The explosion blew out more than six hundred industrial filters and a four-year accumulation of uranium and plutonium nitrate and oxide. The filters had not been replaced since Rocky Flats began operation in 1953.

The blast was thunderous, but the radioactive plume it produced was silent as it floated over the cities of Arvada, Golden, and Wheat Ridge, and then passed on to the north side of Denver and beyond.

There was one lucky break: the freak explosion from the volatile combination of water and plutonium cut off the power and shut down the fans that were fueling and driving the fire. A potentially apocalyptic event for the Denver metro area was avoided.

Official estimates of how much plutonium was burned or released in the 1957 fire varied widely, from 500 grams to as much as 92 pounds of plutonium or more. By comparison, Fat Man, the bomb dropped on Nagasaki, used fewer than 14 pounds of plutonium. Beginning in a production area, the fire had spread through the venting system, destroying most of the filters—flammable filters that were supposed to protect the public. The explosion and resulting plume were caused by volatile gases mixing with plutonium dust caught in the filters. The plume exposed countless people in and around Denver to plutonium.

The government adamantly maintained that residents were not at risk, and that a criticality did not occur.

The day after the fire, a small notice appeared in the newspapers. A spokesman from the AEC stated that "spontaneous combustion" had occurred in a processing line, although he declined to describe exactly what had happened. There was no mention of the destroyed filters and sensors or the deadly plume of smoke. It was the Cold War. No one asked questions.

The AEC repeatedly told the press there was no danger of a nuclear explosion at Rocky Flats. There was no danger to surrounding areas, populations, crops, or livestock from the Rocky Flats plant operations. When pressed for more information by reporters, the AEC said, "Further information regarding the function of the plant would be of value to unfriendly nations, and cannot be disclosed under security regulations."

Elements such as strontium-90 and cesium-135 never occur except in the case of a nuclear chain reaction. Based on soil and water testing completed decades later that detects the presence of these elements, some experts—despite the government's insistence that there has never been a criticality at Rocky Flats—believe that a criticality accident producing various fission products may have occurred on September 11, 1957.

TWELVE YEARS have passed since that terrible, secret fire. Now, on May 11, 1969, it feels like it's about to happen all over again.

"You ready?" Stan asks.

"You and me," Bill says. "Let's go."

When they arrive on the east side of the plutonium processing building, it looks quiet and clean, at least from the outside. There's a loading dock with doors on each side, and a set of double doors that leads into an interior hallway. The men pull on their masks and strap on their air tanks. "CO_2 only," Bill says. "No water."

Stan nods. They open the door, move into the hallway, and enter the main production area.

"Holy cow." Stan stops in his tracks. Usually as bright as a supermarket, the room is nearly pitch black. A few emergency lights glow dully. The only noise comes from the fans, feeding a fire he can feel more than see. "I can't even see my hand in front of my face," he mutters.

Smoke rolls toward them in waves. Bill sees the orange glow and moves closer. It looks like the flames are shooting up over the glove boxes. One, two, three glove boxes—no, all of them. He knows the look of this kind of fire. It reminds him of forest fires he's seen in films—high, fast-moving flames—but the color is different. It's the distinct, unearthly brilliance of burning metal.

"What is that?" Stan yells.

"Plutonium. Probably the magnesium carriers, too."

The heat is intense. Stan feels it through his mask. "It's not just plutonium," he yells. "It's the plastic. The shielding. It's the Benelex around these glove boxes."

"Benelex doesn't burn."

"It's burning! Why is it burning?"

"The Plexiglas, too," Bill shouts. "The Plexiglas is on fire."

It takes a lot of radiant heat to make something like that flammable, Stan thinks. This fire has been going on for a while.

Burning globes crash from the ceiling. It's hard to tell whether they're just light fixtures or pendants, the baskets that carry plutonium nuggets down the production line. "Come on," Stan says. Time is short. He knows this building. Both men have walked it hundreds of times, upstairs and down. The two buildings are connected. The 776 side has two floors; 777 has one. Protecting the roof of 777 is crucial. The plenums—

the filters—stretch across the entire roof area. Stan likes to compare a plenum to the air filter on a car. With a car, you clean the air before you pull it into the engine. In a plutonium processing building, you clean the air in the building before you blow it out into the atmosphere. The flow is reversed, from inside to outside.

If the fire burns through the plenums and the 777 roof, massive amounts of plutonium—as well as other contaminants and radioactive material—will spread over the Denver area and beyond.

Stan opens a cabinet and finds a stack of hard hats. He hands one to Bill and straps one on himself. *Where are the other firefighters?* he wonders. They're unaware of the Jesser team. The men inch into the room until they find the buckets of sand set in corners for extinguishing small fires. They move toward the edge of the fire and throw sand on the flames. It's like throwing grains of rice in the face of an oncoming locomotive. The fire continues to grow.

Bill grabs a CO_2 canister and hands another to Stan. They fire them into the glove boxes. It has little effect. They empty another canister. The air in the room is unbearably hot and the men are breathing heavily—already they're almost out of air. The fire gallops through the line.

"What now?" Stan yells. Bill yells something back, but Stan can't hear it. What are they supposed to do? Who are they supposed to ask? They're alone. The no-water rule is the only rule they've got, but it's useless.

The men bolt out of the building, shaken and gasping. They change tanks and confer briefly, ignoring the radiation monitor who has carefully chalked off a square area. "Don't step outside these lines," he barks. "Keep the contamination inside these lines," as if plutonium could possibly recognize a line of chalk.

"Water?" Bill looks to Stan for confirmation.

"Water." *What the hell*, Stan thinks. He's not a firefighter. He's a guard. He's lived in the country and nearly all he knows about firefighting is how to beat a prairie grass fire with a burlap sack. "You good with these things?" he asks Bill.

"More or less," Bill replies. They wrestle with the nozzle. "Use the fine spray."

"Got it."

"Soft. Gentle-like," Bill says. "Hit the gases from the melting plastic first. See what happens."

"Okay."

They reenter the building, this time with water hoses.

"We'll take turns going forward," Bill says. "I spray you, then you spray me. We need to keep each other cooled down."

"Let's head toward the center," Stan says. "Get under the center beams and see how the plenum looks."

"Okay." Bill turns his hose on Stan, and Stan moves forward into the smoke, trying to follow the emergency lighting on the floor.

"Hey!" Bill yells. Stan looks back.

"Don't blow any of those plutonium pieces together. Keep 'em separated."

"I know." Blue flash. He knows.

FIGHTING THE fire from the other side of the building—two football fields away—Captain Jesser reaches the same decision. Despite the risk, his men decide to use water, too.

Like Bill Dennison, Wayne Jesser fought the 1957 fire. He knows the danger of a criticality. Everyone fears that blue flash. And it's not just about what could happen to him and his crew. If the fire burns through the roof, powdery plutonium ash—toxic radiation—will descend on people living in the Denver area and beyond.

At 2:34 p.m., just five minutes after entering the building with Sweet, Jesser orders his team to bring in fire hoses. They drive a tanker to the north end and hook up a hose to a hydrant.

WORKING IN tandem, Bill and Stan move along the glove-box line, directing a spray of water around the flames and then on each other. They've gone only a few feet when they see where the real fire lies: in the foundry area, where plutonium is melted and cast into pieces that are carried to the production line. The foundry line is one hundred feet long and contains eight furnaces, all held inside glove boxes. The entire line is ablaze.

Bill curses. The men glance at each other. The production area is tight. There's only one way to get to the foundry area: through the underpasses. Some glove boxes have steps beneath them, tiny stairs going down to a miniature basement with steps leading up the other side. This allows workers to get from one side of the production line to the other. The underpasses have no drains. Anything that spills under a glove box is contaminated and has to be cleaned up, not flushed out.

There's no place for the water to go, and the underpasses are filled with water. The water is rising.

"It's like a sheep dip," Bill says, and laughs. He thinks back to all the ranches he worked on as a kid. "One hell of a sheep dip."

"There's our criticality, Bill," Stan says. "We're looking right at it."

"Who's going in?"

Both men stand silent. Bill looks up and sees an elevator flipped upside down, the supporting metal scorched and twisted. People are going to get killed tonight, and he guesses it might be him and Stan. He thinks of his wife, who's pregnant, and his two other kids. Both men were trained for the battlefield, but it didn't prepare them for this. One thing it did teach them was to keep their feelings to themselves—and move.

Bill wades in. The water is up to his knees. He thinks he's moving fast, but it feels slow. He prays they haven't knocked any plutonium pits or pieces into the water, which could lead to the criticality they fear. Then he's up the other side and the foundry fire is so hot, so immediate, that his soaked coveralls dry instantly. His face feels scorched beneath his mask.

Stan is right behind him. They spray the fire until their air runs low—a few short minutes. Then they drop their hoses, wade through the sheep dip again, and fight their way back to the door.

A radiation monitor is waiting when they burst from the building and yank off their masks. He checks their hands.

"You're hot, man," he barks. "Coated." His sharpness can't hide his fear. "You can't go back in."

"We're all right," Bill says.

"No. You're off the chart."

"We gotta go back."

Stan reaches for a fresh tank. "The plenum's about to go. You gonna keep us out here so we can all watch the roof melt?"

"I'm serious. You guys are not going back in that building."

"Who else is there?" Bill asks.

"We're waiting on more guys," another monitor yells. "We don't have anyone else yet. We don't even have enough gear. We're waiting for Boulder and Broomfield to bring more tanks." The only manager on duty that the men are aware of is the guard captain, who's on the phone.

"Is that you, George?" Stan peers into the man's mask. He recognizes him from the lunchroom. They're both model railroad hobbyists.

"I can't let you back in, Stan," George says. "Come on. What the hell are you guys thinking?" He looks toward the road. A van is on the way to take workers to Building 559 for decontamination.

"George," Stan says, "we let the fire get into the plenums and Denver is screwed."

"Give us the tanks." Bill's voice is furious.

"Can't do that."

"Then we'll just take 'em," Stan says. They strap on the tanks and pull on their masks as George, arms folded, blocks their way. Bill shoves past him. Stan follows.

The men duck back into the building. "I'll go first this time," Stan shouts. He runs, crouching, into the production line. He darts back and forth, spraying anything that doesn't look like plutonium. *That man is as quick as a monkey,* Bill thinks. After a few minutes he sprays him down and they switch. Bill can't move quite as quickly as Stan, but it feels like they're making progress. They're both thinking about the roof.

"I'm out!" Bill shouts, and gestures toward his tank. Stan nods. He wonders if the heat and exertion are causing them to go through their air tanks more quickly than usual, or if the tanks are only partially filled.

Abruptly Stan is knocked to the floor. Flat on his back, covered with debris, he can't see anything. He doesn't lose consciousness but it takes him a moment to realize that his body is covered with a heavy material. Ceiling material. His heart pounds. The roof. This is it, he thinks. The roof is gone. It's over.

But nothing happens. He looks up to see Bill still standing. He finds he can move his arms and legs, so he sits up and looks around. He's covered with gunk—messy, sticky gunk—and he pulls a soggy piece off his arm. Bill points to a gap in the ceiling—a false ceiling made of fiber material in two-foot by three-foot sections. They've sprayed it repeatedly with their hoses and the tiles have collapsed from the weight of the water. Stan is covered with nothing more than soaked ceiling tiles. The roof is still intact.

He stands up and Bill cleans him off. He can't read Bill's face.

Outside, they explode with laughter. "I hate to admit it," Stan says, "but I think that's the closest I've ever come to shitting my pants." The statement strikes them both as hilarious and they switch to new tanks. George stands back, watching.

MEANWHILE, JESSER orders several workers up on the roof where fluted steel sheets are fastened to steel ceiling girders and covered by three-quarter-inch Styrofoam, plywood, and a layer of thick rubber. One of the first men on the roof is Jim Kelly, who's been called in for the emergency from a holiday brunch with his family. He doesn't see smoke or detect radiation, and reports that the filters seem to be holding. *I can get back to my family soon,* he thinks. But twenty long hours will pass before he can go home. Soon firemen see smoke trailing along the roof's fluted edge, and then suddenly it begins pouring from the exhaust vents.

A few miles away, drivers on the Denver-Boulder turnpike can see the smoke, but no one understands its significance. The temperature is close to 400 degrees Fahrenheit. The Styrofoam is melting. One area of the roof is soft and beginning to rise like a big bubble. If the bubble bursts, they're in trouble. The entire city is in trouble.

"Water," Jesser shouts. It's against his training, against everything he knows about fighting a plutonium fire. "Get up there with that water!" A couple of firemen climb up with hoses and begin to spray water, cautiously and then more vigorously, thousands of gallons, across the rubber material. It helps slightly. "Keep at it," Jesser orders. He notes that the wind has picked up, not only making it more difficult to fight the fire but dispersing smoke and debris all over the place.

A quick assessment determines that two of the three banks of filters are completely burned out. The third is on fire.

BILL DENNISON's arms and legs are heavy with exhaustion. Stan, too, is tired. A relief crew of firefighters has arrived at the west side of the building, where Jesser and his crew are working, but there's no one to relieve the two guards. They both figure they had the last of their bad luck when the ceiling fell. They survived wading through the sheep dips. The fire, if not diminishing, at least is not growing. The worst is over—must be over—if they can just hang on a little longer.

It's Stan's turn to go ahead. Their tanks are getting low and Bill is misting Stan, keeping him cool. Stan stoops down to pick up his hose. It looks like it's charred on one side but still usable. He turns it on. But it's too hard, too fast, and it shoots out on full stream. No good, he thinks. He wants to wet down the material, not blow it around. He tries to turn the valve and slow it down. But it's still too much. He shuts it all the way off and the hose goes from full pressure to no pressure. Suddenly the backed-up water bursts through the side of the hose. It catches his mask and pulls it off. The burning air hits his face full force. The hose flails around him like a wild snake.

Stan tries to think clearly. *Get the mask back on,* he thinks. *Don't breathe. Where is the strap?* The strap catches on his nose. Everything is out of place. Thirty seconds pass and he's still holding his breath. His fingers fumble through the heavy gloves. Another thirty seconds. He thinks of all the crud they use in there. Plastics, vinyl, rubber, paints. Carbon tetrachloride. Cleaning chemicals. Benelex and Plexiglas. Oh, and plutonium.

He needs air. He can't help it. He knows he shouldn't breathe but he has to. He takes one big gulp and holds it. He keeps holding it, another minute at least, until he gets the mask pulled back on.

Bill pulls him around. "You okay?"

Stan exhales into his mask and nods. "I'm okay."

They go back for new tanks. *One more run,* they think. They can do at least one more. But Bill's bad luck isn't over yet, either. The men finish their air and head for the door. As they're crossing the floor, a blazing fluorescent light fixture crashes from the ceiling, nearly knocking off Bill's hard hat. He staggers, dazed. His oxygen tank is empty. He can't breathe, and he's lost Stan in the smoke and the dark.

A man—another guard—appears. Bill recognizes him. Charlie Perrisi. He's come off the roof to help relieve Stan and Bill. Charlie's a small man, shorter than Bill, but he pulls Bill up on his shoulders and brings him out into the air.

They're all contaminated. Charlie has kept his mask on, but he smells smoke. The same smoke Stan experienced, but somehow it's gotten inside his mask.

The fire isn't out, but it's more or less under control, and the men stagger to the van that will take them to decon.

FOR THE citizens of Colorado, luck plays a big role on the afternoon of Mother's Day, May 11, 1969. There are three lucky breaks, all largely the result of human error.

The first stroke of luck occurred earlier in the week, when workers accidentally left behind a metal plate that blocks the north glove-box line. This plate forces the racing fire to turn from Building 777—a single-story building with an extremely vulnerable roof that probably would have collapsed immediately—to Building 776. Building 776 has a second story and is a little less susceptible. This buys time.

The second lucky break occurs when a member of Jesser's team tries to hose a burning pile of plutonium into a corner. Burning plutonium turns into a heavy sludge of plutonium oxide ash, as heavy as wet cement. The pile won't move. Later an AEC fire investigator will report that if the fireman had been successful in moving the sludge—in pushing pieces of plutonium together—a criticality would have been the inevitable result.

The third piece of luck is the most important, and it is nothing short of déjà vu for Bill Dennison. A flustered fireman inadvertently backs a fire truck into a power pole adjacent to the building. Just as in 1957, an accidental power cutoff occurs—and just in the nick of time. The fans, which have been sucking the fire into the filter bank, feeding it and causing the roof to melt, stop spinning.

The fire still burns, but more slowly. The roof holds.

AFTER FOUR hours of firefighting, Stan and Bill arrive at detox. Both men are crapped up. They strip off their clothes. Everything is contaminated. A radiation monitor begins to check Stan's body. "You'll need to give me that wedding ring," he says.

"That's the one thing you're not getting," Stan says. "No way."

"It's hot."

"It's gold," Stan growls. "You can clean it. I want it back."

The men go to the shower room and scrub themselves with hard brushes, soap, and a sodium hypochlorite solution, or bleach. They scrub and scrub, get measured to see where the plutonium remains, and then scrub again. The goal is to take off the top epidermal layer but not break the skin and let the contamination get in. Their skin feels raw. Stan has no chest hair left. Most of his body hair is scrubbed off. He goes through it again and again. His left leg is still hot. Finally they tape the leg off from the rest of his body with plastic.

He wants to go home. He wants to see his wife. They keep him on site for another sixteen hours, and when they do let him go home, it's with an order not to remove the plastic bag and to come back the next day for another round of detox. Two days later he gets his wedding ring back. Bill scrubs clean, but as in the 1957 fire, most of his contamination will occur in the months of cleanup that follow.

Meanwhile, radioactive smoke continues to billow from the roof and vents, with most of the plutonium release occurring after 4:10 p.m. After burning for nearly six hours, the fire is considered more or less extinguished by eight o'clock.

NEVER MIND that no one is warned. Few people even know the fire happened. A small number of people in Denver notice the smoke on the horizon. I'm home with my siblings. We're having our baths, putting on our pajamas, and getting ready for our Sunday-evening television ritual: *The Wonderful World of Disney*.

The day after the fire, the *Rocky Mountain News* runs a small story on page 28. A plant spokesman states that the fire "released a small amount of radioactive plutonium contamination," all contained on site. The article appears just below a photo of the Pet of the Week.

No one pays much attention. My father skims the paper each day. If he sees the article, he doesn't mention it. My mother, who prefers paperback novels to newspapers, misses it completely. Nothing is mentioned on the five o'clock news, and there is no follow-up in the Sunday paper, which is the only day she bothers to pick it up. None of the neighbor kids say anything.

There are many things to be done for our new house. My mother picks out carpet and orders draperies. The Welcome Wagon ladies pay a visit with a gift basket of cookies and discount coupons from the local bakery and dry cleaner. My father talks to insurance agents and fence companies. He wants to drill a well for our water, but it proves so difficult and expensive that he decides to wait.

When my parents sign the final mortgage and insurance papers, they're also asked to sign some kind of waiver or statement about Rocky Flats from the Department of Housing and Urban Development. Nothing too specific, just some language acknowledging that they've been told about plutonium in the soil, at a level the government says is safe. Everyone in the neighborhood signs it. They wouldn't be building so many houses, my parents reason, if it weren't safe.

Finally, our house is ready.

Every family that moves into Bridledale is fast-tracking on the American dream. It's a blueprint for the perfect place to raise a family: four bedrooms, a gold-and-avocado living room, indoor-outdoor carpet, a kitchen with new appliances, and a formal dining room for show. My mother gets a bay window in the living room. My father gets a wet bar in the basement. Polynesian-style wet bars are all the rage (Trader Vic's is my parents' favorite restaurant), along with big backyard grills and trampolines. My mother hangs a couple of coconut heads from the ceiling behind the bar. Some of the neighbors are building bomb shelters in their basements and stocking them with radios, canned goods, and fold-up cots. "Bunch of paranoids," my dad grumbles.

We christen our new level of affluence with a bearskin rug, a gift from another grateful but strapped-for-cash client at my dad's law practice. There's a hole in the side where it was shot, and fleas hide in the fur. The jaws gape in a vivid plastic pink. My mother lays it in front of the fireplace as our first piece of new furniture.

I'm eleven. My blond hair is long and straight and so fine that my ears stick out. I wear boys' jeans because they have no hips, and I wonder if I will ever wear a bra. My heart belongs to Tonka, but I can't wait to have a boyfriend.

The developer, Rex Haag, is a slender, dark-haired man whose enthusiasm for the new community is contagious. Rex and his family, who live in the neighborhood, occasionally attend the cocktail parties and cookouts hosted by residents to welcome new families. A few months after we move into the neighborhood, Rex and his wife have a new baby. They name her Kristen. It's an unusual name for the time, and my mother is convinced they stole it. But she's willing to forgive. "There's nothing prettier than a Norwegian name," she says.

AT THE end of the summer, the AEC completes a report, classified secret at the time, confirming that forty-one Rocky Flats employees endured substantial doses of radiation during the Mother's Day fire. Both the AEC and Dow Chemical state that no contaminants were released off-site and the public was never at risk. One plutonium conveyor line, two tons of the Plexiglas windows on the glove-box lines, and tons of plastic walls were devastated by the fire. The fire results in more than $50 million in damage, and the cleanup takes almost two years to complete. Dr. Glenn T. Seaborg, chairman of the AEC, notes that despite the fact that "the damage is the most extensive ever incurred in the weapons production complex . . . operations will continue." AEC officials urgently appeal to Congress behind closed doors for $45 million to repair the facility. Production is halted temporarily, but, the Pentagon notes, the stoppage "is not expected to affect presently planned commitments to the Department of Defense."

Due to pressure from concerned scientists, activists, and the media, for the first time the AEC admits publicly that small amounts of plutonium have been released from the plant, but emphasizes that it presents "no risk to the health of employees in the plant or to citizens in the surrounding area."

Bill and Stan are assigned guard duty in the building they saved, patrolling the dark interior, looking for hot spots. Costumed workers in bright white-and-orange hazmat suits scrape out the radioactive mess. The plant begins sending trucks of drums and barrels filled with radioactive waste to a waste site in Idaho, where it's dumped off the trucks like ordinary garbage.

Charlie Perrisi, the guy who carried Bill out of the burning room, got a high body burden. There's so much plutonium in his system that they say he can't take any more. They pull him out of the fire department and make him a janitor. His salary is reduced as a result.

No health warnings are issued to other workers at Rocky Flats. Employees at the plant are not allowed access to their personal health records, and over the years many of those records are lost or misplaced anyway.

Bill will work at Rocky Flats for another thirty-six years. Thirty-six is his lucky number. But he has his problems. No more children, for one thing—he's sterile after the fire. But he's still a lucky guy. He has three children, and he and his wife decide to take in many foster children over the coming years. He survived the two fires at Rocky Flats. He survived the Korean War, one of 36 guys out of a group of 120. Sometimes he still thinks about the war and the faces of the 84 who died. It bothers him that he doesn't know all their names.

Stan Skinger is worried, even though Rocky Flats says he's clean. At first he believes them, but then he changes his mind. He requests a lung scan, even though the plant doesn't feel it's necessary, and the results are inconclusive. Stan's lost a little faith in things, maybe in the plant as a whole. He thinks about going into the tropical fish business—a long-standing hobby—and after another year or so he quits Rocky Flats and opens his own shop. Eventually he becomes a youth counselor. Years later, when he develops mesothelioma, his doctor will attribute it to those few moments when he lost his mask in the 1969 fire.

With a final price tag of $70.7 million, the 1969 fire at Rocky Flats breaks all previous records for any industrial accident in the United States. Roughly $20 million worth of plutonium is consumed in the fire. A congressional investigation later that year reveals that government officials hid behind national security to cover up details of the fire, and it was only the "heroic efforts" of the firefighters [that] "limited the fire and prevented hundreds of square miles [from] radiation and exposure." The report recommends extensive building modifications, and notes that if AEC officials had not disregarded the recommendations following the 1957 fire, there never would have been a fire in 1969. Further, fire investigators note that although they determine that the fire was ignited by oily rags tainted with plutonium, this fact did not appear in the final investigation report, as AEC officials "were afraid it was going to implicate certain individuals."

MY FATHER's law practice grows, and so does our extended family. Guppies and goldfish seem harmless enough, and my mother readily agrees to that. Then come hamsters and gerbils that reproduce at breakneck speed and elude capture by scooting under the heat registers. We progress to pet finches and miniature frogs and crawdads and—for a short time—a baby piranha named Killer.

Tonka, though, is my real passion. If I leave the tip of a carrot sticking out of the back pocket of my jeans, he steals it as smoothly as a pickpocket, and he can eat an apple whole. He licks my neck and face from the collarbone all the way up my cheek. My mother says it's just for the taste of salt, but I know it's love.

Our progress with training is slow. He likes to play hard-to-get. I stand at the gate and call his name and up pops his head. On a good day he casually picks his way around the reeds and tall grasses and moves in my general direction. More often than not we have a chase. I climb through the fence, halter and rope behind my back. He waits until I get within arm's reach, just close enough to offer an open palm with a piece of horse candy. He extends his neck, lips up the treat, and spins around, galloping off to the other side of the pasture.

Often it's dark by the time I catch him. Just in time for his dinner.

He hates saddles. When I tighten the cinch, he sucks in air and swells his belly, waiting to exhale until I get my boot in the stirrup. I swing up on his back and the saddle flips under his belly. He turns to look at me on the ground, pleased with his success.

I decide bareback is best.

A successful mount doesn't mean a successful ride. If Tonka doesn't want to go anywhere, he doesn't. He just stops. All sorts of things capture his interest: dogs, cats, butterflies, weeds. Even bushy purple-topped thistles, which he likes to pluck and eat with bared teeth to avoid pricking his lips. He's a renegade, but he has delicate table manners. If I urge him forward with my heels—or spurs, a poor suggestion from my mother—he rears straight up, again and again, like a mechanical horse, knocking me in the face with the back of his head. His aim is pretty good, and more than once I stagger into the kitchen with blood dripping from my nose.

Glen, who's still working off his debt to my father, offers his advice. "Break a water balloon over the top of his head," he says. "The horse will think it's blood dripping down his face and he'll never rear up again. It never fails."

It fails. Aiming is difficult and a water balloon, it turns out, is hard to break on a moving target. The balloon splashes on the ground. Tonka twists his head around and looks up at me as if to ask, *Are you an idiot?*

A truce is declared. For the moment it seems just fine to sit on Tonka's warm back under a tall cottonwood tree next to the irrigation ditch that runs across the back of our property. I drop the reins and Tonka drops his head and his eyes half-close in a doze. I lie back and rest my head on his haunches, soft as a pillow. We listen to the meadowlarks and horse-flies and the constant drone of construction workers.

ONE AFTERNOON my sister Karin and I notice a car parked on our street with a woman asleep on the backseat. The window is cracked. We peer in to see if she's sleeping or dead. Her jaw is slack and her skin pasty white. The front seat is a jumble of clothes and toiletries. "She's dead," Karin says. A blanket that must have covered her legs has slipped to the floor of the car. "Someone killed her." Karin is the kind of girl who will grow up to love Freddy Krueger movies.

"She's not dead," I say. "There's no blood." By the time we get home and tell our mother, the car has disappeared.

Our mother is reassuring. "She just needs to find a new home, that's all," she says. "Sometimes people get caught in circumstances."

I wonder if she has a husband, or a job.

"Maybe she's an alcoholic," Karin volunteers.

My mother's look is sharp. "How do you know that word?" she asks.

"I don't know," Karin retorts, and the matter is dropped. I add the word *alcoholic* to the list of words we've been strictly instructed never to say, words like *shit* and *fuck* and *damn*. Even *dang* is off-limits. "Just say *uff-da,*" my mother instructs. "That's what good Norwegians say."

We are well-behaved children. We know what not to say. Money, religion, politics, liquor: mum's the word. If you can't say anything nice, don't say anything at all. When I finally start my period, my mother acknowledges the occasion with an incomprehensible pink booklet on my dresser about sperm and eggs.

KURT CELEBRATES his third birthday and we all share a sticky-sweet chocolate cake. My father misses our gathering, as he often does, spending the evening at his office. When he comes home later that night, the house is quiet. I sit in the living room, reading, listening for him. We barely speak. His hours are odd and he spends little time at home, but when he's there, his presence is heavy and I always know where he is in the house. Often he sits brooding in his recliner in the den with the lights out, a big square bottle of bourbon tucked just out of sight.

"Hi, Kris," he says. This night his voice is hoarse. "What's for dinner?"

"I don't know," I say. Dinner was over long ago; the question is strictly rhetorical.

"How's school?" he asks. His shirt is rumpled. His fingers, long and slender, are stained yellow at the tips from smoking. His mind is always on something else. My mind is busy, too, reading every cue and signal, keeping track of all the things that cannot be discussed, that must not be remembered, that have to be erased.

But the scent of tobacco that clings to my father's clothes and skin is familiar and, in a way, comforting.

"Fine," I reply. He doesn't want details. I close my book. *The Carpetbaggers*. A book my mother wouldn't allow me to read that I swiped from under her bed. For hours I've watched a steady glow on the horizon, the lights that burn every night, all night long, against the dark mountain range. They're as predictable as the stars in the sky. "What is that, Dad?" I ask.

He turns and looks out the window. "What?"

"Those lights."

"Oh." He lights a cigarette. "That," he says brusquely but with pride, "is Rocky Flats. The defense of our country."

YEARS LATER, when Kristen Haag is eleven, she comes home with a bump on her knee. Like other children in the neighborhood, she rides horses across the windswept fields, swims in Standley Lake, and plays in her backyard. Bumps and scrapes are common. But this bump won't heal. In May, doctors discover a malignancy and the leg is amputated. By Christmas the cancer has killed her.

Rex Haag is devastated. Small items have begun to appear in the press about Rocky Flats, and a few neighbors quietly express distrust of government assurances that the plant is completely safe. Rex thinks back to the fire in 1969, just about the time he built a new sandbox in the backyard. He wonders what's in the soil and the air. He wonders about the drainage from Rocky Flats that feeds directly into Standley Lake and local streams and ditches. He starts talking about a lawsuit.

The neighbors whisper: It's a terrible thing when a child dies, but looking for a scapegoat won't ease the family's grief.

Besides, my parents agree, the government would tell us if there was any real danger.

After the funeral, Kristen's ashes are sent to be analyzed by three separate labs: a university in New York; a laboratory in Richmond, California; and a primary contractor for Rockwell International, the current operator of the plant. The California lab reports a high level of plutonium-239 in Kristen's ashes, the type of plutonium used in nuclear bombs and routinely released by Rocky Flats. The university in New York—curiously—never performs the analysis. The contracting laboratory for Rockwell International sends back inconclusive results that indicate the presence of plutonium, but at a level within the margin of error for their particular testing procedure.

Kristen's parents do not file a lawsuit. It would be almost impossible to prove a direct connection between Rocky Flats and Kristen's death, and the cost to sue the U.S. government is mind-boggling. Ultimately, Rex comes to believe that matters of life and death are in the hands of God, and that Rocky Flats has nothing to do with it.

2

DRUMS AND BUNNIES

1969

The rumors start with rabbits. One Rocky Flats worker and then another begins to notice long-eared bunnies, lots of them, out by the 903 Pad, an area of 260,000 square feet—larger than four football fields—where more than five thousand rusted oil barrels stand. Barrels stretch nearly as far as the eye can see and have been open to the elements for years. Since 1954, as a matter of fact. There's just no place to put them. Each thirty- to fifty-gallon drum holds waste oil and solvents contaminated with plutonium and uranium, for a total of hundreds of thousands of gallons of liquid waste laced with radioactive material. They can't be shipped, they can't be stored, and no on-site building can hold them all. Weeds poke through the badly corroded bottoms. Contaminants leak into the soil and the groundwater that feeds into Woman Creek and Walnut Creek, and then Standley Lake and Great Western Reservoir. Rocky Flats calls it the 903 Pad, but some workers privately call it the Launching Pad, where all sorts of things are launched into the environment.

Word gets out among the workers. Those bunnies are hot.

But management has known all along. A top-secret memo, with the heading CONTAMINATED RABBIT, dates back to January 1962. The rabbit, dissected for analysis, showed high concentrations of alpha radiation, particularly in the hind feet.

THE SUMMER of 1969 seems apocalyptic. Images of death and destruction in Vietnam are shown on the evening news. The Manson murders stun the country, and Ted Kennedy drives a car off a bridge near Chappaquiddick Island. Apollo 11 lands on the moon and more than 400,000 people show up for three days of peace and music at the Woodstock Festival. In August, several thousand Colorado residents travel to Washington, D.C., to participate in the largest antiwar demonstration to date. The Cold War is in full swing, and Rocky Flats is busy building the heart of every nuclear weapon in the United States. Occasionally demonstrators begin to appear at the west gate of Rocky Flats to protest nuclear war.

Except for my father's sporadic railings about how Nixon can save the country—and possibly the world—my family is largely unaware of anything that's happening outside our immediate cocoon of kin, pets, and general chaos. Vietnam is just imagery on television, and racial tensions are far, far away. The only conflict on the block is when my friend Danny gets teased for being Italian. "I'm a wop," he says. He's proud of it.

As soon as we settle into our new house, we receive more tokens of appreciation from my father's nonpaying clients. My dad says that if I'm to have a horse, every kid in the family should have one. None of them are much good for riding. Chappie is loose-limbed and harebrained, with the golden coat and white-blond mane of a movie starlet. He has a habit of walking into fences and the sides of buildings; we suspect he suffered at the hands of a previous owner. Comanche stands for hours in the same spot, dumbly waiting for dinner. Barney has the sly manner of a mutineer deceptively packaged in the body of a sleek Shetland pony. His favorite trick is to stretch out his neck, grab the bit between his teeth, and tear madly at top speed across the field. In one single brilliant moment, he stops dead in his tracks, plants his front feet, drops his head, and sends the rider flying like a cannonball over his ears. We all take our turns.

Barney has another useful trick. He uses his teeth to slide back the metal lock on the gate, but is polite enough to leave it open for the other horses. He goes to forage and snack on the local lawns and gardens and sometimes on the golf course. "Kris!" my mother hollers up the stairs on early mornings before school. "Go round up the horses!" We've scarcely moved in and the neighbors hate us. Barney seems especially fond of the garden and field of a family on the east end of Standley Lake, not far from our house. The Smiths live off the land. They have horses and cows and grow their own vegetables. Mr. Smith sometimes scolds whoever is unlucky enough to rescue Barney from his garden.

We endure a series of beloved dogs that are almost more trouble than they're worth. Georgy Girl is first, a tall, skittish, red-haired Irish setter named after the character Lynn Redgrave plays in the movie, and the song my mother sings when she folds laundry. My father brings Georgy Girl home from the office with no warning. She's sweet but spectacularly flings herself around the house, unable to sit still for even a minute. "Something happened to that dog," my mother muses, and this seems true of all our dogs—something's happened to them. We never know what exactly. Georgy Girl eventually flings herself out the front door and we never see her again. Thor, my favorite, is a long-lashed Siberian husky with arching brows and inquisitive blue eyes that never admit to the evenings he snacks on the neighbor's chickens. He too disappears, and we suspect neighborly foul play.

Shakespeare is next. My mother hires a young woman to come in one afternoon a week to help her clean house—the mud and dirt tracked in on our shoes is constant—and she has a friend who has a friend who has a dog who needs a home. The following week the woman brings Shakespeare. "He took up the whole backseat of my car!" she exclaims. Shakespeare looks like the sheepdog on the television series *Please Don't Eat the Daisies*. His hair hangs low in his eyes and he has trouble seeing things. Something happened to him, too—he shakes uncontrollably when anyone raises their voice, and it's not long before my mother shortens his name to Shakey. He's sweet and slobbery and we can't bear to discipline him. He takes full run of the house. One afternoon he catches a glimpse of a rabbit in the backyard and bolts through the sliding glass door downstairs, leaving behind a path of shattered glass. He's unfazed, but that's the end of him. My mother calls around until she finds him a new home. "He's a little nervous," she explains. All our animals are a little nervous.

We are, too, I guess. "Your father," my mother likes to say, "is going down the tubes." I like the way she cocks her head and raises her eyebrow as she says this—she makes me laugh even though I have a sinking feeling in my stomach. We hear Dad late at night, stomping around in a series of thuds and crashes as he makes his way down the hall. We watch from our bedroom windows as his car weaves up the long driveway, sometimes barely missing the trees my mother has planted. My siblings and I whisper and worry about what might happen when he's behind the wheel of his car, day or night. Could he hit an animal? A person? Sometimes my mother lets him sleep in the bedroom, but usually he sleeps downstairs in his recliner with his secret bottle.

My mother tells us how she fell in love with my father on a blind date. She likes to tell the story. She wasn't thrilled about the idea of a blind date—or the notion that she needed any help in the dating department—but she took it on a dare. Her best friend, a girl who had been her roommate in nursing school, set them up. After graduation both girls took jobs in public health and worked with the Chippewa tribe at Cass Lake in Minnesota. All week the nurses looked forward to the college boys who came up to visit on weekends. My father was a law student at Drake University in Des Moines, a tall, lanky man with a sweet, open face, who would graduate at the top of his class. She knew at first sight that this was the man she would marry. Then the Air Force sent him to Germany during the Korean War for two years.

Tensions were escalating between the Communist world—primarily the Soviet Union—and the West. In 1951, during the Korean War, the United States conducted its first domestic atomic bomb test since 1945, at the new Nevada Test Site near Las Vegas. Around the same time, a

Los Angeles construction company built the nation's first underground family fallout shelter. Everyone was talking about moving to the suburbs. Some of those new suburban homes would have built-in atomic bomb shelters, including a show home in Allendale Heights, a new housing subdivision in Arvada not far from Rocky Flats. It was one of fifteen "Titan" showcase homes—homes named after the newly developed intercontinental Titan ballistic missile—scattered around the country near other nuclear weapon production sites.

My parents' only date was the one they had before my father left for Germany. They married immediately upon his return.

What was that date like? Photos of my mother at the time show her in her nurse's cap with careful curls and a wide smile. She was proud of her elegant gowns and numerous boyfriends, but she was a good Lutheran girl. "You don't have to go all the way," she confides to me, "to have a little fun."

But this boy was different. His face was sensitive and intelligent and even then he had a slightly wounded look. After my mother's death years later, I look for the letters they exchanged during that two-year period. She saved everything—postcards, photographs, greeting cards, ancient Tupperware containers—but there is no sign of those letters.

They wanted a secret wedding. But an attempt to dodge their parents was unsuccessful; both sets were waiting in cars on the side of the road with engines running. Photographs show a long table of faces with frozen smiles and a towering wedding cake. My mother leans over the cake, regal in a white cocktail dress and hat, an early version of Jackie Kennedy. She balances a long cake knife in her hand. My father has his fingers on her elbow and looks pinched by his tie. Their faces are slightly panicked. My mother hated convention and my father was never comfortable around family.

The families considered it a mixed marriage: my mother was Norwegian, my father Danish. My grandfather said it seemed a shame my mother couldn't find a good Norwegian man. She'd have to make the best of it.

As soon as my father graduated from law school, they packed up the car with a few belongings and a new baby—me—and drove to Colorado. Colorado meant a fresh start: no parents, no farms, no heavy Lutheran traditions. "If I had a wagon I would go to Colorado," my father sang. "Go to Colorado, go to Colorado! If I had a wagon I would go to the state where a man can walk a mile high." His fingers tapped a beat on the steering wheel. People moved to Colorado to start over. Life there was a gamble, but a good one.

The Colorado landscape was brown and dry compared to my mother's beloved Minnesota, but she tried to adjust. Our long drives in the mountains are an attempt to placate her homesickness for trees. She loves Scandinavian cooking and art and all things Nordic, and as we grow, she worries that we might lose our sense of heritage in the homogeneity of American life, in the steady television stream of *Gilligan's Island* and *The Beverly Hillbillies*. She tells us stories about noble Vikings and ugly trolls. She buys lefse, thin potato pancakes that look like tortillas but taste like paste, and fills them with butter and sugar. "You have to love them," she says. "It's in your blood." She adores Henrik Ibsen. She plays Edvard Grieg on the stereo and shows us Edvard Munch's painting *The Scream*. The image haunts our dreams. "That's the Norwegian character," she says. "We keep everything inside." But Norwegians are also warm and generous and welcoming. "Norwegians get along with everyone," she says, "except sometimes the Swedes."

My father, whose Danish heritage requires no small degree of humility in the presence of my mother, rolls his eyes.

NOT ALL of Denver fails to notice the black plume of smoke from Rocky Flats that Mother's Day. By August, larger groups of people begin to stand outside the west gate of Rocky Flats, holding peace signs and waving at passing drivers and the workers coming in to start their shifts. Some protesters have been regularly holding vigils at the gate to commemorate the Hiroshima and Nagasaki bombings of August 1945. They plant small crosses in the ground next to the road bordering Rocky Flats to honor those who died. Others are there to protest the ongoing Vietnam War. But now the group is bigger, louder, more vehement. "Liar,

liar, plant's on fire!" someone begins to chant. It catches on. "Liar, liar, plant's on fire!" the group shouts.

On the west side of town, not far from Rocky Flats, Sister Pat McCormick is about to get arrested for the first time. It won't be the last. She's just returned to Denver after six years in Bolivia and Peru. Those six years changed her life.

One of eleven children, Pat grew up on a farm northwest of Chicago. Her mother was a self-taught musician and singer in the church choir. In high school Pat learned of the Sisters of Loretto, a Catholic religious community of about 1,200 women around the country committed to improving conditions for people who experienced oppression and injustice. In 1953 Pat joined the community and graduated from Webster University in St. Louis. She was working as a teacher in Fort Collins, Colorado, when she was asked if she wanted to go to Latin America.

Pat said yes.

It felt like the world was shifting in important ways. The church was changing in the wake of Vatican II. The life of a nun was no longer as cloistered as it had once been; the sisters were expected to engage in the broader world. They decided to wear regular street clothes. They took jobs in the real world. They became more vocal about how social and government policy affected the lives of the people they were trying to help: the poor, the sick, people marginalized by society.

Pat was living in Bolivia when Robert Kennedy and Martin Luther King Jr. were assassinated. She met Daniel Berrigan, a Jesuit priest who was involved in organizing the peace movement and protesting the Vietnam War. He encouraged her to return to the States to get involved with the peace movement. Pat decided to move back to Denver. Her first action was with the United Farm Workers and the grape strike against growers in California, organized by Cesar Chavez, to protest for fair wages for primarily Mexican-American and Filipino farm workers.

When she's arrested for the first time, Pat McCormick is sitting on the picket line in front of a liquor store in Applewood, near Arvada, protesting the sale of Gallo wine. "Don't buy California wine," she calls to customers. "Don't buy California grapes!"

Across town, in a well-to-do enclave on the east side of Denver called Cherry Creek, Ann White, a young mother of three, is talking on the telephone in her living room.

Her cousin is coming for a visit. That's fine. She doesn't get to see him as often as she would like. It's not just a social visit, though. He's a Quaker and a photographer for the American Friends Service Committee, another organization that works for peace and social justice. He wants to stay with her while he's on assignment—an assignment to photograph a planned antiwar protest at Rocky Flats. "And there's been a fire," he says. "Did you see it in the paper? Some people think a lot of plutonium escaped from the plant."

She can't recall seeing anything in the papers.

It's not that she's naïve about these things. Ann grew up in Santa Fe, New Mexico, and Los Alamos was the biggest secret in town. Ann's father knew some of the physicists and he liked to have them down to the house to talk, although it was illegal for them to leave the closed city. She remembers the driver waiting quietly, ready to take the men back up the mountain before daylight.

Los Alamos is the brains of the U.S. nuclear weapons program. Rocky Flats, she suspects, is the muscle, although she doesn't know for sure. She just knows it's a secret government facility.

"What's going on?" she asks her cousin.

"I'm not sure," he replies. "A lot of groups are starting to take interest. Physicians for Social Responsibility, the Sisters of Loretto, and a group called Citizens Concerned About Radiation Pollution. An antiwar group is going to hike from the Boulder County courthouse to Rocky Flats to begin a five-day protest."

"Yes, come and stay here," Ann says. "I want you to tell me everything you find out."

MY FATHER reads about the protesters in the newspaper. "What do you think of this?" he asks my mother. She laughs and shakes her head. "Hippies and housewives," my father huffs. "Those people should get real jobs."

I rarely read the paper. Early mornings are for tending to our growing brood of critters. The sun is raw and bright when I go out to feed the horses before school. They line up eagerly at the fence, heads up, ears tipped forward, nickering in anticipation. The air is clean and fresh and I can smell everything at once: the sweet hay, the deep musk of the swamp mud, the tall grass, the oddly comforting scent of manure. I pull two flakes of hay from the stack for each horse and pour a cup of oats in each bucket. As they munch, jaws grinding rhythmically, sometimes I try to catch a rabbit. There are rabbits everywhere, hiding in the hay bales, hopping behind the porch, padding quietly around our half-attempt at a garden with their noses twitching. I never come close to catching one, and soon it dawns on me that if I can't catch one, I could buy a tame one for a pet.

My first rabbit is black and white, a commoner, thick and dull and not much interested in socializing. For three dollars I buy a hutch that's been hammered together by a neighborhood boy, two-by-fours and wire mesh, and the rabbit comes with the hutch. Scarcely two weeks pass before the hutch is ravaged and the rabbit gone. I'm pretty sure the neighbor's German shepherd is the culprit, but there's no hard proof. I fortify the hutch and buy two dwarf bunnies, rabbits so tiny they fit in my palm. I adore them but they chew their way to freedom, leaving an escape hole in the side of the cage not much bigger than a half-dollar. I do a little research and seize upon the idea of owning and breeding Siamese Satin rabbits. For ten dollars I buy a pair of rabbits with long fur so shiny it's almost translucent. Like Siamese cats, their coats are a rich gray and their feet and ears look as if they've been dipped in black ink. They put our real cats to shame. And they're crazy about each other. My ten dollars will be the beginning of a dynasty. A local stock show has a rabbit competition. My bunnies are so beautiful, I think, they're sure to win, and with a couple of ribbons, people will be lining up to buy them for pets. I enter the contest and my pair of Siamese Satins takes a second-place ribbon.

When I go to pick them up after the show, the showroom is filled with empty mesh cages. "Where are my rabbits?" I ask the man who is briskly snapping the entry cards off each cage.

"You can pick up your check in the office," he says.

"What check?"

"The check for your rabbits."

"I don't want a check," I say, confused. "I'm here to pick up my rabbits."

He looks at me as if he suddenly sees me for the first time. "Your rabbits are already sold, sweetheart. Packed and frozen. Go talk to the front desk."

Numbly I walk over to where a woman with hair teased up into a fluff the color of orange soda flips through a file and hands me a check. I hold it for a moment and hand it back.

Later my mother explains. "This is the way it is with farm animals," she says. "When I was a girl on the farm, we raised calves and lambs and pigs, and I knew I could never love them very long."

That's the end of my rabbit business. I go back to chasing them around the haystack. I can't bear the thought of one of my lovely Satins ending up on someone's dinner plate.

In our freshly built neighborhood, everyone is new. Each kid has a shot at establishing some social territory, but our parents are all keenly conscious of status. Sons and daughters of the Depression, most of them come from farm or working-class backgrounds. They're eager to show they've arrived, even if it's by the skin of their teeth. Each house in Bridledale has a new boat or snowmobile. Each kid has a pony or dirt bike.

Many of the neighbor kids have fathers or uncles or older brothers who work at Rocky Flats or Coors, the beer factory in Golden. The neighborhood kids regard Coors with equal doses of derision and respect. If you have to work, you might as well work at a beer factory, where everyone is bound to have a pretty good time. On the other hand, it's generally agreed that Coors isn't a real beer, like some of the other brands our parents drink. None of us plan to drink "Colorado Kool-Aid" when we grow up. Budweiser is a real beer, for example. That's what all the older boys drink at the pool hall in downtown Arvada, a strip of shops that holds little more than a hometown bank, an antique store, a dusty bridal shop, and a psychic who keeps irregular hours. Once you turn eighteen you can take a tour of the Coors factory and get two free beers at the end, which is acceptable since it's free. Some of our teachers spend their summers working at Coors as tour guides.

None of the kids seems to know what their fathers actually do at Rocky Flats, though. There's a mystique surrounding the fathers who work at the plutonium factory, if that's what it is—who knows for sure?—and it raises the social standing of their children in the neighborhood considerably. Even secretaries make boatloads of money at Rocky Flats; everyone knows that.

We have big plans for our new house. By early spring our yard is a sea of mud and my mother plants grass and trees and flowers that refuse to take hold. Unlike her beloved Minnesota, nothing grows in Colorado. The neighbors talk about the importance of self-sufficiency and stock their basement bomb shelters with a year's supply of canned goods and a radio. The show home with a built-in bomb shelter in nearby Allendale Heights opened with great fanfare—the governor, two mayors, and two thousand curious residents showed up to see it. Many of our neighbors followed suit and built their own.

This all intrigues me—we have a tornado shelter at my mother's family farm in Iowa, which is terribly exciting—but the only concession my father will make to domestic self-reliance is to try once again to dig a well in the backyard. A man comes out with a rig to dig the well and nothing comes up, no matter how deep he goes. "I don't understand it," my dad says, irritated. "We're right on the Standley Lake water table." We stick to city water that tends to run orange after the pipes freeze and thaw, which happens often.

More and more, our father stays away, and when he's home, he talks to no one. One day Karin and I find a square glass bottle tucked behind the couch, an inch of bourbon in the bottom. We don't dare touch it. A few days later we discover another behind the recliner, half full. Then one in the living room. They seem to be everywhere. After much discussion, my sisters and I wait until our mother is out of the house and then pour the contents down the kitchen drain.

No one seems to notice. The bottle is magically replaced by another, half full, a day or two later. Karin, who's never afraid to take charge of a situation, suggests adding water to the bottles to dilute them. "Sounds good," I say. Karma nods. Karin fearlessly totes the bottles up to the kitchen sink and fills them to the top.

Days pass. No one notices. Nothing is said. Fresh bottles appear in their place. My parents, who rarely see or speak to each other, carry on as usual.

Karin grows bolder. She's the rebel, the one with a temper and the one who can say things the rest of us don't have the guts to say. She wears glittery blue eye shadow and curls her hair like the older girls at school, and her laugh is always the loudest in the room. In her mind there's no time for doubt. She pours out the bottles and puts them in the trash.

It's not long before our mother catches her red-handed. "What are you doing?" she shrieks. "Don't do that. Your father will be furious."

Karin flounces off to her room. I stop inviting friends over to the house. It's too embarrassing to explain the bottle under the sofa or behind the chair—the bottles we're not supposed to know about.

A FEW miles down the road, a nuclear chemist named Ed Martell can't stop thinking about the Mother's Day fire. He's bothered by the bits and pieces of information he's read in the newspapers, and rumors abound.

Martell knows something about radiation. A West Point graduate with a Ph.D. in radiochemistry from the University of Chicago, Martell is a former program director of the Armed Forces Special Weapons Project, when he studied the effects of radiation from U.S. nuclear bomb tests in the 1950s. He saw what happened to life—human and otherwise—in the South Pacific after the detonation of a nuclear bomb. He decided to retire as a lieutenant colonel and go to work at the National Center for Atmospheric Research in Boulder, where he also heads the study group for the Colorado Committee for Environmental Information (CCEI), a nonprofit group of twenty-five people, mostly scientists, that deals with the impact of technology on the environment. The group works independently of the Atomic Energy Commission (AEC).

The CCEI has been in the news lately for their criticism of the Rocky Mountain Arsenal, a former World War II chemical weapons facility on the northeast side of Denver that continues to hold dangerous chemical agents, including mustard gas and nerve gas, for long-term storage. The scientists also opposed a plan by the AEC to conduct a 43-kiloton nuclear test project in Rulison, Colorado, to determine if natural gas could be easily extracted from deep underground levels. Despite the opposition, the AEC went ahead with the test on September 10, 1969. It was ultimately unsuccessful. The natural gas that was extracted proved too radioactive to be sold on the commercial market.

The Rocky Flats Mother's Day fire—particularly compared to the Rocky Mountain Arsenal and the Rulison project—is supposedly no big deal. The AEC reports, "There is no evidence that plutonium was carried beyond the plant boundaries." But Martell is suspicious. He personally saw the smoke billowing from Rocky Flats. Has there been off-site contamination? If so, how much? He contacts other scientists in the CCEI and the group decides to approach Dow Chemical and ask for more details.

The response is swift: Dow denies the request.

Martell calls up an acquaintance from his military days, Major General E. B. Giller of the Air Force, who now helps oversee the nation's nuclear weapons production complex, which includes Rocky Flats. Giller's comments are reassuring: the fire posed no danger to the public. Dow will attempt to answer questions presented by Martell's group of scientists, but of course some information will remain classified, as they can't risk a breach of national security. And, he adds, there will be no off-site testing.

BRIDLEDALE GROWS, and there's a friendly rivalry with the adjacent subdivision, Meadowgate Farms, where houses are springing up just as fast. Summer days are sunny and long and slow. The four of us kids sit up on kitchen stools for cereal each morning—my sisters and I fuss over who gets the latest Bobby Sherman cardboard record, "Bubble Gum and Braces," glued to the back of the Alpha-Bits box—and then our mother boots us outside for the day, with instructions that we're not to come back until suppertime. I have a Bobby Sherman poster secretly taped inside my bedroom closet. Publicly I've sworn to my sisters to having no interest in boys. My life will be devoted to horses. Karma fervently agrees.

A long, fenced bridle path encircles both subdivisions and kids can gallop their ponies and horses around and around. No bikes, no dirt bikes, no motorcycles. A new boy moves into Meadowgate, a boy my age with dark hair and brown eyes, named Randy Sullivan. He's tall, with a quick smile, and he rides a palomino mare. At our kitchen window I hide behind the curtain and watch him gallop past on the bridle path, which goes right behind our house.

Our house is filled with endless places to hide: behind the back patio, in the walk-in closet in the master bedroom, under the basement stairs. Hide-and-seek takes on a new dimension. But the house can be terrifying as well. On top of a hill with no trees to protect it, it's a perfect target for the winds that sweep across Rocky Flats, which hit with the force of an eighteen-wheeler. The windows rattle and buzz and we have to shout to hear one another in the bearskin room. When the chinooks hit, inexplicably hot and fierce, everyone feels on edge. My parents complain that the windows aren't sturdy enough to withstand the weather. The developer, Rex Haag, stands firm: the windows are up to code.

My mother has no qualms about setting aside her nursing degree to look after her own brood, which, she is happy to tell anyone, is more stressful than any job. She finds solace in paperback novels with lurid covers that she hides under her bed and beneath the sofa. "These books are strictly off-limits," she says. "Don't let me catch any of you girls reading these books." Jacqueline Susann, Harold Robbins, Victoria Holt, Mario Puzo—her admonition is better than a library card. I read them all and slip them back into place without bending a page or breaking a spine.

Very few women in the neighborhood have outside jobs. Once a week they meet for coffee, first at one house, then another, ten o'clock sharp after the beds are made and breakfast dishes done. Sometimes I get to sit in when the wives come to our house. My mother pours me milk with a drop of coffee. When I'm grown up, she explains, I'll drink my coffee black like she and my father do, like all good Scandinavians. "I don't miss having a job," she says. "It's impossible with kids. Who would take care of them?" Day-care centers haven't been invented yet and the only baby-sitters around are the fourteen-year-old daughters of her friends, good for one evening a week at best. "They just sit on the phone and empty the refrigerator anyway," she says. She often tells the story of how, when she was pregnant with me and again with my sisters, she wore a tight girdle under her nursing uniform to hide her expanding figure. "You get fired if they know you're pregnant," she says. "I hid it right up to the end."

It's generally agreed that for women there are very few jobs worth having. Although, truth be told, the wives agree, a husband can be a lot of trouble. The only divorced woman in the neighborhood, a good friend of my mother's, comes over by herself on a different day. Everyone whispers about her.

Some of the husbands work at the plant out at Rocky Flats. "I don't know what he does, exactly," one wife says. "He's an engineer. It's too complicated to explain."

"There's nothing to explain," another wife snorts. "It's Dow Chemical. They make bathroom products. What's the glory in that?" The women sit at my mother's kitchen table, smoking and sipping coffee, until someone says she better get busy with her laundry before the kids get home or her grocery list is long enough for the Russian army, and the group disbands until the following week.

But my mother has social aspirations beyond the local coffee klatch. My father's reputation grows, and she's invited to join the Denver Lawyers' Wives Club. This is a great honor and she takes it very seriously, putting on a skirt and heels once a month to attend their afternoon teas. One day she takes me shopping for a new dress so I can accompany her to the Daughters' Tea. We go to the mall and choose a maxiskirt—all the rage—and a matching vest and blouse. The ensemble is stifling hot. At the party I take turns with the other girls pouring steaming tea from a heavy silver pot, and we listen silently to chatter about home decorating and children's soccer games.

Soon it's my mother's turn to host a tea at our house. She frets for weeks—what to serve? what to wear? how to make the house present-able?—until it finally becomes evident to her that there is no way to hide the dust and chaos of our household and the constant parade of half-hidden bottles of bourbon. She quietly drops out of the group and never mentions it again. They were all just a group of silly women, anyway.

NEIGHBORHOOD POLITICS are rough. The girls act like I don't exist and the boys tease me for wearing cowboy boots when go-go boots are all the rage. One girl, Tina, decides to give me a break. She lives up the street and rides a quarter horse mare that can race a motorcycle and win, if the race is short enough. She has a mother who lets her have boy-girl parties and a father who's never home.

Tina's a girl of strong opinions. Cowboy boots—shitkickers, she calls them—are for after school only. She wears short skirts and go-go beads and streaks her hair with spray bleach she buys at the five-and-dime. All the neighbor boys think she's a fox. Her friendship is contingent upon the relentless completion of small rites of passage like ditching biology class, kissing a boy with your tongue, and jumping the pipe.

Every kid in the neighborhood knows about the pipe. Standley Lake is fed on the west side by Woman Creek, the waterway that runs from Rocky Flats. On the east side are several small, open canals that flow from the lake, canals filled with frogs and water skippers and darting glints of minnows. A long corrugated pipe, about four feet in diameter, extends from a high bank and spouts water to a deep pool nearly thirty feet below. Shallow and muddy, the canal meanders at a snail's pace toward the lake, but a small round pool directly beneath the pipe is deep enough to accommodate a cannonball dive. Aim is everything. Speed helps. A quick sprint to the end of the pipe followed by a forceful, frog-like leap works best, although few kids have the guts for that approach. Some kids crawl inside the pipe and jump from its dark cavern. Even if you manage to hit the deep pool and avoid breaking your neck, it's still a mighty task to battle the waist-high mud that makes clambering back to the bank like fighting quicksand.

As a heavy metal, plutonium settles in mud and sediment.

I'm a spectator for weeks. Every afternoon after school a small jury seats itself on the grassy bank. One by one, a kid gingerly inches out to the edge of the pipe, looks down, takes a deep breath, and jumps. Or not. Few do it twice. To look down into the muddy swirl and contemplate a retreat under the vigilant gaze of your schoolmates means not only assuming the role of social outcast but establishing a reputation that will follow you all the way to high school.

Randy Sullivan jumps. Tina does, too. She makes it look easy. She hardly even looks—just eases out to the end of the pipe, closes her eyes, and leaps out into the air as if she could fly. She's dressed herself with an audience in mind: a low-cut, sea-green swimsuit and tattered jean shorts. "Shit!" she breathes, pulling herself up onto the slick bank and flipping back long strands of wet hair. Her legs are covered with mud and tangled weeds. "Your turn," she says.

"Okay." My legs feel like lead. I let two boys go ahead of me—one sprints, to much applause—and I inch out, my legs wrapped around the pipe's rusty ridges, my heart thudding. I can't even dive off the board at the deep end of a swimming pool without serious soul-searching. But failure isn't an option. I'm the first in my family to try to meet the challenge. Our reputation is at stake. And the esteem of my only friend.

I reach the sharp edge of the pipe. I peer over the rusted rim and see the dark blue circle where the water is deep. It looks to be the size of a large dinner plate. I glance over at Tina on the bank, her hair slicked back in triumph.

"Go!" she says.

I can't do it.

"Come on," Tina says. She sounds peeved. A couple of kids laugh. "Holy cow."

My face flushes. My hammering heart seems to have slid into my stomach. After what seems like a decade or two, I inch back off the pipe and climb the bank.

Tina is gone.

☢

THERE IS no further word from Dow Chemical. Ed Martell and the other CCEI scientists, worried that the infamous winds at Rocky Flats might have carried lethal particles of plutonium toward an unsuspecting population, initiate an independent investigation. They take soil samples from two to four miles east of the plant and test for plutonium-239—weapons-grade plutonium—and strontium-90. To account for fallout from atmospheric nuclear weapons testing, they take soil samples from other sites along the Front Range and estimate the background concentration of radionuclides in surface soil. Comparing these samples will allow them to determine if there is an excessive amount of plutonium in the soil—plutonium specifically from Rocky Flats.

Other people are worried, too, and some citizens begin to organize local meetings. Lloyd Mixon, a farmer near Rocky Flats, talks about deformed pigs and how his hens lay eggs that won't hatch. The chicks have beaks so curled and deformed they can't peck their way out of the shells. He wants to know what's happening at Rocky Flats.

Bini Abbott, a local horsewoman, also begins to worry about how Rocky Flats might be affecting her livestock. For nearly ten years she and her husband have owned a large ranch a mile and a half southeast and downwind of the plant. She often shows her horses at local shows and gymkhanas, where she sees some of the kids from Bridledale and Meadowgate. Karma and I envy her horses, her horse trailers, her professional skill. We want to be like Bini when we grow up.

Bini buys several horses each year off the local racetrack and trains them for jumping. Sometimes she buys mares for breeding. She always gets a few crooked foals—it's part of the business—but ever since she moved near Rocky Flats it seems worse than what might be expected. She has high hopes for one mare in particular, a descendant of a Kentucky Derby winner. The mare has two foals before she has to be put down herself due to health problems. The first foal lives for a week. Bini is determined that the second foal, born two years later, will live. But this foal, a little colt, also has problems. For an entire week she camps out in his stall around the clock. March is cold in Colorado—still winter, really—and Bini's husband brings blankets and hot meals out to the barn.

The colt dies.

After hearing about Ed Martell's soil sampling, Bini begins to keep organs of the deformed animals in her freezer. They have misplaced bladders or hearts, sometimes other problems. Bini is a practical and down-to-earth person. She doesn't scare easily. But someday, she thinks, those organs will be tested.

In February 1970, the CCEI completes the report on its investigation. Nearly all the soil samples that were taken show plutonium contamination that originated at Rocky Flats. Plutonium deposits in the top centimeter of soil taken from locations east of the plant are up to four hundred times the average background concentrations from global fallout. In some places, the level is 1,500 times higher than normal. Dr. Niels Schonbeck, a biochemist at the University of Colorado, later notes, "That is the highest ever measured near an urban area, including the city of Nagasaki." Deposits are heaviest in the top surface of the soil, Martell reports, but are present in all levels down to five inches, indicating a long series of leaks from the plant. High readings of plutonium are also found in water along Walnut Creek, which feeds into Great Western Reservoir, the water supply for the city of Broomfield.

Before making the results public, Martell and his associates call a meeting with the Colorado Department of Health and officials from Rocky Flats to report their findings. To their surprise, officials don't dispute the results. Instead, they admit that the plutonium found by Martell—no news to them, it turns out—didn't come from the 1969 fire. There were two previous contamination sources: the 1957 fire, twelve years earlier, and windblown particles of plutonium that have leaked into the soil from the drum storage area. The AEC states that it has never attempted to conceal the fact that "minuscule amounts of plutonium" have been released through ventilation systems during normal operations of the plant. They also concede that there has been inadequate testing of the soil in the areas surrounding Rocky Flats.

The public response to Martell's study, though, spurs them to action. The AEC sends two of its own scientists, P. W. Krey and E. P. Hardy, to take soil samples from the Rocky Flats plant site and its surrounding areas. In August 1970, Krey and Hardy produce a map that shows plutonium from Rocky Flats both on and off the site, covering an area of more than thirty square miles east and southeast of the plant. Nonetheless, the AEC insists local residents are safe. They say that the amount of plutonium released by the plant is "far below the permissible levels" and does not pose a public health hazard. Major General Giller confirms that the commission is in "reasonable agreement" with the CCEI report, but concurs with AEC officials that the amount of plutonium released from the plant does not pose a public health hazard. For plutonium to be truly dangerous, one official notes, people would have to literally "eat the dirt—and large amounts of dirt."

But plutonium is dangerous if ingested *or* inhaled: plutonium particles can lodge in lung tissue and remain active for years or even decades, emitting alpha radiation. In an interview with the *New York Times*, Martell estimates that 200,000 to 300,000 people live immediately downwind from Rocky Flats. He is most concerned about the suburbs of Arvada, Westminster, and Broomfield, but this off-site contamination—which has been found as far away as forty miles from the plant—is not the only problem. A potential nuclear disaster could devastate Denver and possibly all of Colorado. The CCEI report states that "in the not-too-unlikely event of a major plutonium release, the resulting large-scale plutonium contamination could require large-scale evacuation of the affected areas, the leveling of buildings and homes, the deep plowing or removal of topsoil and an unpredictable number of radiation casualties among the people exposed to the initial cloud or the more seriously contaminated areas." There is no emergency response plan to protect the public in the event of a major disaster at Rocky Flats. Rocky Flats is still the biggest secret in town.

MOST PEOPLE in my neighborhood are too busy making mortgage payments and worrying about rising gas prices to pay much attention to Rocky Flats. And they don't want to think about how the situation might affect their property values.

Sister Pam Solo, though, is paying attention. A third-generation

Colorado native, she has deep roots in the state's complicated history. Her father grew up in a coal mining camp in southern Colorado, site of the famous Ludlow Massacre that occurred when coal miners went on strike in 1913. Pam attended St. Mary's Academy in Denver, and after high school, like Pat McCormick, she joined the Sisters of Loretto.

Another young woman, Judy Danielson, a Quaker, has just returned from Vietnam, where she's been doing humanitarian work as a physical therapist. Pam is a serious-minded woman with short dark hair and glasses; Judy, slender with light hair, shares her intensity. They learn of Martell's findings and decide to help organize a group to go door-to-door in neighborhoods east of Rocky Flats. They knock on doors and ask residents if they can scoop up samples of dirt from their backyards to be tested for radiation. The volunteers label the samples with names and addresses and take them to the open public meetings of candidates who are running for Congress, asking to have the soil tested and residents notified of the results. "What," they demand of each candidate, "are you going to do about Rocky Flats?"

WINTER COMES early. In late October the pipes in our laundry room freeze and burst and water spills out into the room in arctic pools. Tonka shivers in the field, his head low and back hunched to the wind. My mother backs the station wagon out of the garage and Karma and I spread straw on the cement floor and bring him in. His chin whiskers are long tentacles of frost and the balls of ice in his hooves make him walk gingerly, as if he's wearing stiletto heels. We cook him a hot oatmeal mash on the stove and serve it on a breakfast tray.

We aren't the only ones concerned about the cold weather. We share our land with a large population of field mice who take up residence in the walls, cupboards, and heat ducts. I don't mind the glimpse of a nimble creature dashing across the kitchen floor in search of a cornflake or two, but my mother is determined to rout them out. She puts small boxes of poison in front of the heat vents. Rather than expire in the open air, the mice climb back up into the warm ducts to take their last breath. We grow accustomed to the smell.

We celebrate Christmas in our new house with a seven-foot tree in the family room. As part of our continuing education in all things Scandinavian, my mother plays Hans Christian Andersen stories on the record player. She directs that each strand of tinsel be hung individually, one by one, on the branches. "You can't let them touch each other!" she orders, and sure enough the tree glitters with flowing streams of silver, at least until the cats start climbing the trunk.

My mother spends weeks deciding on presents for each of us, wrapping them in gold and red foil, tying them with fancy ribbon, and hiding them on the shelf in her closet where we track their location closely. My father mutters about taxes and bills and clients who don't pay, but he never disapproves.

Christmas Eve means church first. My parents reject the staunch brand of Lutheranism my mother grew up with—even cardplaying is a sin—and we sing hymns in a church that looks more like a library than a sanctuary. My mother wears a mink coat and stole my dad accepted from an indebted client: two paws and a stunted nose and tail hang down around her shoulders. On the way home we sing carols and count all the houses with Christmas lights and then have to wait for dinner to be served, lefse and sandbakkels and the threat of lutefisk—dried codfish reconstituted in lye and boiled in saltwater, which tastes like bland fish Jell-O. Fortunately she relents and instead serves a turkey from Jackson's, the local turkey farm out by Rocky Flats, and then we can open the presents we've been eyeing for weeks as the Christmas tree sparkles late into the night. On Christmas morning there are stockings stuffed with rolls of Life Savers, fat chocolate Santas, bookmarks, coins, and color-changing mood rings. We spend the day in pajamas, sitting amid piles of wrapping and tinsel and tape, the record player blaring, dogs bounding around the room in chaotic ecstasy. There is something almost nightmarish in the boxes and paper and ribbon, the plethora of presents we know my father can't afford, and yet we feel loved and spoiled and giddy in our parents' insistence on this heady life of abundance.

Nonetheless, by noon my father reaches his limit of family interaction and heads off in a stony silence to check on things at the office.

☢

THE FIRST time I ride Tonka out to Standley Lake, the wind whips my hair across my face so hard it stings. Tonka is eager to run. I ride bareback with a single leather strap looped around his ears and a rawhide hackamore dropped across his nose, the reins taut, his head tucked and neck arched like a Roman Percheron. He prances and dances—let's run! Let's run! He can gather himself into a ball of muscled energy and shoot across the field like a low-rolling cannonball. I've learned to grip his bare sides with my thighs, crouch low over his neck, and hang on. Maximum contact, minimum control.

I'm alone. That's the best part, to be alone with the horse and the gently rolling hills and the wind bending the tall prairie grass into long ripples of gold. I try to make Tonka walk calmly; my mother has repeated tales she's heard from neighbors about what happens to young riders whose galloping mounts step full speed into groundhog holes. A horse's leg can snap as easily as a slender tree branch, and there's no remedy but a bullet to the head. Like a minefield, the long grass hides hundreds and maybe thousands of potentially lethal mounds and bumps—how many death traps are beneath those dancing hooves? But Tonka dislikes caution. He knows there will come a time on each ride when we will be past the houses, the fences, the roads, and I'll drop the reins, bury my face in his mane, and let him rip.

We sidestep through the metal gate and prance across the wooden bridge arching over the ditch. I try to maintain the illusion of control as long as I'm within range of the neighbors' kitchen windows. We pass the community barn, skitter through another gate, trot past the long swamp—Tonka breaking into a light anticipatory sweat—and canter up a gentle rise to the barbed-wire fence surrounding the lake.

There is a gate, loosely constructed of metal posts and wire. A heavy padlock hangs from the latch. A thoughtful child has neatly clipped the wires below the lock. I slide off, lead Tonka through, and swing back up. He can hardly contain himself.

My vantage point is extraordinary. The lake stretches below us, nearly a mile in diameter. Blue water extends in rows of gentle ripples to a thin line of barely visible cottonwoods on the far side. The wind dies to a whisper and it's quiet, almost perfectly still except for the snap of grasshoppers leaping from the weeds. To the west the mountains rise suddenly, almost violently from the sandy brown of the plains, layered silhouettes of blue and green and gray rising to a turquoise sky. My heart is filled with the beauty of it all.

Tonka will wait no longer. I pull in his head, tuck his nose to his chest, and twist my hands in his mane. "Go!" I shout, and when the reins drop he shoots over the peak of the hill and down the other side, racing to the edge of the lake. His back is slick with sweat, and I barely keep my hold. There is mud, I can see it—should I pull him up? Will he race right into the water? The ground blurs beneath his hooves.

I see the body first. In the split second before Tonka spots it, I ready myself for his response: the sliding stop, the snort of astonishment, and the surge of fear. He knew I had seen it first. He spins around on his back haunches and I pull him up short.

The lower half of the cow's body lies in the water, soggy and swollen. The upper half extends long and rigid across the ground. Her head stretches up achingly, as if she had tried to pull herself out. The eyes bulge.

Has the cow been shot? Drowned? Was she sick? There are no other cows in sight. I look again across the lake, cool, blue, and utterly empty. The mountains feel like a dark, heavy presence, a watching shadow. It's too far to yell for one of my sisters.

I chastise myself fiercely for not having the courage to investigate. We gallop all the way home, Tonka's hooves ringing on the bumpy ground.

IN THE fall I start sixth grade at Juchem Elementary, a small brick school thirty minutes away that stands in the middle of an open, grassy field. My siblings and I ride a yellow school bus down windblown dirt roads that will later become four-lane highways. Randy Sullivan, the boy I've been observing from our kitchen window, rides the same bus. He has a ready smile and more friends than I'll ever dream of. He makes me blush.

My first romance with a boy—not Randy—lasts three entire class periods. He gives me a chunky chain bracelet for my wrist, but by afternoon recess we've broken each other's hearts. I can't wait to go to junior high. I think about all the friends I'll have once I ditch my sisters and brother and the entire sixth grade, which takes the boy's side, not mine. He plays football. I play the clarinet. The chasm between our social circles seems vast.

In junior high I'll be brave enough to talk to boys like Randy.

The wind blows fiercely across the treeless fields. One drowsy afternoon we see a bald eagle settle on the steel post of our playground swing set. The teachers show us films like *Our Friend the Atom*. Once or twice a week we have duck-and-cover drills in case we're bombed by the Russians. We'll be sitting at our desks working out long, dull columns of math and without warning the bell goes off. "Duck and cover!" the teacher yells. "Stay calm!" I crawl under the flat wooden top of my little metal desk and curl up in a ball, forehead to knees, and lock my fingers over the back of my neck as instructed. We've all seen the classroom films of people and buildings instantly disappearing—poof!—in a nuclear blast. I wonder how locking my fingers behind my neck will save me. We huddle until the bell rings the all-clear signal.

"They left out the fourth step," my ex-boyfriend whispers. "Kiss your ass goodbye."

Unbeknownst to us, Rocky Flats is, in fact, a likely Soviet target. Rocky Flats has a sister plant, Mayak, near Chelyabinsk in the Soviet Union. Secretly built in the late 1940s, Mayak manufactures, refines, and machines plutonium for weapons, just like Rocky Flats. Our government knows a little something about them, and their government knows a little something about us. Deeply contaminated, Mayak becomes the site of one of the worst nuclear accidents in history—an explosion in 1957 released fifty to one hundred tons of high-level radioactive waste, contaminating a huge territory including populated areas in the eastern Urals.

But there are no drills on what to do if something goes wrong right down the road at Rocky Flats.

☢

By the time the school bus turns the corner onto 82nd Avenue, I'm thinking about my cowboy boots, scuffed and worn so soft they fit my feet like gloves. My favorite pair of jeans, stained with neatsfoot oil and hoof polish and horse sweat. The way the barn smells of hay and molasses and pungent manure. When the school bus drops us at the top of the hill, I race to the house to change my clothes and go chase Tonka.

I drop my books in the front hallway, pull on my jeans, and I'm back out the door before my mother realizes I'm home. Tonka's waiting at the gate, head up, ears perked. He knows there's horse candy in my pocket.

I fasten the halter strap beneath his jaw and lead him to the grooming post. All my tools are gathered in a plastic bucket like a painter's paint box. I tie him to the rail and set to work. Even now I can recall the pattern of hair growth on his body—the sinewy neck, the whorls at the flank and chest, the straight, wiry texture of his mane. I push my left shoulder into his side and Tonka obligingly lifts his leg. With the hoof pick I clean the underside of each hoof, including the sole and the cleft of the frog, a triangle of dark rubbery flesh.

The face and forelock are last. With a soft cloth I wipe the film from the corners of Tonka's big brown eyes and dab his nostrils. I comb the lick of hair falling down between his eyes. I slip off the halter and pull the hackamore over his ears. We're ready to head for the lake.

Sometimes I don't make it out of the house. Each afternoon my mother rests in her room, which is tidy and quiet and covered with a permanent film of dust. She has a small drawer of pills the doctor prescribes, pills that make her fuzzy. Many of her friends in the neighborhood have the same prescription to help with nerves. She takes them every afternoon. If she hears my step at the door, she calls me and I'm trapped until supper.

She complains bitterly about the dust. The house is filled with it. Dust settles on my mother's dresser and its blurred mirror, on the night table, the windowsill, the lamp with the worn fabric shade. In the living room it settles on the clock and the console record player. Dust swirls and glitters in bars of sunlight striking through the windowpanes. On Saturday mornings, when we do our weekly chores, we spray the dust with lemon furniture polish and rub it in cloudy swirls, but by Sunday it's back. Her bed, with its avocado bedspread, faces two large picture windows that look out to the mountains. I sit on the edge of the bed and watch the dust, suspended in space, floating delicately to the floor. The water tower and tiny square buildings of Rocky Flats sit like toy buildings on the flat plain.

"I love this view," my mother declares. "It makes me feel peaceful." Hours pass and she tells me family stories and hints at dark secrets. "I shouldn't be telling you this!" she exclaims. It's the best first line for any story, and I'm hooked. Her own childhood was troubled; her father drank and the family struggled during the Depression. There are plenty of skeletons in the family closet. But most of her secrets revolve around my father, who seems to completely baffle her. "What's wrong with him? What happened to him?" she asks plaintively. His anger is palpable, even when he's not physically present. He's never hit any of us, but we all fear his threats. She criticizes him, worries about him, and most of all fears him. Her eyes grow wide. "I can't talk to him, you know. I don't dare bring anything up."

My father says the same thing when he shouts and mutters to himself in the dark. "I can't talk to your mother," he announces to the room. Does he intend us to hear this? Everything is said furtively; everything is hush-hush. We don't want the neighbors to know. We have to protect my dad's practice. Keep things within the family, and keep things to yourself. When I feel like exploding inside, or running away, I remind myself that someone needs to hold down the fort. That someone feels like me.

After an hour or two of examining her troubles, I'm itching to get outside, to get away. "I gotta go, Mom," I say.

"Kris," she says. "Don't leave me alone in this room." Her long, cool fingers clasp mine. So I stay until the clock on her dresser shows six o'clock and it's time to brown hamburger for dinner.

☢

I HAVE a few secrets myself. The banks of the old Church irrigation ditch behind our house are lined with wide-trunked cottonwood trees that fill the air with blizzard-like balls of fluff. Wild asparagus grows in thick, sinewy stalks along with plump blackberries and tiny pink straw-berries we pop in our mouths when we can find them. Fat muskrats bur-row in the banks. I find an abandoned baby muskrat the size of a hamster and for three days nurse it secretly but unsuccessfully under my bed. I look for an appropriate burial site. I find a soft spot at the base of the biggest cottonwood and start digging with a spoon I've swiped from the kitchen. I hear a shot, then another. A neighborhood boy is standing at the edge of our property, his brand-new BB gun leveled at my head. I stand, and I realize he's aiming at the horses instead. A loud crack— and then Tonka, who's been nibbling at the short fresh grass near the water, explodes. He leaps into the air, bucking and kicking, astonished with pain. The other horses startle and scatter, but not before the boy has pinged Comanche in the shoulder and sent him galloping heavily around the back of the house. "Get off our property!" I scream. "That's my horse! Leave us alone!"

He turns, shrugs, and slings the gun over his shoulder. He wears combat fatigues and black boots and a black T-shirt with the sleeves cut off. He's fourteen.

That night in the kitchen, I tell my mother what happened while she's making rice and phoning the neighbors to ask if they've seen my brother. "Well, Kris," she says. "I just can't think about this right now."

"But he could shoot their eyes out," I say. "He could really hurt them. We should call somebody." He has lots of brothers and sisters and a mother as belligerent as an army sergeant. I think he should catch hell from somebody.

My mother sighs. The boiling water from the rice seeps over the edge of the pot and drips down onto the stove. "Boys will be boys," she says. "The sooner you realize that, the better."

THE MARTELL study is the first time the public and even the state government learn at least some of the facts about the worst accidents at

Rocky Flats—even though years have passed since they occurred. The AEC and Dow Chemical continue to reassure residents there is nothing to worry about. The amount of plutonium released from Rocky Flats, says Edward Putzier, Dow's health physics manager, is no greater than "a pinch of salt or pepper." He accuses the media of exaggerating the dangers of radiation.

But there is no consensus about what constitutes a "safe" level of plutonium, and even a pinch, according to the AEC itself, can be lethal. Glenn Seaborg, the physicist who isolated and gave plutonium its name in 1941 during the Manhattan Project, said that plutonium "is unique among all of the chemical elements. And it is fiendishly toxic, even in small amounts." Internalized plutonium can be the most deadly. Plutonium emits alpha radiation, which cannot penetrate skin. (Gamma radiation or X-rays can be harmful by hitting the body from the outside.) Alpha emitters have to be inside the body to be dangerous. If plutonium is inhaled or ingested, or if it enters the body through an open wound, tiny particles can lodge in the lungs or migrate to other organs, particularly the liver or the surface or marrow of bone, where they bombard surrounding tissue with radiation. It may take twenty to thirty years for health effects such as cancer, immune deficiencies, or genetic defects to become manifest.

A full gram of plutonium, which is denser than lead, is scarcely bigger than a grain of rice. One microgram—a millionth of a gram—of plutonium, invisible to the human eye, can produce a fatal cancer, according to standards set by the AEC as early as 1945. Plutonium has a 24,000-year half-life, the period of time during which the number of radioactive nuclei decreases by a factor of one-half. This means that every 24,000 years, half of a given amount of plutonium will shed energy, gradually turning into a nonradioactive material. Measured in human lifetimes, 24,000 years is almost unfathomable. And yet, after 24,000 years, half of the material will still be radioactive, and after 24,000 more years, half of that amount will continue to be dangerously radioactive. Even after ten half-lives—that is, 240,000 years—the radioactivity will still not be wholly gone. The physicist Fritjof Capra says plutonium should be contained and isolated for half a million years.

Many scientists believe there is no safe level of exposure to plutonium.

Martell and his colleagues are surprised by the AEC's admission of off-site plutonium migration, but stunned by the news of where it came from. Dow Chemical and the AEC have known about the leaking drums for at least a decade, but they have kept the information from the public.

Over time Rocky Flats removes most of the barrels and covers a portion of the area with asphalt. Some of the barrels are sent to a waste site in Idaho and some are buried on-site (eventually contributing to groundwater contamination). But even after their removal, wind continues to scatter plutonium for miles. Five particularly powerful windstorms in late 1968 and early 1969 suspend a large portion of plutonium as dust and carry it toward Denver. On January 7, 1969, winds reach 125 miles per hour.

There's no reason for concern, officials emphasize. Major General Giller declares that "The AEC is quite convinced that the plant in its present location and operating conditions poses no health and safety hazard either to its own workers or the local population."

But AEC guidelines for worker or citizen exposure to low-level radiation are under dispute. Dr. John Gofman, a professor of molecular and cell biology at the University of California, Berkeley, worked on the Manhattan Project and was an expert on chromosomal abnormalities and cancer. In 1963 he established the Biomedical Research Division for the Lawrence Livermore National Laboratory and began conducting research on the influence of radiation on human chromosomes. Concerned about the lack of data on the health effects of low-level radiation, with other scientists he reviewed health studies of the survivors of Hiroshima and Nagasaki, as well as other epidemiological studies.

In 1969, Gofman and his colleague, Dr. Arthur R. Tamplin, suggest that federal safety guidelines for low-level exposures to radiation be reduced by 90 percent. The AEC immediately disputes the findings.

JUNIOR HIGH proves challenging. In gym I'm issued a blue polyester uniform too tight in the crotch—already I'm long-waisted and taller than most of the boys—and I loathe taking showers with the other girls, who scream and giggle and pinch. They make fun of the gym teacher, a serious-looking woman with muscular arms and legs. Each week we take a test on the climbing rope that hangs from the gym ceiling and each week I make it halfway up, hang just long enough to cringe under the gaze of my schoolmates, and slide back down in defeat. I come home with red welts on my thighs from dodge ball, and a healthy fear of the balance beam, a long wooden rail four inches wide that all the girls have to walk up and back on, our feet dusted in baby powder. The only place I feel graceful is on the back of a horse.

Tina, however—who decided not to abandon me after all—is determined to make me into a real girl. On the day of the first school sock hop, she insists we walk to school rather than ride the bus so we can talk strategy. She wants to give me some advice on dancing with boys. She herself has her eye on a basketball player.

Walking to school is complicated. It involves passing the historic Bunce house, crossing an increasingly busy street, wading through an irrigation ditch (originating, unbeknownst to us, at Rocky Flats) and crossing the railroad tracks. Tina shows up at my door in a blue miniskirt and white fishnet stockings.

"Hurry up," she declares. The morning is cool and her breath hangs in front of her face. "We have to go before everyone starts showing up at the bus stop."

"Hold on." I'm still tugging at the elastic waist of my pantyhose. The crotch is stuck midway down my thighs like a tourniquet.

"Your mother let you buy those?"

"Yeah." The plastic egg-shaped container sits on my dresser. "Finally."

"You have to stretch them up from the bottom," Tina advises. "Don't you know? Like this." She squats down and inches the nylon up past my ankles to my knees. "Now pull."

"They're too tight."

"Just suck everything in. And don't use your nails. Use your fingertips, like this." Tina touches her own stockings delicately, then stands back to watch as I struggle. "That's good enough," she sighs, and slings her purse over her shoulder, leather fringe dangling.

We walk side-by-side through the silent neighborhood, Tina's long brown hair swinging with her stride. The houses look empty. Most of the fathers have already left for work. I wonder what the women do all day, each in her own little fairytale tower. How much cooking and cleaning can a person do before she goes crazy?

We cross 80th Avenue, slink through several backyards, and emerge into the long field before the railroad tracks. Sunlight streams across the dry meadow and the weeds tickle my ankles.

"Lots of thorns out here," Tina says. She wears go-go boots, white vinyl to match her stockings.

"I think I have a snag."

"Don't let it run!" Tina orders. She opens her leather bag and digs around in the bottom. "Here." She holds up a squat bottle of clear fingernail polish. "This will fix it." She paints a ring of wet polish around the hole and blows on my ankle to make it dry.

I feel like a new member of a secret female club.

We reach the railroad tracks, where my sisters and I put pennies on the tracks and wait for the train to come along and flatten them. Every mother in the neighborhood threatens her child with details of what will happen if you get a foot stuck between the rails. It never happens. Once the train roars by, the pennies are gone. The only flat penny I ever get is from a machine at a museum. The real risk is that kids who set off to walk to school in the morning get sidetracked by various attractions along the way and arrive late or not at all. Tina has no time for childhood games, but when we step over the tracks, we stand for a long moment between the rails just to taunt fate before crossing over to the irrigation ditch.

"Damn," Tina says. "The water's up."

"What?" I feel a prickly sensation climbing up my inner thigh as my pantyhose begin to disintegrate.

"Why are they running the irrigation ditches this time of year?" She tiptoes down the bank. "That water's over a foot deep."

I look down the tracks to Simms Road, the narrow two-lane where the school bus passed fifteen minutes earlier. "We could walk down the tracks to the road, cross over, and then walk back," I offer.

"We'll be late for school," Tina says. "We're late already. Today's the one day we don't want to miss, remember?"

"We have a test in French, too," I say.

"Who cares about the test." She bends down and unzips her boots. "It's the dance that matters." She jumps in, right foot first, and makes it in three strides. "*Shit!* That water's cold."

I tug off my new heels and inch into the water. Mud oozes through the nylon and up between my toes. I scramble up the bank and fall in line behind Tina, my feet squishing in my shoes. We're both shivering.

We take our seats just as the bell rings. The teacher speaks with a French accent. Tina and I have agreed it's fake. What would a real French person be doing in Arvada, Colorado? The school has grown so quickly that half our classes are held in the "temps," temporary trailers propped up on cement blocks behind the school that sway with the wind and are always too hot or too cold. Today is a cold day and the teacher wears her coat. She passes the tests down each aisle, and the class sits with heads bent over their papers. I begin filling in verb conjugations. I won't admit it to Tina, but I actually like French.

"This is stupid," Tina hisses. She turns her paper over without filling in any of the blanks. "And look at your legs." A brown water mark crosses each thigh, and my pantyhose, or what's left of them, are a complex network of criss-crossed fibers. "Can I see your notes?"

"No." My ethics are flexible, but this is one test I've studied for.

Tina sets down her pencil in defeat. "What's in your bag?"

"A book," I say. "You can read it."

"What kind of book?"

"I borrowed it from my mother."

"Who wants to read a book you borrowed from your mother?"

"Page nineteen," I say. "Take a look." I pull *The Godfather* out of my backpack.

"I heard about this book." Tina grins.

"Read the wedding scene," I whisper. "The page is bent."

She flips it open and begins reading. "Wow. Your mom gave you this?"

"Sort of."

She begins reading in a low voice. " 'On the landing Sonny grabbed her and pulled her down the hall into an empty bedroom.' "

"Ssshh—"

"Listen to this: 'Her legs went weak as the door closed behind them. She felt Sonny's mouth on hers. At that moment she felt his hand come up beneath her bridesmaid's gown, heard the rustle of material giving way'—"

"Tina?" The teacher stands in the aisle. "Are you finished with your test?"

"No." Tina sits up straight.

"What are you reading?"

"Nothing." She smiles. "Just Kris's book."

"I'll take that, thank you." The teacher holds the book up for the class before setting it on her desk. Then she scoops up our tests. "I think you girls have finished your work for the day," she says.

The class snickers. After what seems like hours, the bell rings. Randy Sullivan sits a few seats ahead. Has he noticed? Students tumble down the makeshift steps and head for the warmth of the main building.

In a metal stall in the girls' restroom, I roll the nylon off each leg, peel off the checkered film of nail polish, and tuck the small brown ball of nylon deep down in the trash. That afternoon I walk home alone and avoid the irrigation ditch. I'd be too shy to dance anyway.

A FEW weeks later Tina announces she's going to have a boy-girl party. The thought makes me feel a little sick. Girl parties are bad enough. My mother, pleased that I am venturing out into the social world, takes me to the mall to buy a new blouse. Suddenly we veer toward the cosmetics counter. "Now is the time to begin thinking about your face," she says.

"Yes indeed," the salesclerk concurs. She's a petite woman with eyelashes as long as her fingernails. Fake, I'm sure. "What you do now

determines what you'll look like years from now. Your face is everything."
She peers at me closely. I realize my face bears serious deficiencies.

"Are you one of our regular customers?"

"She's not. I am. I've used the same products for years," my mother
says with pride.

"What does she use now? Does she exfoliate?" the woman asks.

I have no idea what that means. "I don't use anything," I say.

"Right," my mother says. "Can't you tell? Look at her. She needs the
whole shebang."

"Well, that is certainly not a problem," the saleswoman purrs. In
ten minutes I have a lineup of products that should guarantee my face
for the next twenty years. "Remember, thirty minutes every morning,
thirty minutes every night," the saleswoman says. I can't imagine how
my schedule might accommodate this type of regimen. I roll out of bed
ten minutes before the bus reaches our streetcorner each morning and
sometimes sleep with all my clothes on, I'm so tired from horseback riding. I nod and take the bag.

"Years from now, you'll be glad we talked," the salesclerk chirps as
she hands my mother the receipt. "Come back and see me when you start
to run low!"

We move on to the clothing department. "Are you a misses' or ladies'
size now?" my mother asks. "Let's look in the ladies' section." She smiles.

The ladies' department holds racks and racks of blazers, cuffed
slacks, and polyester pantsuits. If this is what it means to be a lady, I'm
not interested. We finally agree on a lilac blouse.

I take the blouse into the dressing room, my mother fidgeting outside, and slip it on. I face the three-way mirror. The bright light en-
hances every defect. Something isn't right. I turn to the side and catch a
glimpse of my face, newly exfoliated. What is it?

It's my nose. I have my father's nose. The chin, too. In fact, my entire
facial profile is identical to his.

My mother is waiting by the cash register. "Aren't you going to buy
it?" she asks as I put the blouse back on the rack.

"No," I say. "It's not my style."

"You need that blouse. You can't go around all the time wearing T-shirts and cowboy boots." She takes the blouse back off the rack. "This is perfect for you."

Tina's house has a basement as large as the first floor. Dusty wooden stairs descend to a cold cement floor. The room is windowless. A wooden rack with canned goods and a radio stands against the wall—a pint-sized bomb shelter. For the party, her mother sets up a couple of card tables and chairs and tapes paper Chinese lanterns over the hanging lightbulbs. Tina brings down her eight-track tape player and sets out bags of corn nuts and party mix. She invites four boys and four girls.

"Are you inviting Randy?" I ask.

"Not this time," Tina says. "I think you should get to like David." She prefers Bridledale boys to Meadowgate boys.

"I don't even know David." I wear the lilac polyester blouse, which itches.

"Well, you won't talk to Randy, so what does it matter?" The closest I've come to getting up the guts to talk to Randy is riding Tonka past his house after school, hoping to catch a glimpse of him. If someone's in the yard I turn around and go the other way.

We start out by playing hearts, then Twister, arms and legs tangling. "Let's play spin the bottle!" Tina exclaims, and I know it's been her plan all along. My heart sinks. Truth be told, I've never been kissed except by Tonka and our string of dogs. We sit cross-legged in a circle. My heart sinks when the Coke bottle spins to me and David, a sullen boy with long blond bangs who rarely speaks at school. He glances over at Tina, then leans over and pecks my cheek.

"Oh, come on," Tina says. "The rule is you have to go to the back of the room and kiss at least a minute." She's keeping time on her watch

"Over here, then," David mutters, and pulls me away from the group. His mouth is wet and cool and tastes of cigarettes. He stops kissing when I don't kiss back. Tina spins the bottle again and it flips to her and her basketball player. They're gone longer than a minute.

The next morning on the school bus, I wince as we approach David's stop. He stands waiting, expressionless. "There he is!" Tina says.

David strides up the steps. "Hey, Kris," he says.

I stare at the floor, cringing. I hate Tina.

"Kris?" It takes some effort on his part to repeat my name. The other kids pile up behind him, waiting to get on the bus. "Come on!" someone shouts.

I feel the tips of my ears turning pink.

"Ah, hell," David says. He saunters to the back of the bus.

Tina turns to me. "Well, that was stupid," she says. We ride the rest of the way in silence.

I TURN thirteen. I feel electric.

I love Joni Mitchell and Carole King. I know all the lyrics to "Big Yellow Taxi" and "A Natural Woman." I embroider the bottoms of my bell-bottom jeans in brightly colored thread and love the way they flop over my cowboy boots. My skin and clothes smell like hay, horses, and sunshine, and my face and the back of my neck are tender with sunburn. I love the way my hair whips in the wind when I'm riding, and the way the hair rises on my forearms just before lightning arches across my backyard sky. I like the sound of a meadowlark's call on a hot afternoon under a cottonwood as I lie back on Tonka's broad, warm haunches, and the dry wheat-and-tobacco smell of the tall prairie grass and the occasional stink of skunk.

It's a fierce love, a love for small things. The scent of cigarettes on my father's clothes. The way he settles back into his big recliner with a smile, his belly rounding over his belt. His technique of making hamburgers by rolling them into golf balls and scorching them on the grill. The way he looks in the morning on his way to work, his shirt pressed and hair slicked back, clear-minded and serious. I think of him as a kind of dark genius, buffeted by the world but bravely setting forth.

And my mother: the way her closet smells, sweet and musty with a touch of Joy perfume. The way her hair curls up in an elaborate blond beehive and how she taps it with her fingertips or searches for an itch with a long knitting needle. The way she sits at the kitchen counter and paints her nails a deep pink. The way she sings "Moon River" while she makes dinner and tells us long stories that make us laugh. I love the way she says, "I gave up everything for you kids," or "I would do anything for you kids." She is a displaced queen, unseated, usurped, somehow denied what the world promised her, always waiting for her ship to come in. I love the way she tells me I'm her best friend.

I hate the small things, too. I'm too thin. I'm not thin enough. My hair is too fine and straight. I'm awkward and clunky and can't speak in class. I'm painfully shy, scared of boys, and blush crimson at a moment's notice. Tall as a giraffe, I wear moccasins to school to make myself seem shorter. I can't bear to look at myself in a mirror.

I hate the way my father smolders with anger and how I can tell he's been drinking by his eyes, hard and dark. The way he drives, erratic, distracted, his cold eyes blurry, barely watching the road. I hate the way he disapproves of me. I'm a straight-A student, but it seems nothing I do is good enough. He criticizes my clothes, my hair, my weight, my friends.

I hate the way my mother simmers with fear. The way she keeps up appearances and covers things up. The way she slips off to her room at any sign of trouble and lies on the bed with her eyes closed, saying prayers to herself. The way she says, "I gave up everything for you kids. Everything."

Rocky Flats is also interested in keeping up appearances, but Dow Chemical is finding it increasingly hard to do. At the 1970 congressional hearing on the Mother's Day fire, behind closed doors, Major General Edward Giller described Rocky Flats as "old, outmoded and increasingly hazardous . . . old, crowded, the equipment corroded, and it has reached the state of [words censored as top secret]." He talked about tanks that were open and vented to the environment, ongoing leakages from cracks in the concrete holding ponds, and other problems. The AEC report following the 1969 fire called for substantial improvements. But in the three years since the fire, safety and fire protection work on some of the key facilities has slowed or stopped, particularly in the plutonium recovery building and liquid plutonium waste treatment facility.

In January 1972, James Hanes becomes the fourth general manager of Dow's Rocky Flats Division. In a statement to the press, Hanes says that plans are in the works for a new plutonium recovery facility, and ongoing improvements are designed to make Rocky Flats "as safe as is technically feasible." He emphasizes, however, that "the improvements being made shouldn't imply that Rocky Flats is unsafe."

Supported by tax dollars, Rocky Flats has cost the federal government and American taxpayers more than $620 million over the plant's twenty years of operation.

MY SISTER Karma loves to ride as much as I do. I'm on Tonka, who's still not quite trained, and Karma's on the witless but fun-loving Chappie. We ride as far as we can go and still get back by dark. We gallop bareback across the fields, flying until the horses are slick with sweat: down Alkire and Indiana streets, past the scattered houses and barns and the barbed-wire Rocky Flats fences with the Keep Out signs. We slow down, finally, to a trot back along the creek and stop to cool off at Standley Lake. We twist our hands in their manes, and Tonka and Chappie splash down eagerly into the water until it reaches their shoulders. Suddenly the underwater shelf ends and we all plunge into the cold, deep water, Tonka striking out with his front legs, galloping and then swimming hard and fast, his head just above water, nose high, nostrils extended, eyes ringed white with excitement. The lake water is frigid. I look over at my sister and she's laughing, gulping air. The water reaches my neck, and my lungs feel frozen. "Let's turn back!" I yell, and we turn the horses around, paddling, swimming steadily now, beginning to tire. We emerge, dripping, and wait to let the horses have a good shake. "Race you back!" Karma calls, and it's a dead heat, hooves pounding, until we get to our pasture gate. We're dry from the wind and aching with exhaustion when we slide off and take the horses into the barn for their supper.

There was no sign of the dead cow.

In Colorado the air turns cold at night, even in the summer. One evening, just as the sun is setting, I head down to call Tonka to his dinner. There is a large pasture shared by all the neighbors, where we can let our horses graze and enjoy a little more room to gallop around. Several horses stand idly in the field, but Tonka's not there. I walk out into the pasture and find him in the swamp, a wide swath of mud and reeds fed by the irrigation ditch.

Stupid horse. He likes to nibble on the fine, thin grass that grows at the edge of the swamp, and the cool mud keeps the flies off.

But this time he doesn't come when I call. He stands as stiff as a statue, almost completely submerged. His head and neck strain high above the black muck and he's shivering. His eyes are glazed and he doesn't recognize me.

The sun dips below the mountains and the temperature drops a little more.

I'm not sure what to do. There's no one to help. I think of Walt, the neighbor up the street. Walt has a big white pickup with a winch on the front. He's the first to go out and rescue cars stuck in driveways from mud or snow. The black swamp mud that oozes across the pasture toward Standley Lake is as treacherous as quicksand and as sticky as glue.

It takes me ten minutes to run as fast as I can to Walt's house—will Tonka sink by then?—and pound on his door. He's home, having supper with his wife and boys. "How long has Tonka been in there?" he asks.

"I don't know," I say. "He's numb."

Walt throws a rope in the back of the truck and we bounce down the dirt road to the pasture in a flurry of dust and rock. He backs up to the swamp and anchors one end of the rope to the truck. He leaves the headlights on. Without hesitation he wades into the mud, clothes and everything. It's as thick as wet cement. Tonka doesn't move. Walt throws a lasso and loops the end around Tonka's neck and shoulders. Black mud spatters up on Walt's neck and he wipes it from his eyes with his thumb. He pulls himself out, slams into the cab, and puts the truck in gear.

The rope strains. Nothing happens. Walt blares the horn and I nearly jump out of my skin. He pulls again. Nothing. He hops out. "He's pretty far gone," he says. "He's lost the will." He wades heavily back into the mud and starts slapping Tonka around the head and neck. He's a muscular man and the blows are hard.

"What are you doing?" I yell.

"Waking him up," he shouts. He slaps a hand across Tonka's haunches. "Come on, you old thing!" He reaches over to Tonka's head and pulls down an ear, bellowing into the soft cavity. "Wake up!" he yells. "Move it!"

Tonka startles. His head jerks up.

"That's it," Walt shouts. "Wake up! Wake up, old buddy!" He slaps him again. Tonka shudders, hard, and tries to shake himself.

Walt shoulders his way out of the mud. "Yell at him, Kris. Yell!"

I feel frozen, too, but I make myself react. "Tonka!" I shout.

The truck roars to life. Walt blares the horn again and Tonka's eyes startle and clear. He puts the truck in gear again and pulls the rope tight, tighter, and finally Tonka scrambles out like a spider tugging from a web, the mud sucking his legs and clinging like filmy glue to his skin.

"Take him up to the house," Walt says as he unties the rope. "Wash him down with warm water, give him a hot mash. Watch him. Sometimes they get pneumonia after that."

"Pneumonia?" Hard shivers move up and down Tonka's body.

"Don't worry," he says. "It looks like he'll be fine."

"Thanks," I say. "I mean, thanks a lot."

"No problem," Walt says. "You could probably use some supper yourself. Go tell your mom the horse is okay." He puts his hand on my shoulder.

Tonka spends the night in our warm garage and is fine the next morning. I think of Walt from time to time, although I rarely see him and never get to know him very well. Some years later, when I learn he has died of cancer, I feel a sharp sense of loss. What kind of cancer? No one seems to know. It's not polite to ask questions.

There seem to be many cancer illnesses and deaths in our neighborhood. Karma, Karin, and I whisper about it from time to time. It scares us. "There are all these big houses with sick people inside of them!" Karma says. "Scary cancers, too. Like tumors and things." But no one mentions Rocky Flats. After all, you can get cancer from just about anything. Teflon, for example. Or overcooked hamburgers. Everyone knows that.

☢

HARVEY NICHOLS has been following the news at Rocky Flats closely. A new assistant biology professor at the University of Colorado at Boulder, Dr. Nichols has just completed his postdoc work in the biology department at Yale. His wife is a nuclear physicist. In 1974 Nichols is working on a National Science Foundation grant to study long-distance pollen transport in the Arctic when he's asked by a biologist from the government if he'd like to work on a project tracking pollen at Rocky Flats. Nichols is interested. Is it possible, he wonders, that particles of radioactive material might attach to pollen? The question turns out to be almost moot. In the summer of 1975 he reports that the research team has found that "there [is] so much radioactive particulate matter out there, it [doesn't] matter whether pollen [is] transporting it or not." Years ago Dow Chemical had set up twenty-five air samplers, ten on-site and fifteen off-site, to detect radiation releases from the plant. Nichols's team tests the filters from the samplers and discovers that they do not effectively detect smaller, lightweight airborne particles of plutonium, the size that can be easily inhaled.

Nichols takes it a step further. Given the heavy snowfall in the area, he's curious about whether radioactive particles might attach to snow. He receives a grant from the Department of Energy (DOE) to test snow for radioactive particulate during the winter of 1975–76. The research team collects snow samples from the plant boundary all the way to Indiana Street to the east, a distance of nearly three miles. They collect freshly fallen snow, taking samples close to the soil and other samples from the top level of snow after a heavy snowfall of six to nine inches. The team then melts the snow and passes it through a series of very fine "millipore" filters for testing.

Researchers find radioactive particles at all levels.

Nichols finds it odd that the snow, which presumably is picking up plutonium contamination from the soil, also tests positive for radioactivity on the top layer after a heavy snowfall. He wonders if this could be new plutonium coming from ongoing radioactive emissions at Rocky

Flats, and he asks an official directly. He is surprised when the answer is yes.

"Do you admit," he repeats, "that you emit tiny amounts of plutonium out through your stacks as part of your routine operation?"

The official admits that this is the case.

"And do you regard that as dangerous?" Nichols asks.

In a surprisingly unguarded moment, the answer, again, is yes. But the official explains that there is no cause for alarm regarding public health.

But there should be concern, Nichols exclaims. Not only is there a great deal of radioactive particulate out there, but it's moving constantly, like pieces of cork floating in water. It doesn't "settle out," he emphasizes, and the material will continue to be suspended in wind or snow, moving around and far beyond the plant site.

But no action is taken, and the DOE decides not to pursue further testing.

ONE AFTERNOON after school I reach the bus stop to find Adam waiting for me. Adam lives up the street and we're in the same grade, but he hangs mostly with older kids. With longish brown hair and brown eyes, he's as cute as David Cassidy. A pale mustache shadows his upper lip, and he has a dirt bike. "Hey," he says. I look the other way. Some kids call me stuck-up behind my back, but the truth is I'm petrified. Adam occupies the unsteady territory between cool and geek, tending a little more toward cool. I tend a little more toward geek.

"Let's race," he says.

"What?"

"I challenge you. Me on my bike, you on your horse. Race you to the lake and back."

"Today?" I need time to prepare and raise a little support. Boy versus girl, cool versus geek, dirt bike versus pinto pony.

"No," he says. "Tomorrow. Three o'clock, right after school."

"Fine." A race is one thing, but a challenge is neighborhood jargon for a full-on formal battle, with spectators. You can't turn it down. "Where?"

"At the barn gate." This was the standard starting point for lake races.

"From the barn gate to Standley Lake?"

"Yeah. Past the pond, over the hill, first one to touch water."

"You're on."

That night I give Tonka an extra helping of Omolene, an oat mix with extra molasses and vitamins.

At two forty-five the next afternoon, Tonka and I clop down the driveway and over the wooden bridge arching over the old Church irrigation ditch. Adam appears at the top of the hill on his bike and immediately slows, cutting the noise of his engine so as not to spook the other horses. The cottonwood trees are just starting to leaf out, and the air smells sweet. Tonka raises his nose in the air appreciatively and his front hooves dance. By the time we reach the barn gate, a small group of kids has gathered. We line up nose to fender.

"I'm the one who gets to say go," I say.

Adam looks dubious. "Someone else should be the starter. A neutral party." He glances toward our audience. The dirt bike purrs.

"You made the challenge, I make the rules."

He laughs. "All right." He wears no helmet, no gloves, just a T-shirt, jeans, and boots. He looks over at the kids on the fence and revs his engine for their approval.

I clap my heels into Tonka's sides. "Go!" I shout. We've got a split-second advantage. He half-rears in surprise and leaps out front. Behind us the bike roars. Tonka hates dirt bikes. We catapult down the beaten path. From the corner of my eye I see Adam's back tire spin briefly in the mud before it catches the track. I crouch low over Tonka's neck, my face in his mane, eyes filled with tears from the wind. Tonka stretches his body as long as a greyhound. It sounds like the bike is right on our heels. We sweep across the field, around the edge of the pond, and up the hill. I see the lake. "Go, go, go!" I scream. Tonka's ears press back flat against his head, catching my words, and he thunders to the top.

At the crest of the hill, Adam passes us in a blur of dust and oily smoke. The tall weeds slow him as he comes to the edge of the lake, but by then it's beyond hope. I pull Tonka up in a long slide as Adam reaches down and splashes his hand in the water.

"Hey!" he says, smiling. He revs his engine.

"You won," I say.

"You cheated."

"You're spooking my horse."

"Sorry." He cuts his engine and dips his head in mock apology. "No hard feelings, eh?"

The next afternoon Adam appears in my driveway, his bike purring, chrome and red enamel with an intoxicating scent of heat and grease and all that is forbidden.

"Want to go for a ride?" he asks.

"Where?"

"Out by Rocky Flats."

I glance up at my parents' window. No one's home anyway.

"Okay."

Adam unbuckles the helmet hanging from the back of the bike and fits it over my head. "Keep your legs away from the exhaust pipe or you'll get burned," he instructs. "And hang on tight. Lean into the curves." He doesn't wear a helmet. He steps on the bike and strips off his T-shirt. "I like to feel the wind," he says.

I climb on the back and grip the sides of the seat as we motor down the driveway. We rumble slowly through the neighborhood and just as we reach 82nd Avenue, Adam reaches back, grabs my hands, and links them around his waist. "You're gonna fall off unless you hang on the right way!" he yells, the wind taking his words.

We roar down the road. Suddenly I feel certain it is my last day on earth. The turn onto Alkire is sharp, but Adam doesn't slow until we reach the east gate of the Rocky Flats plant.

"Was that fun?" he yells.

"Too fast!"

He can't hear me. The wind is fierce.

"You're going too fast!"

The next afternoon is a repeat, except we stop by the side of Walnut Creek at Indiana Street to kiss for a while under a cottonwood tree. Adam's mouth tastes like sunshine and wheat and sweat. This time I kiss back, and my heart races. The wind grows cool as the sun sets, and he shakes out his T-shirt and puts it around my shoulders.

OUR FAMILY Sunday drives continue, although somewhat more sporadically than before. My mother is a master of logistics. We gave up going to church as a family long ago, and now Sunday mornings are meant for sleeping. She has to work to get everyone out of bed and committed to the project. "Let's go, kids!" she calls, standing at the bottom of the stairs in her pedal pushers and Keds, hair and makeup perfect. One by one we shuffle into the bathroom and tug combs through our tangled locks. My mother tries to fix our long, straight hair with frizzy perms that we endure once a month at the Arvada Beauty School, but they never take very well. Once we're dressed she packs us girls into the backseat of the station wagon, side by side, and then puts down blankets and pillows in the back of the wagon for Kurt so he can lie flat, his leg in a long plaster cast from being bucked off by Comanche. In our household, someone is always falling off a horse. The cast is covered with autographs and illustrations from his horde of friends; he won't let his sisters touch it.

"I guess you'll have to look at the ceiling, honey," Mom jokes.

"I can see fine," he says. He props his pillow against the backseat and braces his foot against the rear window.

Mom slides behind the wheel and starts the engine. "Where's your father?" she mutters. She taps the horn. A minute or two passes and she honks three short bursts. She rolls down the window and lights a cigarette. "I get so sick of this," she says. She blows a thin stream of smoke and honks again.

"The neighbors probably don't like honking on Sunday morning," Karin says. Karin is never afraid to speak up. Sometimes she refuses to go on Sunday drives at all.

"He's coming," Mom says. She glances at her watch, counts under her breath, and then lays on the horn for a good long one.

Dad bursts from the house. "I'm coming, Marilyn. Christ." She smiles and slides over so he can get behind the wheel. He hasn't showered, but his shirt is clean and he smells minty, like mouthwash.

"Hey, guys!" he says, grinning into the rearview mirror. He's best in the morning. He puts the car in reverse and backs down the driveway.

"You're going too fast," my mother says. "You'll hit one of the dogs."

He ignores her. "Where are we going today, guys? Where do you want to drive? Golden Gate Canyon? Rocky Mountain National Park?"

We don't want to drive anywhere. It's hard to see from the backseat, even with my mother pointing out the scenery, and the winding roads make us all carsick.

"Let's drive up to Rocky Mountain National Park," Mom says. "We haven't been there in ages."

"That's a long drive," Karin pipes up. "I have homework."

"This is a family day," Mom admonishes. "Maybe we'll see some elk. Don't you kids want to see the elk?"

The silence is taken as agreement. "There's that nice little restaurant in Estes Park," Mom says. "We can have lunch."

"All these kids deserve is a hot dog stand!" Dad jokes.

"I like hot dogs," Kurt says.

"We'll have a very nice lunch," Mom says, and gives my father a dark look. "It's the least you can do for your family."

It will be years before I learn to share my parents' appreciation for mountains, and it will take a backpack and a pair of cross-country skis for me to really begin to understand what wilderness means. In those days we never got out of the car. "Look, kids!" Mom exclaims. "There's snow on top of that peak—in July!"

"If we're lucky we'll see a mountain lion," Dad warns, though we never do. Maybe it's the flat cornfields of Iowa that makes them so ecstatic about the Colorado mountains, but they never tire of driving and looking, the car mostly quiet except for the hum of the tires. It's a sin to doze off. Sometimes, if Dad's in a gregarious mood, we sing John Denver songs. We know all the words. Occasionally we see deer standing alone or in groups at the edge of the road or a hawk floating on air.

We stop for a late lunch at a restaurant in Estes Park, and as soon as we've cleaned our plates, our parents dismiss us while they finish their Manhattans and have a cigarette. We stand outside and feel the cool wind on our faces. If we're lucky we'll get to stop at the fudge shop, a tourist trap. Already the sun is beginning to slip behind the mountain.

"Let's go," Mom calls, emerging with Dad, smiling, and we take our seats. Kurt braces himself in the back of the station wagon. It will take a couple of hours to get home; the road is narrow and winding. No one sees the deer until it's too late.

"Damn!" Dad yells, and slams on the brakes. For a moment I glimpse the deer in the windshield and then it vanishes. The car stops, shuddering, the hood hissing, and the deer appears again, dazed, and staggers into a ditch by the side of the road.

Karin, Dad, and I jump out of the car. The front of the car is pushed in like the nose of a bulldog and the deer lies bleeding, one eye calmly looking up. "We need a veterinarian," I say.

"No," Dad says. "There's nothing a vet can do." His voice is tense.

The road is dark, the car steaming, and what happens next seems unreal. My father walks into the forest and comes back with a large rock. "I'm sorry about this, kids," he says, and in one swift motion he crushes the deer's skull. He drops the rock to the ground and speaks in a low tone. "He was suffering," he says. I feel the sadness in his voice, the compassion. "I'm sorry you had to see that."

But it isn't my father who kills the deer. Years later I will learn that no one else in the family remembers the incident the way I do. The truth is that another man kills the deer, a stranger who drives up behind us in a pickup truck with big headlights. He wears a flannel shirt and heavy boots—a real mountain man, Karma says later. He looks things over, finds a rock, pulls the deer's carcass farther off the road, and then leaves to call the cops so we will be sure to get home safe.

I want to believe it was my father.

The sheriff arrives and he and Dad examine the car. "It's fine!" Dad barks. "The car is fine!" It looks bad, but it's drivable. We climb back in to begin the slow journey home. But all is not fine. The cast on Kurt's leg has been crushed in the jolt, and when we get home my mother takes him back to the emergency room. He returns with a fresh cast—the leg has been broken again—and he has new doctor's orders of six more weeks of no soccer.

"He'll be all right," Mom says. "He's tough."

IMMUNE TO outside regulation, Rocky Flats has always policed itself. Following the 1969 Mother's Day fire, however, the Colorado Department of Health begins sampling air, water, and soil around the plant. One afternoon in 1973, Al Hazle of the Colorado Department of Health is driving back from the Western Slope, where he's been collecting water samples to test for tritium, a radioactive isotope of hydrogen, in the aftermath of the Rulison Project, the 43-kiloton nuclear test project that was intended to extract natural gas from deep underground levels. Tritium, usually a by-product of a nuclear explosion, is a health hazard when inhaled, ingested through food or water, or absorbed through the skin.

Al Hazle decides to stop by Rocky Flats to pick up some samples for another test, and on a whim decides to take additional water samples from one of the canals to use as a background test for the Rulison water.

When he gets the results, he's not surprised to find tritium in the water, but he is surprised to discover where it comes from. It's not the Rulison water but the runoff from Rocky Flats that contains tritium. *What the hell?* Hazle thinks. *How did that get in there?* Unless there was a criticality or nuclear explosion at Rocky Flats, there should be no tritium anywhere near the plant.

TINA GIVES me one more chance to redeem myself.

Halloween is a big night. Jackson's Turkey Farm, on Indiana Street near Rocky Flats, is a family-run business with four teenage boys. My mother knows the family; for years she's bought our holiday turkey at Jackson's, where she knows the meat is fresh. "Those boys, though, are wild," she says disapprovingly. The Jackson brothers are famous for the haunted house they set up each year in one of the farm's outbuildings.

It's bitter cold when Tina's mother drops us off at the turkey farm with a promise to come back in an hour. A snowstorm is on the way. "She'll be late," Tina notes, but that's fine with us. We walk up the gravel drive. The night is pitch-black except for the lights of Rocky Flats just down the road. There's a line of kids in puffy down jackets standing at the door.

"I wonder if this is the slaughterhouse?" Tina muses.

"I don't think I want to go," I say. Tina has a pool of strength garnered from years of watching stuff like *Night of the Living Dead* and *Rosemary's Baby*.

"Wimp," she says. "Come on. It'll be fun."

The door opens and a shrouded figure gestures for us to come in. The shedlike space is dark. A long maze has been created by taping together large black plastic garbage bags and hanging them from the ceiling. The effect is suffocating and creepy.

"Begin here," the shroud moans. I can't tell if it's one of the Jackson boys or maybe Randy or one of his friends. "These are the parts of the witch," the shroud intones. "Eyeballs, hair, tongue. Give me your hand." He takes Tina's hand and puts it in one bowl, then a second and third.

"Peeled grapes," Tina says. "Cold spaghetti. Peeled green chilies. Come on. We did this in Brownies in second grade."

I keep my hands to myself.

"It gets better," the shroud assures us. We turn a corner and a guy in a gorilla suit rocks behind the bars of a rubber cage and leaps out to grab us as we pass.

"Bor-ing," Tina comments.

In the next scene, a girl lies dead on a table with a tall vampire leaning over her, fake blood dripping from his fangs. He looks up as we approach. "Are you next?" he hisses, reaching out to touch our jackets.

"Yeah, right," Tina says.

The vampire looks a little crushed. "Come on, Tina," I say. "At least play along with it."

We come to a long table lit with candles and another hovering, shroudlike creature. "This," the creature says, "is the real body."

"Don Jackson, is that you?" Tina asks.

The shroud ignores her. He points to the first bowl. "Here are the fingernails."

"Pretty skinny fingernails," I say. "Those don't even look human."

"Next, we have the innards."

I lean down to take a look. It doesn't look like spaghetti. I'm not sure what it looks like.

"Touch it," the shroud suggests.

"No."

"Oh, for Pete's sake." Tina puts her fingers in the bowl. "Touch, touch, touch. Okay?"

Whatever feelings the shroud might be having, they're masked by the costume. He continues unfazed. "Here, on this plate," he says, "is the skin." A pile of pale greasy casing or membrane is layered on the plate.

"It kind of looks like skin," I say.

"It's not skin," Tina declares. "Cooked lasagna noodles, I would guess."

"And here," the shroud declares with a small note of triumph, "is the heart." He gestures toward the final bowl. A purplish, thumb-size organ lies on the white porcelain, a ventricle extending like a tiny straw.

"Must have been a pretty small witch," Tina says.

"I think that's real, Tina."

"Just stopped pumping a minute ago," says the shroud.

"That's disgusting," I say. I've never felt strongly about turkeys, but I'm about to lose my dinner. I turn away.

"You can't get out that way," the shroud says. He pushes me toward a kid standing with a white sheet over his head, two holes cut for eyes. "You have to go all the way through."

"Fine." I run. I push past the other kids and run. I can hear Tina scrambling behind me, screaming with laughter. Just as I reach the door, I hear my name.

"Kris!" A masculine voice.

I stop. A masked monster comes up behind me, grabs my shoulder, and shoves a fistful of vaseline into my hair. I fly out into the dark night, furious. "What the hell!" I'm learning to cuss like Tina.

"It's just a boy," Tina says. "Don't worry about it. He likes you."

I'll never be cool. I don't like boys. I don't care. My flirtation with Adam broke up when I refused to let him go past first base.

"The turkeys are radioactive anyway," Tina laughs. "Those parts probably glow in the dark."

WALKING HOME from the bus stop one afternoon, I notice that Adam's house—nearly identical to ours—is suddenly empty. The curtains are gone from the windows and there's a Realtor's sign in the yard. I arrive home to find my mother in the kitchen making chili. She's humming "In the Mood."

"Why is Adam's house for sale?" I ask.

"Oh." She turns to me. "I thought you knew."

"Knew what?" I haven't spoken to Adam in weeks. He has a new girlfriend one grade up.

"They moved to California. His father got a new job."

"Oh," I say. I climb up on a kitchen stool and set my elbows on the counter. I wish I had been able to say good-bye. Adam was scary and sweet, all mixed up together.

"You knew he was sick, didn't you?" My mother is in her housewife attire: pedal pushers, a long shirt tied around her waist, and a scarf to hold her hair up. I like the way she looks: calm and efficient, as if she could set the whole world straight.

"No," I say. "I didn't know." I'm surprised. "What was wrong?"

She turns to face me and her voice drops, as if someone might overhear. "He has testicular cancer. He had surgery right before they left." She looks away. "I think he must have been embarrassed about it. His mother said he didn't want anyone to know."

Later, in a whispered conversation with Karma, I learn that Adam is not alone in his experience. Karma's had a crush, too, on Scott, a tall, blue-eyed, athletic boy at school. She's been too shy about it to tell me or Karin. Scott, too, has testicular cancer. "Don't tell anyone," she says. "He would die if anyone knew."

A few years down the road, in 1981, a scientist by the name of Dr.

Carl Johnson will publish a study on high cancer rates in three exposed areas around Rocky Flats, including our neighborhood. "The remarkably higher incidence of cancer of the testis in the three exposed areas merits special attention," he'll report. "One possible explanation is the demonstrated propensity of plutonium to concentrate in the gonads."

But no one will believe him.

Dow Chemical and the AEC don't bother doing water samples before releasing a statement that the Rocky Flats plant has no source that could possibly account for tritium contamination. The Colorado Health Department tests the water again and confirms that radioactive tritium, released from Rocky Flats between April 1969 and September 1974, has entered Walnut Creek and flowed into Great Western Reservoir, the primary water supply for the city of Broomfield. The Environmental Protection Agency—which had just been established in 1970—confirms the sharp increase in tritium levels. Tritium emits low-level radioactivity and passes through the body over a period of days or months, but if left in the water supply it continues to be replaced in the body, which can lead to health problems. Nearby families are asked for urine samples, including a couple with a new baby. Several residents, including the new mother and baby, test high for tritium—seven times higher than what officials consider to be "normal"—but the woman's husband, oddly, tests negative. Officials are stumped until the man admits he drinks a six-pack of Coors every night when he comes home from work and never drinks water at home.

AEC officials are slow to acknowledge that the tritium leak has occurred, and when they finally do, months after the initial discovery, residents are told by state officials that the tritium levels are far below what might be "judged to be harmful." Dr. Ed Martell once again disagrees. Broomfield's water reservoir tests at 23,000 picocuries of tritium per liter of water, and the AEC itself considers normal background radiation in Colorado to be approximately 1,200 picocuries per liter.

Even the latter number, Martell claims, is unsafe. Radiation is measured in curies, which quantify its rate of decay or disintegration. A picocurie is one-trillionth of a curie. Scientists estimate that 50 to 100 curies of tritium, or 50 to 100 trillion picocuries, eventually reach Great Western Reservoir.

Rocky Flats maintains there is no threat to residents, and current and past discharges of radioactive material are in "very low quantities." The Environmental Protection Agency sidesteps the controversy by concluding that the public health impact of these radiation doses is "considered to be minimal based on established criteria."

Laverne Abraham, a resident of Broomfield, isn't taking any chances. Every Monday morning, two five-gallon jugs of bottled water are delivered to the Abrahams' front porch. Laverne doesn't want her family, including her six-year-old daughter, Jennifer, to have one sip from Broomfield's "plutonium-lined" reservoir, never mind tritium. Plutonium is heavier than water, and residents are told that the plutonium in the reservoir is harmless as long as it remains where it is—at the bottom of the lake. Rocky Flats officials stress that plutonium in Great Western Reservoir and Standley Lake will stabilize into lake sediment and not create a hazard. Plutonium, residents are told, isn't dangerous unless it's inhaled into the lungs. But Laverne is concerned that even the tiniest amount of plutonium could cause cancer, whether it's ingested or inhaled. "They keep saying it isn't dangerous," she says to a reporter as she shops for groceries with her daughter. "Well, even if it was, I don't think they'd tell us. We won't drink the water here until we get a new reservoir."

But where did the tritium come from? Had there been a criticality at Rocky Flats? Rocky Flats officials explain that the tritium is not their fault; it was apparently brought on-site via scrap material shipped to the plant from the Lawrence Livermore Laboratory in California.

The tritium incident is a public relations nightmare. A storm of publicity eventually forces the plant to reveal that over a period of seventeen years, hundreds of tons of contaminated material were buried in seven trenches and at five other sites at the plant. The mixed waste included asphalt, soil, sewage, and radioactive materials including plutonium and uranium. Rocky Flats officials insist that the buried wastes pose "no hazard." Tests by the Colorado Department of Health will later confirm that plutonium, americium, and strontium-90, a by-product of a nuclear explosion, exist in areas off-site. Strontium causes particular concern, as it can be readily absorbed in the body and deposits in the bones of humans and animals. The presence of strontium strengthens the suspicion that a criticality occurred, perhaps during the 1957 fire.

Al Hazle, who's worked with the Colorado Department of Health for years, is beginning to feel like a detective. As an aside to Broomfield's worried city manager, he jokes that "Broomfield has its mouth over the plant's anus." That's just the way it is, he says. There's a direct connection.

HAZARDOUS OR not, Rocky Flats is a boon to the Denver economy. In 1972, Rocky Flats employs 3,700 people working in three shifts, seven days a week. Plutonium triggers are rolling off the assembly line. Hundreds of millions of federal dollars are being pumped into local communities through salaries and commercial contracts. Real estate is booming. Jefferson County, which includes Rocky Flats, is the county with the second-highest population in Colorado and is growing fast. In 1973 the Health Department of Jefferson County needs a new director.

Carl Johnson didn't get an easy start in life. Diagnosed with tuberculosis at age twelve, he changed his diet and began a strict weightlifting regimen to work himself back to health. In 1946 he joined the army, and after serving three years, he decided to go to medical school. An epidemiologist and radiation specialist, he was hired at the University of Colorado as an associate clinical professor, and in the fall of 1973 Johnson is appointed director of the Health Department of Jefferson County.

Johnson is familiar with some of the problems at Rocky Flats. Still, it's a surprise when a newspaper article crosses his desk revealing that yet another radioactive element is quietly in use at the plant: curium, which is three hundred times more toxic than plutonium. Al Hazle notes in the article that "curium is hazardous, and when they have significant amounts of it at Rocky Flats, we would like to know about it. We're kind of upset when we find out about things [in the newspaper], without the plant letting us know."

One day Johnson is approached by the Jefferson County commissioners, who seek his approval for a new housing development about to break ground just three miles from Rocky Flats, expected to house approximately ten thousand new residents. Johnson checks the state's radiation surveys and discovers the land is contaminated with plutonium. He's shocked that anyone would want to build houses on contaminated land, but he proceeds cautiously. He tells the county commissioners that further study must be done before any development begins, and he gains their approval to go ahead with a study to be conducted by himself and soil scientists from the U.S. Geological Survey, working with scientists from the Colorado Department of Health and the Colorado School of Mines. The study will measure levels of radioactivity in breathable dust on the surface of the soil.

The results are worse than he anticipated. Tests show plutonium concentrations forty-four times greater than what had been measured at the same locations by the Colorado Department of Health method of sampling whole soil, not surface dust. Several of the readings exceed earlier ones by one hundred times or more, one by a remarkable 285 times. The readings are much higher than what Martell and the CCEI found in their study.

Further, he takes issue with the state standard for soil contaminated with plutonium (two disintegrations per minute per gram of soil), which does not take into consideration the size of particles that can be suspended in the air and inhaled into the lungs. Developers typically plow contaminated soil beneath the surface, and while this may bury the soil, it also creates breathable dust. Studies show that concentrations of plutonium are much higher in dust than in soil and that the particles are easily carried and dispersed by wind. Johnson feels this creates a potential hazard for children playing outdoors and that, even for adults, ordinary activities like gardening could be risky.

Johnson presents the results to the local planning board. The board vetoes the proposed development. It's decided that no more subdivisions will be approved near Rocky Flats until further studies are done.

The response from Rocky Flats—and local homebuilders—is swift. In an interview with the *New York Times* the following day, Dr. Robert

Yoder, in charge of safety at Rocky Flats, states that Dr. Johnson's plutonium sampling techniques are questionable and that he has vastly over-
stated the amount of plutonium in the soil. "We don't think he's shown
an increased hazard," Yoder says. "He is just measuring it [plutonium]
differently."

Harold Anderson, chairman of the Jefferson County commissioners
who originally approved the housing subdivision, sides with local home
builders and believes there is no hard evidence that the plant is a hazard.
"If it were," he says, "I'd be the first to get it moved."

Others aren't so sure. Local rancher Marcus Church owns some of
the land for the proposed development. Ranching is the family business.
His family homesteaded the land, bringing the first Hereford cattle to
Colorado in 1869. They grew hay, raised cattle, and gradually expanded
their landholdings. Church's Crossing was a popular stop on the Over-
land Stage Route for bullwhackers and stagecoaches on the two-day ride
between Denver and Boulder.

Marcus Church's nephew, Charlie McKay, wants to be a rancher like
his uncle and the two generations that came before him. The family feels
deeply connected to the land. Charlie's been coming to the family ranch
at Rocky Flats since he was two, spending summers riding out to Standley Lake to count calves and check fences.

In 1951, when Charlie was nine, the federal government approached
the Church family to buy roughly 1,400 acres of their land at Rocky
Flats, along with land owned by two other families. Only the price was
negotiable: the federal government can take private land for roads, dams,
or national security through eminent domain. Landowners were originally offered eighteen dollars an acre and refused the price. Four years
later the government agreed to fifty-six dollars an acre. Marcus Church
wasn't happy with the price, and a deal was reached only when the government threatened to condemn the property.

The government moved in with guards and fences, and Marcus
fought to maintain access to the irrigation and mineral rights he still
owned. He grew accustomed to throwing hay to his cattle under the
close scrutiny of armed guards with binoculars.

Marcus knew little about what went on inside Rocky Flats. But as the years went by and information about the plant's rumored activities began to surface in the media, he started to worry about the value of his property. The family had big plans for eventual development of homes, business parks, and a shopping mall, but after reports of plutonium in the soil and no clear determination of how much plutonium was "safe," Marcus could no longer get building permits. Charlie watched as his uncle grew increasingly frustrated.

In 1973, Marcus Church decides to sue the government. He contacts the head of a local law firm, who puts a young attorney, Howard Holme, on the case. "It's really gone too far here," Marcus tells Howard.

Church family members aren't the only ones who are unhappy. Builders and developers are angry about falling land values. The city of Broomfield wants the AEC to divert Walnut Creek, which is contaminating the city drinking water with tritium, and wants a new wastewater reprocessing plant to keep radioactive material out of the city water supply.

The same year Church files his lawsuit for property contamination, Dow Chemical awards its employees cash rewards in recognition of "superior performance in safety, environmental control, production and energy use reduction." Soon, however, the situation changes. Dow Chemical has been the AEC's contractor at Rocky Flats for twenty-two years. To the early employees of Rocky Flats, the corporation was known fondly as "Mother Dow." To the early activists, Dow Chemical was known not only as the producer of plutonium detonators, but as the manufacturer of napalm during the Vietnam War. But Dow has grown tired of ongoing accidents, leaks, protests, media scrutiny, and its relationship to the AEC. And its reputation in Washington is suffering. Employee Jim Kelly, now president of the Steelworkers Union at Rocky Flats, tells a congressional committee that Dow is not operating the facility safely. The company decides not to renew its contract, and the AEC requests bids from private contractors for a new company to run the plant.

Dr. Carl Johnson finds himself facing a growing storm.

☢

ONE DAY, years in the future, Randy Sullivan will tell his children how lucky he is to have grown up in Meadowgate Farms. The world feels wide open. Kids run free in the neighborhood, and most boys and girls have a horse or dirt bike. In the winter all the kids skate on the frozen pond. Sometimes there are as many as thirty or forty kids on skates, twirling and chasing hockey pucks, and some of the older kids like to spin donuts on the ice with their motorcycles and dirt bikes, even though it's been strictly forbidden. Summer days are for floating on inner tubes along the network of streams and irrigation canals that flow to and around Standley Lake. He has a host of pets—turtles, hamsters, dogs, and whatever he can trap and domesticate. Randy remembers days in high school when he and his friends would ditch their afternoon classes and drive down to the local grocery store and dare each other to shoplift a six-pack of beer. They would then head out to Standley Lake and take turns jumping off the pipe.

The cycles of snow in Colorado are punctuated with days of clear, clean sunshine, when the snow melts and the air turns as warm as summer. The setting sun burns the sky peach and then brilliant orange and the mountains turn from gray to cobalt blue. In his house in Meadowgate, Randy looks out the window of his room to a house on a hill a few blocks away in Bridledale. He has a straight view. He watches a second-floor window to see if the lights come on. They do. He goes out back to the horse pen behind the house and bridles Cocoa, his creamy-colored palomino. He loves to ride. His father has given him a beautiful saddle, but he prefers bareback. He calls to his Labrador and the three of them clop along in the half-dark. He rides down the quiet streets, turns left on 82nd Avenue, and trots to the end of her driveway. He pauses for a long moment. If his friends found out what he was about to do, they would tease him mercilessly. But he does it anyway. "Hey, Kris!" he yells. He waits for a response. There is none. He yells again. "I love you!" Silence. Suddenly embarrassed, he turns and sets his heels to Cocoa and gallops back up the street.

☢

Iᴛ's ᴀ Saturday morning, early, and Karma and I are getting the horses ready to go to the county fairgrounds for a gymkhana, one of the first equestrian competitions of the summer. Karma loves pole bending, a timed race that involves weaving your horse in and out of a line of tall poles without knocking one down. My love is the barrel race. Tonka loves it, too. My knees are scarred from taking the barrels too close—you can skim them as long as you don't knock them over—and more than once I've had a rein break or slip from my fingers as Tonka races across the finish line on his own with my hands twisted in his mane. Our best time is just over seventeen seconds, nearly as good as the pros. There are other contests as well: the keyhole, the goat rope, calf wrestling. Sometimes there's a greased-pig contest. It costs two dollars to enter each event, and if you're lucky you get a ribbon or a plastic trophy.

With a round wire brush I rhythmically rake Tonka's coat to tease out the last of his winter hair, and then use a soft brush to stroke his coat into a high sheen. Karma works over Comanche in the same manner. The bridles and saddles are oiled. My hair is in pigtails to keep it out of my eyes, and my white cowboy hat waits on the backseat of my dad's Blazer. Karma, never one for frills, doesn't wear a cowboy hat. None of those fancy cowgirl shirts for her, either, just boots and jeans and a quiet determination that serves her well in the arena.

My mother usually drives us to the fairgrounds, but this morning my father emerges from the house, looking unshaven and unkempt from spending the night in his recliner. He's heavier now and seems disconnected from his body, his clothes loose and flapping. Let's go, he says. The familiar scent of bourbon is on his breath, and I find it almost as comforting as the scent of cigarettes and aftershave that clings to his body. He'll drive us to the fairgrounds and drop us off, leaving the horse trailer, and then head to his office until the end of the day, when he will return to pick us up. He backs the Blazer up to the trailer and drops the hitch over the ball. We load the horses, first Comanche, then Tonka. They've had their breakfast, but there's a cup or two of Omolene in the feed bin to keep them busy. Tonka presses his face to the little window at the front of the trailer. He likes to watch the road. Trains and loud noises make him nervous. We don't tie either horse to the front of the trailer because sometimes, when spooked by a locomotive or motorcycle, Tonka will jump feet-first into the hay bin and tangle his feet in the rope.

We pull out of the driveway. Karma sits in the backseat and rolls the window up. The air smells like rain. I sit in the front, my legs tucked up, hugging my ankles. None of us wears a seatbelt.

The neighborhood is quiet. There's no one on the road. We don't speak, but it's a comfortable silence. Or comfortable enough. We turn onto 80th Avenue and then left on Simms, to go past the junior high school. We pick up speed.

"Dad," I say cautiously. It's not uncommon for him to drive fast, but this is faster than usual. He doesn't like backseat drivers.

The road rises slightly and we hit the railroad tracks, the same tracks where my sisters and I laid pennies. Then the road dips. Suddenly the Blazer feels like it's flying. The trailer begins to fishtail, and the car swerves sharply to compensate. My thoughts freeze as my body seems to rise in slow motion, up toward the roof of the car.

Later my father will say he swerved to avoid an oncoming car. He will mention the rain and the slick pavement. He will say he saved us from a head-on collision.

My sister and I never saw another car.

The car and the horse trailer separate after the first roll. My father swears. The car rolls again and again and then everything tumbles into an explosion of glass and grass and bodies. The rain comes in. I feel a sharp, crushing blow to the top of my head. Then I find myself lying flat in the back of the Blazer, surrounded by metal and glass, facing the back window, curiously open. My head and neck feel wrong.

"We have to get out," Karma says. Her voice is calm and seems to be coming from a long distance away. "Are you okay?"

"I think I have glass in my eyes." I'm afraid to turn my head.

"Can you see?"

"Yes."

We crawl across what used to be the roof of the car and out the back window. The window itself is gone. The Blazer is upside down in a dry irrigation ditch, flattened, the roof and hood pressed into the ground and the belly of the car facing the sky. My father's door is partially open. I can see his shoulder pressed against the window glass. After a moment the gap in the door widens and he pushes out. There is blood on his cheek, and his voice is thick. "I'll flag someone down." He stands. "I think I've hurt my back," he says.

I feel the cool rain on my cheek. There are no tears. Karma and I walk over to the horse trailer, which has rolled and then lodged in the ditch a few yards back from the Blazer. Tonka and Comanche lie on the floor of the trailer. They look like they're asleep. *I'm glad we didn't tie their heads,* I think. Unlike the car, the interior of the trailer is nearly intact.

A man appears, slim, blue-jeaned, wearing a straw cowboy hat. Who is this angel? A passerby or a neighbor. He stoops down and puts a hand on Tonka's warm flank. There is a shudder in the flesh. I look over to Karma, who has moved to Comanche's head. "They're alive," she says.

"They've just been knocked unconscious," the man says. "It's probably what saved them." He tells us to sit on the grass. He crouches next to their heads, stroking their cheeks, and then he rises and kicks open the trailer door. The horses scramble awkwardly to their feet and he backs them out, blinking and shivering, into the rain. "They seem to be okay," he says. He ties them to the fence. "Do you live far?"

"No," we say.

"You should walk them home," he says. "They probably won't want to get in a trailer again."

"Okay," Karma says.

"Are you two sure you're all right?"

"We're fine," we say in unison. We never see him again. We brush the crumbled glass and bits of dirt and grass from our arms and faces. No blood. No tears. I can't stop shaking. It's the cold rain, now steadily streaming.

An officer arrives and agrees to drive up to our house and inform our mother. "Tell her everything's okay," we say. My father agrees. "Tell her we're all fine."

My mother takes the news calmly—it's not the first accident my father has been involved in. She hopes the neighbors don't notice the police car in the driveway.

There is minimal fuss and no ambulance. The Blazer is towed away. Karma and I walk the horses home and I lie on the couch until I stop shaking. I have a headache and I feel afraid to turn my head. My parents do not take us to the doctor. "I'm fine," I say. Karma says she's fine, too. We're Norwegian. Norwegians are tough. Or so we think.

Besides, we all agree this is something the neighbors don't need to know, especially since my father's practice has been a little shaky lately. He needs his clients—many of whom, ironically, are fighting DUIs. He's the best in the business. Neighbors, clients, family, friends: no one needs to know about the accident.

Weeks later, when my father can no longer sit in his office chair, he goes to a doctor and discovers he has a fractured vertebra in his lower back. He wears a brace for a few weeks and then tires of it and throws it out. Years later, after ongoing headaches and pain, I learn I have a broken neck, two fractured vertebrae that have fused together over time. A quarter-inch higher or lower and I would have been severely disabled. You're lucky you didn't end up like Christopher Reeve, a doctor will say.

We never speak of the accident again. Silence is an easy habit for a family or a community. This is just for us to know. Eventually we'll forget this ever happened.

Before long the entire incident is, indeed, forgotten.

A painting hangs in our living room, a blur of blue and gray with a woman's face in the center, her body mostly obscured. A long lock of dark hair falls across her eyes and only her mouth is visible, a pouty smear that's sullen or seductive, it's hard to tell. My mother doesn't like the painting—our uncertain budget has curtailed her home decorating impulses—but it's a gift from one of my dad's clients, a payment, and he says it's art.

My father's law practice is, as my mother likes to say, going down the tubes. It's feast or famine. From time to time the electricity at our house is shut off. My grandfather, a stern man who worked as a banker during the Depression—traveling from bank to bank, looking at balance statements, and determining whether or not an institution should remain open—swoops in to oversee my dad's office. He and my grandmother Claire sell their home, drive west, and move into an apartment in Arvada. Grandma Claire is a retired schoolteacher. She dotes on my father, her only child, and now she dotes on us kids. She wears round, dark-rimmed glasses that make her look intellectual despite her flowery dresses and plump arms. My grandfather is as impenetrable as my father.

Both my parents feel a little panicked.

I've shot up like a weed, Grandma Claire says. I hunch down in my desk at school. I drink coffee on the sly when she scolds that it will stunt my growth.

Tonka suddenly seems like a pony. At fourteen and a half hands, he's on the small side as far as horses go. But my father knows a man at the local racetrack—the same place where our neighbor Bini Abbott buys many of her horses—and the man has a horse. A mare, a tall Thoroughbred mix, too old and used-up at four years of age to make it at the racetrack but maybe a good kid's horse or broodmare. Her bloodlines are good, he says. Her name is Sassy Cowboy. I call her Sassy.

On a startling fall day we hitch up the trailer to pick her up. The sun burns bright and the aspen leaves glitter like gold coins. Sassy is a tall, elegant sorrel with a white stripe down her nose. She's jumpy and nervous and we have to coax her into the horse trailer with horse candy.

"Is she broke to ride?" my dad asks. "Or just broke to race?"

"Not saddle broke exactly." The man stands back with a rope of tobacco chew tucked under his lower lip. "Just go slow. She's a good gal. The track makes 'em a little crazy."

That's all right with me. We don't mind a little crazy at our house.

Sassy is old for a racehorse, but in real horse years she's a youngster. She gallops back and forth across our back pasture at full speed, bucking and farting in bursts of energy and general high jinks that cause the other horses to snort and stamp their feet. She loves it when the chinook winds blow down strong and hard from the mountains. All our animals act a little crazy when the chinooks hit. She stretches her nose out like a greyhound, shaking her mane back and forth like supermodel Christie Brinkley.

We fall in love slowly. She lips up horse candy from my palm and the back pocket of my jeans and rests her head on my shoulder, but she's skittish about having a rider. It's not long, though, before I can swing up on her back and sit while she finishes her dinner mash. We gradually work up to a walk, then a trot, then a slow gallop. I curry her coat into a high sheen and varnish her hooves. She's the most beautiful horse in the neighborhood. Sometimes, when we go for long rides around Standley Lake, I bring Tonka along on a halter lead. He takes two steps for every one of Sassy's, but he always enjoys the trip.

It's getting harder to find places to ride. When we first moved to Bridledale, Karma and I could gallop down dirt roads and across fields and ride all the way to the Tastee Freeze in Broomfield for chocolate dip cones. Now there are houses in the way.

RABBITS AREN'T the only animals people are starting to wonder about. Cattle graze on land near Rocky Flats, cattle that are eventually shipped to the stockyards and slaughterhouse on the far side of Denver. In early December 1974, residents wake up to a shocking headline in the *Rocky Mountain News*: CATTLE NEAR ROCKY FLATS SHOW HIGH PLUTONIUM LEVEL. An Enviromental Protection Agency (EPA) study has found that cattle in a pasture just east of Rocky Flats have more plutonium in their lungs than cattle grazing on land at the Nevada Test Site, where the United States conducted hundreds of aboveground nuclear explosions in the 1950s and 1960s. Plutonium, uranium, americium, tritium, and strontium are found in measurable quantities in the cows' bodies, and levels of plutonium in the lungs and tracheo-bronchial lymph nodes of the cows are especially high.

My family's not too worried. "They're always finding something out there," my mother says. "The turkeys from Jackson's always taste fine."

Marcus Church is outraged. "People don't want to buy my cattle," he tells Howard Holme, the young attorney who's just started work on his case. "They think they're contaminated. I've cooperated with the plant in the past, but something needs to be done."

The DOE and officials at Rocky Flats challenge the test results, and a month later the EPA changes its mind and says that the conclusions are based on too little data. Nonetheless, Church, along with other landowners, continues to press his suit against the federal government and the operators of Rocky Flats based on the allegation that the plant has contaminated surrounding properties, driving property values down.

Attorney Howard Holme grew up in Denver, attended Stanford University and then Yale Law School, and was a conscientious objector during the Vietnam War. He jumps at the chance to work on the Rocky Flats case. He begins reading piles of documents about the 1957 and 1969 fires, the accidents, the leaks, and the ongoing problems with soil and water contamination. He knows plutonium is dangerous. The question for the court is, how much plutonium constitutes actual risk?

Holme employs a number of scientists and physicists to examine the data, and hires physicist Steven Chinn specifically to evaluate health risks. Multiple regression analysis is considered one of the most effective means of doing epidemiological analysis to determine what might cause an increase in cancer or other health effects. Using computers, Chinn conducts a multiple regression analysis examining the residents of each census block around Rocky Flats in comparison to more than a hundred criteria, including socioeconomic class, education, age, gender, income, and—most important—where they live with respect to previously measured levels of plutonium in the soil, correlated with soil measurements following the two major fires.

The pretrial statement painstakingly prepared by Howard Holme and Steven Chinn is five hundred pages long. A review of all the studies, consultation with national and international experts, and the regression analysis show that there is, indeed, a rise in cancer that can be attributed to Rocky Flats. Not only is off-site land generally contaminated, but there are also isolated "hot spots." The Holme Report, as it comes to be called, concludes that the plant has been highly negligent, with results including fires, accidents, and poor storage of waste plutonium. In addition to the thousands of leaking oil drums left out in the open, plutonium, uranium, americium, curium, and neptunium have seeped from waste ponds. The report concludes, however, that most of the contamination resulted from the 1957 fire, which burned through the entire plenum of filters and completely melted the top of the ten-story smokestack, releasing an unknown amount of plutonium and other materials into the air. Church's attorneys claim that even though facts regarding the plant's contamination are still forthcoming, negative publicity alone has stripped the remaining Church land of its value. The Denver area is growing rapidly and development in other areas has soared, but homes and properties in the Arvada area are not as valuable. Less cancer was found in Boulder, which is upwind of Rocky Flats, and property values are much higher. Could all of that be chance? Holme asks.

Church and his attorneys eventually submit a formal claim against the Energy Department, asking for millions of dollars in damages. "It is as if the government had moved a tiger in a papier-mâché cage onto land adjoining the plaintiffs' lands," Holme writes. "Maybe the paper cage will hold, but no man in his right mind is going to move next door and find out."

But many residents are unaware of the tiger next door, or they refuse to take it very seriously. Houses continue to creep up to the edge of the buffer zone, an area surrounded by three strands of barbed wire and an occasional No Trespassing sign. The three-string barbed wire around the buffer zone is a dare. Karma and I ride along on Sassy and Chappie and take turns kicking the metal signs with the toes of our cowboy boots, daring each other to crawl under the wire.

One afternoon Randy Sullivan is riding in a car with two friends. They drive along Indiana Street, just north of the east gate of Rocky Flats. Neither he nor his friends know much about Rocky Flats—they've heard, like the rest of us, that the plant makes cleaning supplies. Randy lives in Meadowgate, but his family owns open land near Rocky Flats and the boys often hang out there or host late-night keggers with other kids from school.

There's an argument in the car and the boys pull over to the side of the road. Randy stands back as his two friends argue. No fists, nothing serious. Suddenly a guard in full uniform appears.

"What are you guys doing?"

The boys stop and stare. The man is armed.

"Nothing, sir," Randy says.

"Do you know this is Rocky Flats?"

"Yes."

"Can't you read this sign?" The guard points to a sign hanging from the barbed-wire fence.

"It says 'No Trespassing,'" Randy's friend says. "We weren't trespassing."

"Can't you read the sign?"

The boys know what else the sign says: "Use of Deadly Force Is Authorized by Presidential Order."

What they don't know is that due to "increasing terrorist activity," guards at fourteen nuclear sites around the country—including Rocky Flats—are now under orders to shoot to kill. M-16 rapid-fire rifles have replaced .38-caliber revolvers and all Energy Department facilities have new armored personnel carriers.

"Get in that car," the guard orders. "Don't come back."

Despite the voluminous documentation, the judge assigned to the Church lawsuit refuses to set a trial date. The case drags on and on. Marcus Church doesn't live to see the results. Just before his death in 1979, he tells his nephew, Charlie Church McKay, to keep fighting the lawsuit. Hang on to the land, he says. Keep fighting.

OUR NEIGHBORHOOD is transient. The oil business, the airline industry, and the changing workforce at Rocky Flats mean that people move in and out. Tina moves away. I miss her, but not without some small sense of relief. I've become a solitary sort. My mother's reading tastes have changed, and so have mine. She still reads racy thrillers with bosomy heroines on the covers, but once in a while something else sneaks in. *The Gulag Archipelago. Love Story. Emerson's Essays. Jonathan Livingston Seagull.* One afternoon I find a copy of John Updike's *Couples* under her bed and I'm thrown into a new realm of reading. From *Couples* I move on to John Barth and Virginia Woolf, Sylvia Plath and John Irving, and then back to Updike again, books I buy with my own money and hide under my own bed. I want to be a writer. It seems impossible.

Arvada grows so quickly that the high school can't accommodate all the incoming students. The school board decides to do a split session until a new high school, Pomona, is built near all the new neighbor-hoods east of Rocky Flats. Half the students will attend school from early morning until noon, and the other half from noon to early evening. I pull the first shift, which means riding to school in the dark and having chemistry class so early in the morning it makes my head ache.

Our house has changed. The sod never really takes, and the front lawn, which slopes sharply to the driveway, begins to look dry and rough. The split-rail fence we've plastered with linseed oil splinters and cracks in the dry heat and the horses get out. Barney still visits the Smiths' garden. Their daughter, Tamara, rides with a local riding club called the Westernaires. Karma and I ride with the same club.

Mr. Smith has few kind words for our wayward pony, who can open any gate. Sassy doesn't wait for a fallen rail; with a good enough start she jumps the fence with no problem. She's more sociable than naughty. I can always find her touching noses with the horses up the street.

One Sunday afternoon my father declares he's going riding. He pays for the horse feed, he declares, so he might as well get some benefit from it. Karma and I exchange glances. As far as we know, he's never ridden a horse in his life.

Comanche seems to be the sturdiest bet. We call him to the gate and saddle him up. Comanche's eyes are wide, ears pricked, and he sidesteps nervously when dad swings up into the saddle. Karma hands him the reins. "You need to pull the reins up shorter," she says. "You can't let them hang like that. Do you know how to make him turn?"

"Yup," Dad says. His speech is a little slurred and his breath is strong. He settles into the saddle, looks up to the house to see if our mother is watching—she is—and claps both legs hard against Comanche's sides. Comanche bolts.

The gate is open and they fly out of our pasture, galloping heavily across the wooden bridge over the irrigation ditch and down the path that leads to Standley Lake. My father leans far back in the saddle, one hand on the saddle horn, the other arm in the air, reins flying out in either direction, a loose, flapping version of a Teddy Roosevelt Rough Rider. Comanche is not an elegant animal and he gallops forcefully, running hard and spooked like he's got the devil on his back. Karma looks stunned. I can't breathe. They gallop around and around the field, barely missing trees, barbed wire, and fenceposts.

They are completely out of control.

I can't tell for sure, but I think my dad is grinning.

When they return, they're both spent. Comanche is covered with sweat, his flanks heaving. My father's hair is whipped by the wind and he, too, is breathless, his face shiny with perspiration. But he dismounts with the air of a conqueror.

"Thanks," he says as he hands the reins to Karma.

"I think he wants to kill himself," my mother mutters that evening over hamburger casserole. "What did he think he was doing?"

I say nothing. I adore my mother, but I fear for her. She seems helpless, caught in the vortex of my father's dark moods and unpredictable behavior. I try never to displease her. I love the scent of Juicy Fruit gum on her breath and the hint of Joy perfume on her neck, the crisp crinkle of her hair stiff with aerosol spray and the chipped pink polish on her nails. I study her closely, just as closely as I watch my father, vigilant for the slightest hint of a crack, a slip, a breakdown. We are always on the edge of catastrophe.

"Everything will be all right," I say. I look around at my siblings at the table, sunburned and hungry. Kurt grins. We'll be all right.

MANY THINGS occur in my father's life that we're not supposed to know about. Car accidents. Missed appointments. Nights in jail. One afternoon my mother drives us all to the courthouse, where we sit in a row on the front bench and watch as our father comes out in a jumpsuit. "He has to be able to get to work, Your Honor," the attorney says. "He has to support this wonderful family." My mother nudges us and we stand. "Hold your heads up straight," she whispers. The judge takes pity, but for weeks we don't see our dad. He goes to work during the day and returns to the jailhouse at night. "Is he in jail?" Kurt asks. My mother won't answer directly. "He loves you," she says. "He loves you all very much, but he just can't be here right now."

Sometimes I find myself counting the days until I graduate from high school. I have to get away. But then I feel guilty. "You kids are spoiled," my mother reminds us. "Pets and horses, and this big house!" She complains about the house—it's too much work, inside and out, and the mortgage sometimes goes unpaid—but she loves it all the same. "Your father and I would do anything for you," she says.

The court lets Dad come home for good, though we still rarely see him. Some nights, when he comes home very late, my mother will ask one of us to stay with her in the bedroom and be there to help block the door. She keeps a baseball bat in the front closet. "Just in case," she says. All bark, no bite, my brother says, but we all stand ready to rescue her. Together we hide under the stairs with the baseball bat.

"Why don't you leave him?" Karin demands. "Why don't you stand up to him?"

"I can't say anything to your dad," she says. "He has a hard enough time as it is." Besides, she says, she made a commitment. "That's what marriage is. A commitment."

My sisters and I whisper that we will never marry.

"And your father loves you," she reminds us. "He just can't show it."

EACH OF us thinks about running away at one time or another. We live in a house where the rules are inconsistent or nonexistent, and the contours of our lives constantly change. Some nights, just to get out of the house, I sleep with a blanket on the trampoline in the backyard, gazing up at the stars, wishing the time would pass more quickly and I could leave for college, for a job, for a boy, for anything.

One afternoon, after a long night of hearing our parents argue, Karma stands at the top of the stairs, listening to our mother on the phone. Suddenly she is filled with a muted sense of rage. She returns to her room, takes some clothes out of her dresser drawer, and quietly slips down the stairs. She saddles up Tonka—her new hand-me-down, now that I have Sassy—and puts the clothes in her leather saddlebag. She's accustomed to long rides, usually by herself, and she loves to gallop over the hills and fields. She feels most comfortable when she's out in the landscape. It's dusk when she and Tonka reach the box canyon at Leyden, a secluded spot near the train tracks just before the foothills where high school kids sometimes go to drink and make out. The spot is empty now except for broken beer bottles.

With her jacket bunched under her head for a pillow, she lies on the ground and looks up at the black sky riddled with stars. The night is cool but not cold, and Tonka stands peacefully over her. No one will find them here, she thinks, and she's right. No one comes looking.

She wakes when the sun strikes her face. The ground is rocky and her knees are stiff with cold. Suddenly she realizes she's hungry. How foolish not to bring food, or even a sleeping bag! Tonka is hungry, too, and nuzzles her pockets for horse candy. But what to do? She has no money—she's eleven and too young for a job. She can't stay here, but she doesn't have anywhere else to go.

Feeling like a whipped puppy with its tail between its legs, she turns Tonka toward home. She'll have to face the music. It won't be pleasant.

But when she gets home, no one says anything. No one's even noticed her absence. Or Tonka's. They haven't been missed.

SOMETIMES I wonder how my father feels about having three daughters. He complains about women drivers and dislikes female attorneys. My mother says he's just mad at the world.

My sisters, my brother, and I all share a fierce, if silent, loyalty. But we begin to splinter away from one another. I move into the bedroom in the basement, where I paint the walls vivid purple and keep the door closed. Karma is never home; no one knows where she goes. Karin's rarely around, and when she is, her bedroom door is closed, too. Kurt spends time at the homes of friends. We tiptoe around the house, hoping not to be called up to our mother's bedroom during her afternoon siestas. She has a row of little orange bottles in her medicine cabinet, and she keeps a broom next to her bed to pound on the floor when she hears we're home from school. "Kris!" she calls. "Who's down there? Karma? Karin? Come up here!"

But she wants to keep the family together. "We have to do what we can to help your dad," she says.

One Saturday afternoon she piles us all in the station wagon, just like old times, for a drive in the mountains. She slides in behind the wheel and starts the car and we wait. Dad doesn't come out. Finally she goes into the house and gets him. His mouth is set in a hard line. We drive silently through the foothills to a pretty house surrounded by pine trees, the house of a psychiatrist. "This will be fun," Mom says. "Lots of people do this sort of thing now."

We all nod grimly.

A man in a sweater walks out to the driveway to meet us. "Welcome," he says, and extends a hand to each of us. "This must be your wonderful family," he says to my father. He doesn't answer.

The man shows us to his living room. My parents each get an easy chair and the kids are instructed to sit cross-legged on the carpet. Alcoholism is a disease, we're told. But it's a disease of the family and not just the person.

My heart sinks. My siblings look glum. It looks like it's our fault after all.

The psychiatrist takes out some construction paper and crayons and hands them around. "I'd like each of you to draw something for me," he says. "Draw me a picture of your house. Of your family home."

We clutch our papers close so no one can see what's being drawn, except for Kurt, who never follows the rules. He lies on his stomach on the carpet and draws a square little house with a door and two windows. A puff of smoke comes from the chimney and a stone path curves up to the door. He draws a tree on each side of the house. "Good," the psychiatrist says. "You did a nice job there."

I can't decide what kind of house to draw, so I draw a horse instead. Karin and Karma are reluctant to show their drawings. "How about you?" the man says, turning to our parents. My mother's drawing is detailed and precise—she took art classes before her father talked her into going to nursing school. Her house looks like a bare-bones version of our house in Bridledale. "And you?" the man asks. My father turns up his paper for us to see.

The paper is solid black, colored in hard crayon from corner to corner.

The psychiatrist nods. When we get back to the car, Dad mutters something about a very expensive drawing lesson.

We don't bother to stop for dinner on the drive home. "Well, that was a good first step," my mother says cheerily as we file back into the house. We don't go back.

IN LATE 1974, seven companies respond to the AEC's request for bids from private contractors. On November 21, the AEC selects Rockwell, one of the nation's largest industrial corporations, as the new contractor at Rocky Flats. It awards Rockwell the typical cost-plus contract that will provide the company immunity from most lawsuits should problems occur.

Rockwell is well-known to activists like Judy Danielson, Pam Solo, and other volunteers from the Denver office of the American Friends Service Committee, a Quaker-affiliated organization working for social justice and human rights. One of their projects is to curb production of the B-1 bomber, which is built by Rockwell.

And there are still many unanswered questions about the effects of the 1969 Mother's Day fire. No one—not Dow, Rocky Flats, the Colorado Health Department, or the Jefferson County Health Department—can or will provide clear answers. With Maury Wolfson of Environmental Action of Colorado, activists form a coalition of citizens and grassroots groups called the Rocky Flats Action Group. The group begins a public information campaign with the slogan "Local Hazard, Global Threat," and bumper stickers bearing the slogan begin to appear on cars. Meetings are set up with incoming Governor Dick Lamm and Congressman Tim Wirth. One of the people in attendance at these early meetings is my sister Karma.

Denver has changed from a cow town to a burgeoning city of oil and gas companies and successful sports franchises. Colorado is riding an early wave of environmental enthusiasm. Citizens are finally able to stop the Rulison Project, the blasting of underground nuclear bombs to stimulate production of natural gas. But no one's sure what, if anything, to do about Rocky Flats.

Governor Lamm and Representative Wirth respond to citizen pressure by creating the Lamm-Wirth Task Force to study Rocky Flats and make recommendations for its future. In late 1974 the task force conducts a series of public hearings to seek answers to citizens' questions about Rocky Flats. More than fifty people testify, including experts from around the country. Farmer Lloyd Mixon of Broomfield brings along Scooter, a piglet with deformed ears and hind legs. Mixon testifies that Scooter is only one of many animals with deformities born on his ranch southeast of Rocky Flats, beginning as early as 1965.

Rocky Flats officials contend that the plant makes a "vital and substantial contribution to freedom" by manufacturing plutonium triggers, and that the plant's $70 million operating budget is necessary to the region's economy. They dismiss claims of health problems and sinking property values. The Colorado Health Department, often caught in the middle between Rocky Flats and the public, attributes the problems with Mixon's animals to unsanitary conditions and inadequate nutrition, an allegation Mixon vigorously denies.

When the Lamm-Wirth Task Force publishes its report after the hearing, it concludes that there are not only serious safety issues at Rocky Flats but also the potential for a catastrophic nuclear accident. The risk is too great for a weapons plant to be located near a large population center, and the report recommends that the nuclear work done at Rocky Flats be closed or relocated. In a press interview, task force member Patrick Kelly, a Rocky Flats worker and United Steelworkers of America official, states that Dow Chemical is "neither responsible nor responsive" to the public or to Rocky Flats workers, and that the secrecy at the company is supported by the AEC.

The report recommends the establishment of a permanent citizen monitoring committee. It also criticizes the Price-Anderson Act, passed in 1957, and renewed in 1967 and 1977, which indemnifies the nuclear industry against nuclear accidents and exempts corporations from penalties associated with their actions, even in the case of gross corporate negligence. (In 2003, the Price-Anderson Nuclear Industries Indemnity Act was extended until 2017.) Companies like Dow and Rockwell can pollute without penalty, and taxpayers bear the cost. "The Price-Anderson Act should be repealed and replaced with a nuclear industry liability act," the task force states, "which requires contractors and licensees to bear the risk of doing business in the industry."

Contradicting its own recommendation, however, the report also emphasizes that "strong consideration should be given to maintaining the economic integrity of the plant, its employees, and the surrounding communities."

Some critics claim that the Lamm-Wirth Report is "a masterpiece of compromise." It compromises the health of local citizens with the competing interests of a government that wants to make bombs, developers that want to sell houses, and workers who need jobs.

Dow Chemical has left Rocky Flats after two decades of accidents, plutonium releases, and safety problems, most of which are still hidden under the cloak of Cold War secrecy. Now that Rockwell has stepped in, it's business as usual.

WHAT LITTLE popularity I enjoy in junior high disappears by the time we move into the new high school. I'm not a stoner or a jock or even a proper redneck. I play clarinet in pep band but I don't hang out with the band geeks. I go to Rodeo Club meetings but I'd rather listen to Led Zeppelin than Tammy Wynette. I don't smoke pot, I think beer tastes like soap, and I'm still painfully shy. I persuade the principal to let me take auto mechanics instead of home economics—who wants to be a housewife?—but the teacher won't let me actually work on a car like the guys in the class.

Spirit Day arrives. Classes end early and all the students and teachers gather in the new gym for a special assembly, a pep rally in the gym just before our first basketball game of the season. I sit with the band in the bleachers, wearing my Pomona Panthers T-shirt, and we belt out a fumbled but deafening version of "Rock Around the Clock." Our theme for Spirit Day is "Happy Days," from the popular television show.

The assistant principal walks up to the microphone. "Panthers!" he cries. "Are you ready for a little spirit?" The crowd roars. We're a big school. We stamp our feet on the bleachers until the whole gym rocks.

Suddenly there's the sound of a motorcycle in the hall, revving its engine. The gym falls silent. How can this be? Motorcycles aren't even allowed in the parking lot.

The doors swing open and the motorcycle roars in. The rider wears a white T-shirt, black leather jacket, and black boots. His dark hair is shiny and slicked back. Before he can take the bike full-speed around the gym floor, the assistant principal waves him down. "Boys and girls!" he yells into the mike. "I give you . . . The Fonz!"

The crowd erupts. Girls start screaming. I'm shocked. It's Randy Sullivan on his big brother's motorcycle.

Randy smiles and gives a big thumbs-up. All the girls love him, and he gets a full page in the yearbook. There's no hope for me now.

THE SUMMER of my sixteenth year, my mother goes back to work, my grandfather starts to have heart trouble, and bill collectors call every night at suppertime. My father is adored by his clients. He gets them out of their divorces and lawsuits, accidents and DUIs—some of the neighborhood kids have begun to fall into the latter category—and they all love him for it. He wins a pro-bono award from the local bar association. "Your father is wonderful," people say. "He's so smart, and he's fun to talk to."

I think they're nuts. Who are they talking about? He must be living a double life. Nothing has changed, and even our mother seems stumped. "I thought I knew him when we married," she says. "But I often wonder what happened to him." She sits on her stool in the kitchen and smokes one cigarette after another. She's not convinced that all men aren't like this anyway, to some degree or another. Unreliable, unknowable, impossible to trust. And they can't even do their own laundry. But she likes men—she's always been a big flirt—and she firmly believes in marriage. Everyone in the world should be married.

"Look what I have, despite everything," she says. "A wonderful family and four beautiful children." This has become a mantra: a-wonderful-family-and-four-beautiful-children.

My mother starts working a few days a week on the swing shift at the local nursing home. She hides the money from Dad so she can buy groceries. In our whispered bedroom conversations, she tells me that the money he earns seems to go into his sinking law practice or the cash drawer of Triangle Liquor. She never sees it. Our household grows a little leaner. We give Chappie to a local 4-H kid—he's a little addled in the head anyway. There's little money for hay and horse brushes and riding lessons, never mind clothes and books and tennis shoes.

When school ends I look for a summer job. I'm old enough to get a real one, but pickings are slim. I skim the classifieds and find nothing. Then I see that the truck stop a short drive from our house is hiring. I fill out an application for a waitress position. "Thank goodness you came in today, honey," the woman behind the cash register purrs. "I had two girls quit just this morning." Her copper-red hair is piled up in a beehive as stiff as a football helmet. It's 1974. Even my mother has stopped wearing beehives.

I place my application on the counter. A long U-shaped bar stretches around the edges of the room, with truckers on bar stools hunched over their meatloaf and mashed potatoes. A few tables stand empty in the middle. A black-and-white TV roars in a corner and the room is filled with smoke. "When will I hear back?" I ask.

The woman rings up a gas purchase and hands the receipt to the man waiting. Several truckers stand impatiently behind him. It's a busy place, and she looks at me as if she's already forgotten me.

"Do you have any experience, darling?"

"No." This is my first job, except for the long Saturday afternoons spent hanging out at my dad's office.

She looks me up and down. "That's all right. Just show up tomorrow morning at five thirty. We'll give you one week on the early shift, six to two, and if that works out we'll put you on the late shift. Girls make a little more money on the late shift."

I watch as a waitress takes a pot of coffee up the line of truckers at the bar, filling each cup.

"Anything else, dear?" The words are not kind.

"What does it pay?" I blurt.

"A dollar twenty plus tips. And you'll get a uniform. The uniform is free. You need to get yourself a pair of pantyhose, honey. Neutral color." She shoos me away with her long fingernails. "Tomorrow morning. Five thirty. Girls get fired for being late."

I don't tell anyone at home about my job. It feels new and exciting and a little dangerous. The best thing, though, is the paycheck. A paycheck that's all mine.

EVERYONE—THE governor, the mayors, Rockwell, and the Energy Department—seems to like the idea of a citizen watchdog group. Following the recommendation of the Lamm-Wirth Report, Governor Dick Lamm establishes the Rocky Flats Monitoring Committee, probably the world's first and only group of citizens formally tasked with monitoring a nuclear weapons facility. Sister Pam Solo is the sole woman appointee. The group meets regularly at the Rocky Flats plant, where they're treated like VIPs. They drive through the checkpoint and put on booties and respirators for formal tours of the plant. They shake hands with managers and watch films produced by ERDA.

They have no real authority.

Few women work at the plant except as secretaries, and Pam has a hard time finding a respirator that will fit her face. She's also one of the few people who are skeptical of the intentions of Rockwell and ERDA. For one thing, being inside the plant makes her uncomfortable. She feels like she's entering a world she's feared and hated since she was a child, when she had to dive under her school desk for drills during the Cuban missile crisis. And she's struck by the language the workers and managers use to distance themselves from the product they manufacture. A nuclear bomb explosion is called an "excursion"—as if, Pam thinks, one were going for a mountain hike on a nice summer day.

It begins to dawn on Pam that the ultimate recommendation of the Lamm-Wirth Report—the phasing out and closing of the plant—may never be addressed. She worries that things will continue as usual, with the committee merely gathering data and indecipherable information as a substitute for actually carrying out the real recommendations of the report.

But small groups of activists now appear regularly at the gates. Pat McCormick and fellow nuns from the Sisters of Loretto meet with a prayer group every Sunday morning just outside the plant. Housewife Ann White drives up from her home in Cherry Creek to march with people carrying signs and beating drums from Boulder to Rocky Flats, a twelve-mile hike along a busy highway.

I SHOW up for my first day at the truck stop and I'm handed a short-skirted uniform with a white bib, a pocket for a leaky blue pen, and an order pad. I feel like I've become a stereotype. "No runs in the panty-hose," another waitress warns. "They'll send you home. You need to have a spare pair in your purse."

Through the window I can see trucks moving in and out, filling up with gas, their drivers jumping down from the cabs to light cigarettes and stand and talk. The trucks carry beer and milk and gas and who knows what—I wonder if some of them come from Rocky Flats. Later I will learn that they do. The truck bay is brightly lit, but inside it's dim and smoky all day long. Plastic globe lighting hangs from the ceiling, and the carpet is gray and stained. By 10:00 a.m., I'm exhausted. The cook yells and the busboy is slow and the truckers—a constant stream of them—smile as if they expect something that's not on the menu. Speed is everything. Two girls work the counter and one girl—the new girl, me—gets the family dining area, a small room set off from the counter area where no one actually sits down to eat. You have to pay your dues and prove you're willing to work hard for nothing before you get moved to the counter, where tips are thick. The coffee is scalded and the food all looks the same, some version of an open-faced beef sandwich and mashed potatoes coated with gravy and a slice of microwaved cherry pie for dessert.

If I'm going to be a truck-stop waitress, I'm going to be a good one. Before long I can carry four plates on each arm just like any other girl. The manager lets me work the counter. Tips are good. One waitress befriends me, a heavy woman in her forties named Shelley who lights a cigarette while still finishing off the last one. She's raising two kids on her own. "Don't stay here too long," she laughs. "It grows on you."

I count my tips each night and keep them in a shoebox under my bed. It's rumored that some girls make extra money at night after their shifts end, climbing up and tapping on truckers' windows. I begin to receive letters from places as far away as Nevada and Utah from a trucker not much older than I am who stops at the diner every other week or so. I barely know him and he writes three or four pages at a time. "Is this boy in love with you?" my mother beams.

One evening I'm making a chocolate malt and Shelley grabs my arm. "I think someone wants you in the family dining area," she says. *Oh no, I think.* It's my long-distance admirer, who's taking a new approach. But no. It's my mother. She waves as if I'm on the other side of the planet. The other waitress—the new girl—is trying to pour her a cup of coffee. "No, no, no!" my mother exclaims. "I want Kris. I want my daughter!"

I take the coffee pot from her hand. "I'll get it," I say. The girl is pissed. There are few tips to be had in the family dining area, and later—when my mother has left—I give her the five-dollar bill my mother leaves under the plate. Karma and Kurt sit on either side of the table and order grilled cheese sandwiches. My mother has the beef. "Sit down and join us!" she says, pulling my arm. "I'm so proud of you."

"I'm working, Mom," I say, crimson. I retreat to the back room, where Shelley is smoking just as fast as my mother is out front.

"That's your family?" she asks.

I nod.

"What a nice family," she says, and stubs out her cigarette in the last remains of her patty melt.

At the end of the summer I quit. It's not the long hours or burned coffee or even the girls with questionable morals. It's the palpable feel of loneliness, the edge of desperation. It's the way the waitresses smoke in the back and the men joke out front and the drawer of the cash register bangs constantly.

WITH FUNDING from the National Institutes of Health, and in the face of growing opposition from nearly every side, Jefferson County health director Dr. Carl Johnson continues his studies on areas downwind from Rocky Flats. In 1975 and 1976 he and his colleagues find an average of forty-four times more plutonium in soil near the plant than had been reported by the State Health Department at the same locations, with some "hot" spots even higher, reinforcing earlier independent studies. Concentrations in the air and drinking water are also high.

Using data collected by the National Cancer Institute from 1969 to 1971, Johnson examines the relation between cancer rates and exposures to plutonium for people living near Rocky Flats. The study involves 154,170 people in the most contaminated suburban area and 423,870 in an unexposed suburb. He finds higher-than-average rates of cancer, primarily leukemia and lung cancer, in areas downwind of Rocky Flats. He also finds that in zones of increased contamination there is increased cancer. Johnson estimates 491 additional cancer cases, where the Energy Department (then called ERDA) had estimated only one. The excess cancers are mainly the same types found in excess among the survivors of Hiroshima and Nagasaki. He also believes that plutonium from the plant could increase the potential for birth defects in future generations.

Johnson prepares his findings for formal publication. A panel of international peers in his field reviews and confirms his results.

The government responds swiftly. A report by ERDA, the Energy Research and Development Administration, questions Dr. Johnson's soil sampling method, claiming that the soil samples are too shallow. Although Martell praises the method as innovative, ERDA argues that samples must be obtained by scooping deep into the ground to get a "meaningful" measure of plutonium contamination. Johnson stands by his results, emphasizing that surface particles of plutonium are more significant, as they are more easily inhaled. Plutonium typically moves around on surface levels. Studies have confirmed a concentration of 50 picocuries per gram of airborne soil in the Rocky Flats area. "In this arid, windy climate," he writes, "a cubic meter of air may contain one or more grams of suspended dust." Further, "in contrast to the Rocky Flats plant workers, who wear protective clothing, breathe carefully filtered air, are monitored frequently for radiation exposure, and have medical supervision, families downwind from the plant have no such protection."

Some scientists question Johnson's study. One complains that Johnson did not look at whether other factories in the area might also have released dangerous materials, though there are no other factories near Rocky Flats.

The debate is lost in a flurry of negative publicity. Before Johnson's article reaches print, officials at Rocky Flats and the Colorado Department of Health publish a scathing editorial attacking Johnson in the *Denver Post*. The editorial claims Johnson's methodology is "largely useless" and runs the risk of being "a cruel joke." The carefully worded op-ed doesn't deny the existence of plutonium in residential areas, noting that "[our] analysis does not say there isn't cancer danger from Rocky Flats. Plutonium admittedly has drifted beyond the plant boundaries. What [our] analysis does say is that Dr. Johnson has not proved scientifically any connection with cancer."

Years later, in 1990, a reporter from the *Denver Post* will uncover the fact that the Department of Energy paid Rockwell a cash bonus for persuading the *Post* to publish the derisive editorial.

Further, ERDA's own report reveals several previously unpublished facts: approximately 11,000 acres of land, including 7,413 acres outside the plant area, are contaminated with more plutonium than is considered safe by the Colorado Department of Health, and almost all operations at the plant create "small releases" of radioactive material leading to measurable doses in "all segments" of the environment.

The report also states that trucks carrying about five hundred shipments of radioactive cargo to and from the plant each year leave measurable amounts of radiation along Colorado roads. Hundreds of radioactive shipments have traveled by plane as well. Over the past twenty-five years, plutonium and other hazardous materials have quietly been flown to and from Rocky Flats through Denver's Stapleton International Airport and Jefferson County Airport on government aircraft, cargo planes, and—occasionally—passenger planes. Records going back six years show shipments of up to 4.5 pounds of plutonium were not always inspected by the Federal Aviation Administration to determine if they were safe for travel. Earlier records were lost or destroyed. When this news hits the papers, Felix Owen, Rockwell's new director of information services at Rocky Flats, downplays it. "I believe it's irrelevant," he says. "A rehash of history isn't too productive. All the flights took off and landed [safely]."

Citizen outrage is slowly growing. But in Bridledale and Meadowgate, all the neighbors agree: that Johnson guy must be a kook.

THE SMITH garden is still a constant magnet for our pony Barney, but we rarely see the Smith children. Tamara Smith was four years old when she moved to Standley Lake in 1978. She'd spent her first years on Rainbow Ridge near the old part of Arvada, not far from our first house, and when her family moves to Standley Lake, just down the road from us, her parents are as thrilled as mine were when we moved to Bridledale. It's the home they've always dreamed of, the perfect place to raise kids. Just over the ridge from our house, Tamara's house faces the lake with a full view of the mountains, and with eight acres there's room for a big garden and a pasture for horses and cattle. The family is devout Mormon and they live off the land, growing their own vegetables and raising their own meat. One of the first things her father does is dig a well.

Tamara is like a lot of kids who live by the lake. She rides her horse around the water, loves to swim, and rides with the Westernaires. She likes to spend time outdoors, but she has to be careful. Like her four brothers and sisters, she has allergies, and it's hard for her to be outside for any length of time. Tamara seems to have it much worse than the others. It upsets her that she's allergic to the things she loves most. Her ears itch and her eyes water. If she tries to help load hay or feed the horses, her skin breaks out and she can't breathe. When she comes home from her riding club, she has to stand in a steaming-hot shower to get her lungs to open up. She's allergic to various foods and can never find a diet that suits her. She gets sick a lot. Her siblings tease her: if there's something to get, Tamara will get it.

Jonathan Smith, Tamara's father, is a tall, slender man with an authoritative presence. His voice carries across the field when Barney gets in through the fence. Doreen Smith is quiet and determined. They make an efficient pair. They believe in God and family and taking care of themselves, raising their children the way they see fit. They pay little attention to local newspapers or television, and they don't like going to doctors. "We are not a doctor-going family," Jonathan likes to say.

The Smiths intend to breed their cows and they put a bull out in the field. But none of the cows get pregnant. The same thing happens with the horses. The mares can't conceive or carry a foal.

Tamara and her siblings eventually go to Pomona High School, the same school I attend with my sisters. She dreams of going to college and becoming a teacher.

Tamara's older cousin Carol sometimes brings her children to visit the Smiths and their animals. Carol never lets her children drink the water because, she says, it might be contaminated. The Smiths don't believe her. Tamara thinks it's because well water tastes different from city water. City water always tastes a little better.

THE PRESENCE of my father takes on gargantuan proportions in our house, although he's rarely there. I feel weighed down by his dark despair and the sense that his life is completely out of control, and by his glowering disapproval of us and our mother, of himself, probably. I can never see far enough into him to know what he really thinks. My mother, on the other hand, displays a forced daily optimism that seems to have little to do with reality. She is the queen of high drama, ruling from her olive-green bedspread with moans and sighs and rolling eyes. "What am

I going to do?" she laments, clutching my arm or stroking my fingers. "Everything, *everything* is going down the tubes."

But she knows exactly what she will do. Once the afternoon sun has filtered down to the bottom of her window blind, she'll rise, put on some lipstick, and go down and make supper.

Caught in the crossfire of my parents' war, I live for the afternoons when I can gallop Sassy out to Standley Lake, the wind blowing hard against my face as the ground blurs beneath her hooves. We gallop up to the dam and across to the other side of the lake. Sassy extends her neck and stretches out her body like the racehorse she's always been and we run, run, run until her neck is soaked with sweat and foam and we're both breathing hard.

I skip homecoming and prom. When my mother asks if I'd like to have a big party for my high school graduation, I feel a flash of anger. I'm a serious kid, and I live between the pages of books. In some ways it's simpler with my father: I just avoid him. I can sidestep all the anger and fear by pretending he's a strange, dark satellite to our odd little family. My mother, though, is at the center of everything. She needs me. She needs my siblings. What would happen to her if she were on her own? She tells us stories of how she sacrificed her childhood to take care of her mother during the Depression, and how she came home after school each day to boil potatoes and carrots for supper. She tells us how she went off to college—paid for by her spinster aunt—and had boyfriends. So many boyfriends, and look who she ended up with. So many missed chances! "I gave up everything to be with your dad," she sighs.

There's no money for me to go to college but I apply anyway. "Things will work out," my mother says. "They always do. You never know what's going to happen." She believes in signs. She loves palm readers, fortune-tellers, and omens of all kinds. She reads her daily horoscope and goes to psychics and buys lottery tickets. Every night she prays that her fortune will change. "Someday," she says, "my ship will come in. And when it does, I'm going to share it with you kids."

The outdoors is our salvation. Karma is always on a horse. Kurt spends afternoons with his friends at Standley Lake, jumping off the pipe and hitching waterskiing rides off the neighbors' boats. Karin is happy to join him and, when she can get away with it at the local Stop-N-Go, brings a six-pack of beer to share. Sometimes late at night, when our dad has fallen asleep in his chair, Karin steals the keys to his beleaguered two-door Maverick and drives out to Rocky Flats with a boy and a six-pack. They're both too young to drink; she's barely old enough to drive. They sit on the old orange hood and crack open warm cans of Coors and make out for hours. The wide ribbon of the Milky Way spreads thickly across the sky. Sometimes the moon is nothing more than a thin curl of ribbon and other nights it's as round and full as an orange. The other beacon in the night is Rocky Flats, whose lights shimmer on the silhouette of land almost as beautifully as the stars above it.

One day Kurt comes home sick from school. He complains of a headache, fever, and being tired. My mother, always the nurse, presses her fingers into his neck. "Your lymph nodes are swollen," she says.

"At least it's not a concussion," he quips. He stays home from school but doesn't get better. When she takes him to see our family doctor, the news is not good. "Kurt has all the symptoms of leukemia," he says.

My mother is devastated. Kurt takes up permanent residence on the family sofa with his pillow and blanket. The medical tests are inconclusive. It doesn't seem to be leukemia, but doctors can't determine exactly what it is. He misses months of school, watching endless episodes of *Charlie's Angels* and *Love Boat*. When Kurt finally returns to school, no one's quite sure what he's been suffering from.

Soon he's back to his old high jinks. Kurt, too, swipes the keys for the Maverick—an uncertain prize, with its tattered seats and full ashtray, the tires as bald as bowling balls—and picks up a friend or two before driving out to the big hill on Ward Road where they bounce over the bumps at high speed in true *Dukes of Hazzard* style. Our father is furious when he finds out, and the keys never leave his pocket again. Kurt is resourceful; he learns how to hot-wire the ignition.

It becomes a pattern for all us kids as we grow older: chronic exhaustion, fever, and swollen lymph nodes, symptoms that no one can diagnose, symptoms that never really go away.

☢

LOTS OF kids from the local high schools end up working at Rocky Flats.

Debby Clark is one of them. The Monday following her graduation from Golden High School, in 1973, will be her first day of work at Rocky Flats. She starts in the cafeteria, serving sandwiches on the line for $2.75 an hour. Most of the workers are older than Debby and they refer to one another as family. It's a friendly place.

Debby lives at home with her parents, two brothers, and two sisters. When she arrives home from her first day of work, her mother meets her at the door. "What have you done?" she asks. "Did you do something wrong?"

"What?" Debby's shocked. She never gets into trouble.

"One of the teachers called. They said the FBI showed up at school and they want to interview your teachers." Her mother's face is pale.

"I haven't done anything," Debby says, "except fill out paperwork." The paperwork didn't seem out of the ordinary for a new job. She's not sure what goes on at Rocky Flats—until now she wasn't even aware that it was a government facility. One agent asked a teacher if Debby was the kind of girl who partied a lot. The FBI interviewed the neighbors as well.

It's not long before Debby is promoted to janitor, cleaning the main floor in the 881 building, including the computer rooms, offices, and labs. The labs are a mystery. She doesn't know what they do at Rocky Flats. Everything is on a "need to know" basis. She's instructed not to talk about her work, not to talk about what she does or sees. Her parents never ask.

A job at Rocky Flats is lucrative. Debby keeps an eye on the job boards and begins to think about becoming a radiological tester or a security guard, which pays more than being a janitor. She has friends who are security guards. It's fun, they tell her. It pays well and you get a lot of overtime.

When Debby begins her security guard training, she's one of only three women on the guard force. At five feet six inches tall and 125 pounds, with a long mane of red hair, Debby is the last person someone would peg as an armed security guard. But she's learned to be tough. Some of the men give her a hard time at first. She doesn't mind a catcall or two, but unlike some of the other women, she never runs to the boss. She can stick up for herself. She doesn't think twice about getting in some guy's face and making sure he understands her point of view. She learns how to shoot a gun and how to tear down and reassemble an M-16 in the dark. She's trained on explosives. She learns about the various buildings at Rocky Flats and the routes the guards are expected to walk in each particular building. Depending on the time of day, they walk it one way and the next hour they walk it in reverse. They're instructed to look for things that are unusual or out of the ordinary, and to check if anyone's left out classified documents or forgotten to lock a safe.

As part of their final test for graduation, each class of new guards is expected to "pull an exercise" on the regular guards—that is, pose as terrorists and try to break into the plant. The regular guards never know when it will happen. Debby's team waits until midnight to cross Indiana Street and sneak up to the border, where they split into groups and try to enter the plant from different places along the fence. One guy gets snagged on the barbed wire and is caught by a patrol car. But a member of Debby's group gets all the way to the guard shack, wrestles the key from the guard, and opens the gate to let everyone in. For the fake terrorists, it's a successful exercise. They all graduate.

When Debby figures out that Rocky Flats is part of the nuclear defense network, she's proud. Her grandfather was a sailor in World War II and her father was a Marine. She believes in a strong defense.

Pulling fake terrorist exercises, though, eventually comes to a halt when one night a guard gets a little too excited and almost shoots a guard-in-training.

MY HIGH school graduation is a heady mix of pride and relief, my feelings of freedom tempered by the weight of my family. It feels as though the roots of our cottonwood trees are twisted around my ankles. I want to get as far away from Arvada as I can.

Two weeks after my graduation, a small item appears in the back pages of the *Denver Post* reporting that the radioactive element thorium has been found in the gonads, or sex glands, of three horses in Jefferson County. The horses belong to rancher Lloyd Mixon, on his ranch just east of Rocky Flats, not far from our property. Lloyd Mixon is the rancher who brought Scooter the pig to the Lamm-Wirth hearings.

I lose track of Randy Sullivan and his entourage of friends and the throngs of adoring girls that no doubt follow him everywhere. Undecided about college, Randy takes a job at the Ralston Purina pet food factory down by the Denver stockyards. The pay is good, but the job includes cleaning up sticky cat food when the conveyor belt breaks, and he only lasts a year. He feels he's dodged a bullet with the military—the Vietnam draft ends a year before he becomes eligible—but he's not sure what he wants to do with his life.

With the help of student loans I start classes at Colorado State University and move to Fort Collins, where I take a job making doughnuts at an all-night doughnut shop. The college is two hours away and I drive home on the nights my mother calls, frantic, with my father banging at the door. He wants in. She's afraid of him but won't call the police. "He knows all the cops and judges," she explains. "It wouldn't do any good."

My father sporadically sends small checks. Often they bounce. I pin them up on my bulletin board. "He loves you," my mother says. "He's just trying to show his support for you." It's not unusual for the two of them to go out for a nice dinner after one of their fights. When my mother calls to tell me about one of their make-up dinners, she reports that she's filled him in on all the tiny details of my life: my classes, my job, my half-attempts at a dating life. "You kids don't talk to your dad, so I'm the one who has to tell him everything," she says. I feel like a splayed fish, split right down the middle.

I finish my final exams and move home for the summer to work and save money, but also to help my mother. "I need you here," she says. She seems on the verge of divorce, but can't quite do it. Divorce is a disgrace. "And I love your father," she says. Then her eyes grow sad. "Doesn't he realize what he's doing to me? What he's doing to you kids?" *How easily love and hate lie together,* I think. *Side by side.*

I need a job. The truck stop is out of the question. I comb the classifieds and find a job as a driver on a roach coach, a lunch wagon that winds around construction sites and office parking lots all day. Construction is booming in Arvada and business is good.

I love my roach coach. The front cab is like an old taxi. The seat is worn and the dashboard is dusty and bleached from the sun, but the back of the truck is as shiny as a big kitchen, opening up on both sides to reveal rows of cold soda pop and racks of wrapped sandwiches and honey buns. Each morning at five I report to the loading station and shovel chipped ice into the soda bins, stock the shelves, and fill the tank with gas. The owner is a caustic man who watches over his fleet of three trucks like a hawk, carefully counting the remaining stock each evening to make sure his drivers aren't snacking on his profits. I'm given a list of stops and a timetable—no more than five minutes at each stop—and I'm promised a twenty-five-cent raise if I can handle it for a week. He has a hard time keeping drivers.

I meet Mark Robertson on my first day. He comes up to the side of the truck shyly, bare to the waist and wearing a tool belt, worn jeans, and steel-toed construction boots. He has hazel eyes and his hair is light but his skin is dark from the sun. He buys a flattened bologna and cheese sandwich and a soda.

"You the new girl?"

"Yeah." I'm nervous, mostly because I have a line of construction workers three bodies thick standing around the truck. I stand at the corner near the bumper, frantically trying to tally up chips and honey buns and give correct change. The owner has warned me about shoplifters— I'll have to pay for anything that comes up short.

"We get a new girl almost every week," Mark says. "They never last."

"You won't last either," another worker adds. He buys Twinkies and corn nuts. "Let's bet on how long she lasts."

"I'm not going anywhere," I say. "At least not until the fall."

"College girl?" Mark asks.

"Trying to be."

"Hang in there, then." He gestures toward three guys sauntering away from the truck. "You better yell at those guys."

"Hey! You need to pay for that stuff!" I holler. I slam down the sides of the truck and run after them, my change belt banging against my thigh.

The men turn and spread their empty palms. "Just testing you," one laughs. He reaches into his pocket and tosses me a five. "Keep the change, sweetie."

When I get back to the truck, Mark is gone.

IN APRIL 1977 the Rocky Flats Action Group holds a rally at the plant that attracts several thousand people, including the well-known activist Helen Caldicott. Thanks to the work of Sister Pam Solo and Judy Danielson and others, it's now a good-sized coalition of citizens, scientists, churches, and environmental and pacifist groups. As Rocky Flats and DOE officials hide behind the wall of Cold War security, and scientists and politicians vehemently debate the potential health effects of exposure to plutonium, the Rocky Flats Action Group wants to pull back the veil that's hidden Rocky Flats from the public for more than twenty-five years. Members of the action group begin talking about organizing a big rally that would gain national attention.

When she's in high school, my sister Karma, who's taken an interest in Greenpeace and Rocky Flats, begins to attend meetings. Always a quiet, serious girl, she no longer seems to be the "sweet little bird" our grandmother Opal used to call her. She spends time at meetings and rallies, talking about feminism and environmentalism. She doesn't hide her anger and disgust. She refuses to talk to our dad and she avoids our mother. "Karma's so angry," our mother sighs. "Why is Karma so angry?"

I FIND myself thinking about the boy who comes each day to the roach coach to buy a bologna sandwich. He seems almost as shy as I am. "Sandwich and soda, two fifty," I say. He hands me three bucks and I drop fifty cents into his callused palm. "Thanks. See you tomorrow," he says. My cheeks burn.

It's three weeks before Mark asks me out. He picks me up from our house in Bridledale in a two-door orange Honda with a peace sticker in the back window. We go to an art cinema and watch an Italian film with subtitles, sipping Cokes silently in the dark. He drives me home afterward. "I want to get you back on time," he says. I walk into an empty house. My father's gone, my mother's started working again, and my siblings are out. There's no benefit to being a good girl in this household.

On our second date, we eat pizza in the garden of a tiny Italian restaurant with the moon shining through the trees. Mark orders a beer and shares it. The patio is nothing more than two round tables and folding chairs set outside the restaurant's back door, but it feels like we're in Venice. Once again he brings me home to an empty house.

The next night I go to see Mark perform at the Denver Folklore Center, where he works as a stagehand and occasionally plays guitar. His shyness is gone as soon as he steps on the stage. He plays a six-string and then a twelve-string guitar and sings with a rich, warm voice. I wait for him in the lobby and when he greets me with a kiss on the cheek, I hate myself for blushing so furiously. It's late. "Should I take you home?" he asks.

"It doesn't matter," I say.

We go to his place, a tiny shotgun house next to the railroad tracks in the oldest part of Arvada, white clapboard with a jumble of spider plants on the front porch. There's a kitchen with a single burner and a big room with a mattress in the corner. Books on planks supported by cinder blocks line the walls. A cat with smooth gray fur and eyes as green as grass sidles up next to me. "His name is Arlo, for Arlo Guthrie," Mark explains. We sit cross-legged on the floor and Mark plays records: Arlo and Woody Guthrie, the Nitty Gritty Dirt Band, Joan Baez, and a local bluegrass group called Timberline. He tells me about his father, a machinist, and a brother who works in a molybdenum mine near Leadville. His father built the house Mark grew up in, brick by brick.

He pours me half a cup of red wine in a jelly glass and we sit on the mattress and listen to the trains go by, one after the other, the rocking of the wheels making the little house tremble.

"Shall I take you home now?" he asks.

"No," I say.

THE SUMMER of 1977 is about sitting on Mark's porch, listening to the trains, waking up to Arlo's paw on my cheek. Mark is the gentlest person I've ever met. When the summer ends, I briefly consider forgetting about college, at least for a while. Mark has no desire to go to college. But when September rolls around, I pack up and return to school in Fort Collins, renting an old carriage house with three other female students. One has a wealthy father, one works as a waitress like me, and the third makes more money than the rest of us put together working as a stripper at a local club. My room is at the top of the stairs and was once a hayloft. It's a tiny room filled with plants and books and a poster I tack on the wall that reads "War is not healthy for children and other living things." I sell my yellow Rambler for fifty bucks and buy a used red Volkswagen bug with a rainbow in the back window and a ski rack on top. I wear jean skirts and hiking boots and my hair hangs in a long soft braid down my back.

Mark drives up on weekends. "You'll be a writer," he says, "and I'll be a musician." It all seems possible. "We just need a little faith," he says. "And time."

One afternoon Mark says he wants to take me rock climbing. For years he's climbed local peaks, and he occasionally teaches beginning rock-climbing classes for extra money. I've done some bouldering and rock scrambling with friends on long hikes in the mountains, but I've never done any serious climbing.

"It's not hard," he says. "I'll show you."

The next weekend we pack up his little Honda with ropes and sandwiches and head up to Castle Rock, a well-known spot about twelve miles up Boulder Canyon. A three-hundred-foot tower of rock near the west end of the canyon, Castle Rock is a challenging climb, but typical fare for newcomers to the sport. "I'll lead the way," he says, "and you follow me up on belay." He snaps a harness around my hips and hooks the end of my rope to his own harness with a metal clip called a carabiner. "That's a beaner. See? We'll be on the same line. Just watch me and put your hands where I put my hands. Watch my feet, too. I'll put in some pitons as we go." He gestures to the small steel spikes attached to his belt. "These act as anchors, so if one of us falls, we won't fall very far."

I've never been afraid of heights but my hands are shaking. "Here," Mark says. He hooks a small bag onto my belt. "This has chalk inside. If your hands get sweaty, dip your fingertips in this."

It takes us a long time to get to the top. I move slowly, frozen in those long seconds when I cling to the rock while searching with a single hand or foot for a secure hold. "You're fine!" Mark yells over his shoulder. "Just keep going!"

At the peak we share a flattened sandwich and warm beer. The wind is clean and fresh and smells of pine. The view on all sides is breathtaking.

"We could hike down the backside," Mark says. "But I want to teach you how to rappel. It's a way of going down the face of the rock on a rope. I'll control your line from the top."

"I don't mind hiking down the back side," I say. "Honestly."

"You've got to try this. It's the most incredible feeling in the world."

Mark checks the harnesses and ropes and anchors himself to a large boulder. "Nothing to worry about. Piece of cake," he says. "There's me, and then there's the mountain. We're both holding you. Down you go."

I want to turn around and back away from the rock face, but I slide my legs over the edge. "How can I go down when I can't see where I'm putting my hands and feet?"

"Find your first point." He grips my hand until my right foot finds a hold. "Now the next foot. Now put one hand here." He gestures to a little crevice.

"Okay."

"Got it?" he asks.

"Yeah."

He lets go of my hand and I grip the rock. "Now put your weight in the harness."

"What do you mean?"

"Put the weight of your body in your butt. Relax into it. Let the rope pull tight."

I feel my body hanging in the air, barely touching the rock.

"That's it. Now I'll lower you down a little."

I swing out from the rock and slide down a few feet until I find a new toehold.

"That's it!" Mark yells, exhilarated. "Keep going."

I feel like a spider, moving weightlessly, point by point, down the face of the mountain.

"Now I want you to try something," Mark yells. I can no longer see him but can hear his voice. "Kick out from the rock. Bounce against it."

"How do I do that?"

"Give yourself a little jump! Push out from the rock!"

"Are you crazy?"

"Trust me, Kris! I won't let you fall."

"Are you sure?"

"Trust me. Trust yourself."

I push my legs against the rock and swing out and down. It feels like flying. One, two, three jumps and I'm nearly at the bottom.

"I love this!" I yell.

"You did it!" he shouts back. Ten minutes later we're both on the ground.

"Not bad for your first try," he says. "You're going to be fine."

A DATE is set for the first big national protest at Rocky Flats: April 29, 1978.

To ensure a peaceful protest, the Rocky Flats Action Group spends months negotiating with dozens of participating local and national organizations as well as Rocky Flats, the Jefferson County Sheriff's Department, and the DOE. The rally will include a speech by Daniel Ellsberg, who is famous for having released the Pentagon Papers. Demonstrators will be taught how to follow well-established methods of nonviolent resistance, and, following the rally, more than one hundred people plan to occupy the railroad tracks in a symbolic blockade. Although Rockwell has agreed to the rally and to a short demonstration on the tracks, no one knows for sure what will happen. If demonstrators remain on the tracks until Monday morning when the trains start running again, they'll likely be arrested. It's a risk they're willing to take.

Dr. Carl Johnson, the county's health director, is constantly on the defensive. And yet another radioactive element has popped up. Soil and dust testing reveals an "excess amount" of cesium-137 on the west side of Rocky Flats, which Johnson believes is indicative of a more serious cover-up at the plant. Subpoenaed in connection with the suit filed by unhappy landowners, Johnson testifies that the presence of plutonium and cesium could be the result of a single criticality incident at the plant. Rocky Flats' official position has always been—and continues to be—that there has never been a criticality at the plant. But Johnson feels strongly about his hypothesis, as do other scientists and engineers. Radioactive cesium and strontium are produced only during fission—the actual splitting of atoms—and there is no nuclear reactor at Rocky Flats. Johnson recommends additional testing of soil samples and analysis of forage, vegetables, and other fresh produce in the contaminated areas; evaluation of livestock and milk; and testing of air, ground water, and water in reservoirs.

That's not what developers want to hear. On cross-examination, the special assistant U.S. attorney Jake Chavez reminds Johnson of a recent Golden District Court decision that determined county officials could not block residential housing development based on land contamination— that power belonged only to the state board of health. Johnson says he's aware of that, but his job is to investigate radiation hazards for residents and make the results known, and the county board of health has approved his activities.

But the chairman of the Jefferson County Board of Health, Dr. Otto Bebber, suddenly has a change of heart. Some members of the board are unhappy that the latest report on radioactivity near Rocky Flats was released to the press. A Jefferson County commissioner says that he's been receiving phone calls from people concerned that the cesium report will make their property values go down.

Bebber tells the press that Dr. Johnson was instructed to release the results of his research only to specific agencies and not to the public. Johnson denies that such a limitation existed. Bebber responds that the board will consider a censure motion against Johnson at their next meeting. However, he says, "Even if it [a vote to censure Johnson] does happen, it would be just a censureship. It wouldn't mean his job or anything." The motion to censure Johnson is defeated.

APRIL 1978 will go down in Colorado history as the beginning of the Year of Disobedience and the Summer of Protests. But except for Karma, no one in our family—or in our neighborhood—is paying the slightest bit of attention to Rocky Flats.

In the middle of my sophomore year I decide to transfer to the University of Colorado at Boulder and, against the advice of everyone except Mark, major in English and creative writing. I'll be closer to home—just half an hour from my mother, who calls every day. She's a confiding sister, a clinging daughter, my biggest fan, and—sometimes—a stone around my neck. She is as vulnerable as a child and as stubborn as a bulldog. Mark is my escape.

I pay a local rancher fifty bucks a month to board Sassy in a field so I can ride on weekends. I take classes in the mornings and work two waitress jobs: afternoons at the New York Deli on the Boulder pedestrian mall, where I get a free bowl of matzo-ball soup, and nights at the Oasis Diner, a fifties-style restaurant where I serve burgers and shakes to the tunes of Frankie Avalon and early Beach Boys.

It's tough to make a living as a musician and rock climber, at least in the short run, and Mark takes a job in a hardware store. When he comes over, our attempts at cooking together are creative: we try combinations of tofu, bean sprouts, and brown rice. But we're halfhearted hippies and poor vegetarians. "Try this," Mark says one night. He flips a slice of bologna into a frying pan.

"That's disgusting."

"No, it's good," he says. "And cheap. My mother used to make it for us when things were tight." He lets it sizzle until the edges curl and then tucks it between two slices of white bread. He takes a bite. "Want some?"

I grimace. "No thanks."

The kitchen is so small there's no room to sit and we head to my cramped bedroom. Mark studies the flower poster on the wall.

"So you were against the war?"

"Of course. Weren't you?"

"I was nearly drafted," he says. "Missed it by a hair."

"Would you have gone?"

"No. I would have been a conscientious objector."

"I can't imagine you carrying a gun or even in a uniform. In fact," I say, "I can't imagine you wearing anything else than what you've got on right now." He wears a thin cotton shirt tucked into faded jeans, a brown leather belt that hits just above my hipbone when we hug, and leather boots with thick heels.

"I don't want to go to Vietnam. Except maybe as a tourist someday," he says.

We sit on my Indian-print bedspread and he finishes his sandwich. "So you must be against Rocky Flats," he says.

"You know about Rocky Flats?" I'm surprised.

"Everyone knows about Rocky Flats."

Not in my neighborhood, I think. "No one talks about it. What's there to know?"

"There's a lot to know." He shakes his head. "You live right next to it. Don't you know what they do out there?"

"We used to think they made cleaning supplies."

"You're kidding me."

"No, I'm not kidding. Scrubbing Bubbles. Dish soap. And who knows what else?"

"They make bombs, Kris. How can you be antiwar and not know about Rocky Flats?"

"You don't know that for sure."

"You really think they're making cleaning supplies?" He sits up. "They make plutonium detonators, but that's not what they call them. Disks or buttons or pits. I don't know what they're called. But they make a lot of them. Besides," he adds, "there's all sorts of crap out there. Plutonium and who knows what. Probably right where you live. Isn't your house out by the lake?"

"They would tell us if there was anything really bad, right?"

"You think they would?"

"Stop it, Mark." I've never seen him get so upset.

"Don't you know about the protests?"

"Yes. Of course." I try to pull him back down on the bed. "But those protesters are just people without jobs, you know, people with nothing better to do. Like students and housewives. They just want to get their names in the papers." I'm quoting my dad, but suddenly it occurs to me that even though I have negative feelings about my father, I've never really disagreed with him, and I've never confronted him. In fact, I've never disagreed with or questioned anything about Rocky Flats, either. Maybe I just don't want to know. I drive past the plant all the time and see protesters out there every weekend.

"You should listen to what they're saying, Kris. Read one of their pamphlets."

"I don't have time for that. I'm trying to put myself through school, remember?" I hear the bitterness in my voice. "Besides," I say, "nothing they say can be proven." My own vehemence surprises me. "Don't you think the government would tell us if it wasn't safe?"

"The government!" Mark explodes. "The government is busy covering things up!"

"How do you know that?" Suddenly it dawns on me. "You're going out there, aren't you?" I recognize my father's critical tone in my voice. "You're one of those people!" One of those people like Karma and her angry hippie friends. Or maybe like Kurt and Karin, who just want an excuse to have a good time.

"Yeah," he says, standing above the bed. "You're right. I go out there.

Someone's got to pay attention. And you should, too—that is, if you give a damn about anything."

He walks out the front door.

It's our first and only argument.

Half an hour later, Mark comes back. I'm still lying on the bed.

"You okay?" he asks. "Sorry."

"I'm sorry, too." I feel shaken.

"Good." He grins. "I just want to say one thing. There's a big rally coming up. Do you want to go?"

"No."

We drop the subject.

ON THE morning of Saturday, April 28, 1978, as many as six thousand people gather at the west gate of Rocky Flats to demand that the plant be moved or shut down. The spring sun is bright and the day starts out warm. The crowd is peaceful. The press is there. After the speeches and music, clusters of wild doves are released to the sky as a symbol of peace.

Somewhere in the crowd stands Mark Robertson. My sister Karma is also there with her friends. One of the security guards is Debby Clark.

As the day goes on, the wind picks up and the temperature drops. Rocky Flats is known for extreme weather, especially in the spring, and it begins to rain. As the protesters walk to their cars to pack up their belongings and head home, the rain grows cold and turns to snow. A group of about 120 protesters, including Daniel Ellsberg, stays behind. They take their signs and sleeping bags and pitch camp on a rail spur on the south side of the plant, where the railroad tracks cross Route 72. In front of the camp they post an American flag and the solar energy flag— ironically, the DOE's Solar Energy Research Institute is just down the road. As darkness falls they pull together their sleeping bags and sing the familiar Vietnam-era slogan "Hell no, we won't go" with a slight twist: "Hell no, we won't glow."

As the night wears on, the weather turns fierce, with high gusts of wind, sleet, and snow. The protesters put up a makeshift shelter of plastic sheeting over the tracks, but it's almost impossible to stay warm and dry. When the sky finally begins to lighten and turn pink over the dark blue mountains, Daniel Ellsberg emerges from his sleeping bag to discover two wild doves perched on a tent frame and sixty-five soaked, freezing people—more than half of the original group—still huddled on the tracks.

After coffee on camp stoves, the group discusses strategy. Ellsberg and many others want to continue the occupation on the tracks. President Jimmy Carter is scheduled to speak at the Solar Energy Research Institute. Since he took office in 1977, one of Carter's major foreign policy objectives has been to reduce the production and number of nuclear weapons in the world and establish a second Strategic Arms Limitation Treaty (SALT II), work that began with SALT I. SALT II would further limit the number of nuclear arms in the United States and the Soviet Union. In his inaugural address, Carter spelled out his hope for a world without nuclear weapons.

"The world is still engaged in a massive armaments race designed to ensure continuing equivalent strength among potential adversaries," he said. "We pledge perseverance and wisdom in our efforts to limit the world's armaments to those necessary for each nation's own domestic safety. And we will move this year a step toward our ultimate goal—the elimination of all nuclear weapons from this Earth. We urge all other people to join us, for success can mean life instead of death."

With the world's press focused on the protesters and on Carter's visit, it seems a perfect opportunity to increase public awareness of Rocky Flats. The activists believe they are effectively obstructing the flow of radioactive materials in and out of the plant. Some want to stay.

Others disagree. The protest has already been successful, they note, and they don't want to jeopardize the positive effects of Saturday's nonviolent action and the careful negotiations that helped achieve it. But Ellsberg's group stands its ground. They decide to create a new name to distinguish themselves from other groups associated with Rocky Flats. One activist suggests *satyagraha*, Mahatma Gandhi's term for the philosophy of nonviolent resistance, a concept that influenced Martin

Luther King Jr. The literal translation of *satyagraha* is "truth force," and the group settles on the Rocky Flats Truth Force. They hold a press conference to announce that they plan to remain on the tracks. Someone hangs a mailbox on a fencepost printed with the words ROCKY FLATS TRUTH FORCE, and supporters bring food and supplies.

The word gets out. Even people from other states and countries send food and supplies and letters of support.

Rockwell's first response is to appear unconcerned. They announce that they don't use that particular spur of track anyway. Making note of the ongoing blizzard conditions, one Rockwell official says, "All I can say is that I hope they don't catch cold."

On Friday, May 5, however, their patience runs out. Rockwell and the Jefferson County Sheriff's Department order the Truth Force to leave. The activists refuse. When the first train approaches, they sit down on the frozen railroad tracks. The train slowly chugs to a stop, and officers step forward to make arrests. As the group is led to waiting school buses that will take them to Golden for legal processing and then on to the Jefferson County jail, Daniel Ellsberg flashes a victory sign to reporters.

Marian Doub is with that group. At seventeen, she's the youngest member of the Rocky Flats Truth Force and a senior at Boulder High. She calls her parents that night, after her arrest, to tell them she intends to go back out again. Her mother is upset. She wants her daughter back in school the next day, studying for final exams. "One arrest is plenty for one school term," she says. Ellsberg sees Marian on the phone in tears and pulls her aside. He says he has a daughter just a little older and he understands how her parents feel. "Protect your relationship with your parents," he says. "It's important to maintain their trust."

Marian decides to go home.

Two days later the Rocky Flats Truth Force is back on the tracks. Wearing down jackets and wool caps, the members trek through snow carrying a huge American flag. They set up a new campsite and pack walls of snow to shelter it from the wind. People on cross-country skis and snowshoes bring supplies, but it's not long before security officers in black riot gear show up to warn them of their impending arrest.

The train comes again: a brown diesel locomotive pulling several cars, including a white car with the familiar sign that indicates radioactive cargo and the words FISSILE MATERIAL—RADIOACTIVE. Once again the protesters sit in the path of the oncoming train, some lying across the tracks with their heads on one rail, feet on the other. The train continues to chug toward them. For a moment it seems like a crazy game of chicken. Finally the train slows and stops. There is silence. Suddenly the doors slide open and two rows of officers in black helmets emerge from each side of the caboose, moving as if in silent slow motion through the deep snow. The protesters offer no resistance, letting their bodies go limp, and sing "We Shall Not Be Moved." The officers lift the bodies from the rails and carry them to the waiting bus. This time twenty-three people are arrested.

"We'll be back," Ellsberg tells the press. "We are effective here, as the arrest shows. If they keep us in jail, it doesn't matter, because there are so many more ready to take our places."

The booking process lasts into the evening. Later that night, Ellsberg finds himself in a room with two women who've just been arrested. One is seventeen-year-old Marian Doub. Ellsberg is surprised to see her. She tells him she spent the day in classes at Boulder High as promised, but saw the arrest of the Truth Force on the evening news. She saw that no one was sitting on the tracks. "So the two of us," she says, nodding to her companion, "went out there." The camp was deserted. Carrying flashlights in the dark, the two women held hands and began walking down the tracks toward the plant, singing "We Shall Overcome." When they saw the train coming toward them, they sat down in the snow between the rails. The train chugged to a stop, and a solitary security officer emerged and tried to pull them off the tracks. They went limp. He called for backup and they were arrested.

"The two of you stopped the train again?" Ellsberg asks. "Alone?"

Marian nods.

"Who is your friend?" Ellsberg asks.

"My mother," Marian answers.

✪

THE PROTESTS at Rocky Flats continue. Many of the workers at Rocky Flats are unhappy about the negative publicity. About seven hundred plant employees sign petitions condemning the DOE and Rockwell for "taking the criticisms [by the antinuclear activists] passively."

Karma attends many of the protests, sometimes taking along Karin or Kurt. The rallies turn into long afternoons of sunshine, music, and impassioned speeches. Sometimes they talk about whether it's safe to stand out there on supposedly contaminated soil. "I guess it doesn't really matter," Karma says. "Whatever there is to get, we probably already got it in Bridledale."

"That's why we all have such glowing personalities," Karin says wryly.

Mark and I spend weekends together at my place in Boulder. When he drives past the activists on the way to visit his parents, he honks in support. When I drive down to Arvada to visit my mother, I take the long way around so I don't have to see the protesters. They seem a little crazy with their signs and handouts. They scare me. I look at the rolling hills, the fields where I galloped my horse. The wind is fresh, the sky turquoise blue. How can there be contamination here? It looks so pristine.

One evening Mark picks me up after work and we drive to a local park. He takes a six-pack from the trunk and we sit on a bench in the dark, cupping our cans of beer, sipping the cold foam and watching the stars above the black line of mountains. "We should get married," Mark says suddenly.

I'm stunned.

"There's an artist I know, a jeweler, who's making a special ring. I designed it just for you. It's not ready yet, but he's working—"

"Mark—" I feel a weight in my stomach. "Don't say this now."

"What do you mean?"

"It can't be now."

He dips his head between his knees.

"I'm not even twenty-one. I'm in school and working like hell to stay there." I feel terrible. Guilty. "I can't get married now."

"You could stay in school."

"I don't need your permission."

"That's not what I mean."

"I don't—" I feel choked. "I'm sorry. I don't even know where I'm going yet." Everything feels ill-timed and off-kilter. Maybe this is the only good man I'll ever meet. Maybe this is my only shot at love. But I need to finish school. I don't want to end up like a housewife in Bridle-dale, cleaning house and folding laundry. "I just want a chance at things, I guess."

"Well," he says. "All right." But it's not all right.

"Maybe we can wait a while. I just need time. I love you. But marriage . . ."

"Okay," he says. "Let's give each other time." He doesn't look at me as we walk back to the car.

Later that night, alone in my room, I can't sleep.

My mother is upset when she hears I've turned Mark down. "You could settle down and start looking for a house. All I've ever wanted is for you girls to get married and bring me grandchildren. I'm ready for grandchildren." College, for me or for any potential husband, is less important to her than seeing me married. She's proud of her daughters, but a woman's primary role is to be a wife and mother.

I don't ask her how we might afford a house and children when I can't pay my tuition and Mark can barely pay his rent. "Mom, I'm not ready."

She sighs and grasps my arm. "You never know what's going to happen," she says. "These things don't always come again."

But that summer is wonderful. Mark and I go rock climbing up Boulder Canyon. We sit on the grassy banks of Standley Lake and watch the sunset. He plays softball with a local team and I rarely miss a game. One weekend we drive all night to Las Vegas in his tiny orange Honda to see Burt Bacharach in concert. We spend the night on the old part of the strip in a cheap hotel with tattered furniture and glitter on the ceiling. Mark buys a cassette tape and we sing Bacharach songs all the way home. "That man," Mark says, "sure knows how to write a love song."

THROUGHOUT THE summer, contentious public debate over arms control and SALT II fills the media. President Carter refuses to comment on Rocky Flats during his visit to Golden. Many activists are crushed—he's become a hero of sorts. Although Carter and Leonid Brezhnev, the leader of the Soviet Union, will reach an agreement in the coming year, SALT II is met with opposition in Congress, as many believe it will weaken U.S. defenses. (Signed by both Carter and Brezhnev, the treaty is never ratified. Both sides, however, honor the commitments laid out in SALT II, until 1986, when the Reagan administration withdraws from the treaty.)

In 1978 the world is watching what's happening at Rocky Flats. Groups stand with signs at both the east and west gates, handing out flyers to workers and people who drive by. A white canvas tepee becomes the symbol for protests at Rocky Flats. Erected by members of the Rocky Flats Truth Force, the tepee that straddles the tracks just outside the plant boundary becomes a familiar landmark to drivers on Highway 93. Two flags mark the tepee: the American flag and the solar energy flag, green with a golden sun. Some drivers honk and wave or stop to drop off food and supplies. Others make threats and obscene gestures. Protesters come and go, some staying for days or weeks. Patrick Malone is one of the most steadfast members. His face is dark from the sun. In his black beard, bandanna, and gold earrings, and wearing the fingerless wool gloves of a mountain climber, he looks more like a pirate than a peace activist. The protest goes on month after month, but Patrick remains. When the train approaches once a week, Patrick and the others quickly dismantle the tepee to save it from being taken away by the authorities. Then they sit on or lay their bodies across the railroad tracks. The train stops, police approach, and they're arrested. Each week the scenario is repeated. As the protest continues, Patrick is arrested ten times.

Across the rock-strewn field of amber grass lies the Rocky Flats complex, surrounded by towers and lights and patrolled by security guards.

One of those guards is Debby Clark, and she's no longer a rookie. She's been trained to deal with terrorists, but sometimes she has to yell at neighbor kids who hop over the boundary fence for a lark. The protesters, though, are something else. They make her angry. How could anyone be against defending the United States? They're misguided, she thinks. They don't understand what they're doing. She doesn't believe what the so-called environmentalists are saying. At Rocky Flats, everyone does their job, and workers depend on each other. They're working to protect the country, to protect the very people who feel the need to protest.

During the weeks and months that the Truth Force occupies the tracks, part of Debby's job is to sit in a converted truck nearby—nothing more than a heated railcar, really—and watch the protesters. She and the other guards have equipment, including night-vision goggles, to help them keep a close eye on what's going on. Some of the activists are students and she wonders if they're smoking pot or exactly what they're doing out there. There's no place to pee or take a shower. There's no privacy. She talks to them from time to time, but she has no patience for students who should be in class.

One of the protesters is poet Allen Ginsberg, a founder of the Naropa Institute writing program in Boulder. Ginsberg is a frequent visitor at the home of Ann White, who now marches regularly with the protesters. Ann and her husband have a home in one of Denver's more prestigious neighborhoods, and Ann enjoys watching Ginsberg in the mornings on their front lawn, doing t'ai chi in his beard and business suit and astonishing the neighbors.

In an effort to "calm fear among local residents and to clear up the mystery about the work done at the plant," Rocky Flats begins conducting public tours. The tour is intended to demonstrate how safe Rocky Flats is, and includes a look at Building 707, a "component manufacturing facility." Building 707, the replacement facility for the building that was destroyed in the 1969 Mother's Day fire, holds a storage vault containing several tons of plutonium. Tour participants, mostly curious local residents and members of the press, are outfitted with laboratory smocks and elastic-topped booties, and each person receives a respirator. The tour guide, a Rockwell employee, instructs the group to put on their respirators if an alarm sounds while they're walking through the pressure-sealed corridors and air locks because a particle of plutonium the size of a grain of pollen, if inhaled or taken into the body, can cause bone or lung cancer, leukemia, or genetic defects.

Despite the new openness, the public is not reassured. On June 16, 1978, when Rocky Flats is conducting a tour, Ginsberg sits on the railroad tracks with several others and reads his poem "Plutonian Ode" as a train approaches. Members of the local and national press are present when Ginsberg is arrested. As he is led off by an officer, the officer jokes, "We're equipped to deal with terrorists, but we're not equipped to deal with you people."

There are repeated arrests for trespass and obstruction of justice, and many activists, including Daniel Ellsberg, are arrested several times. A trial date is set.

Rocky Flats decides to suspend public tours of the plant.

THE ACTIVISTS' trial takes place in Golden, in the old courthouse just down the road from the Colorado School of Mines. Golden, an old mining town, is a mix of college students, Denver yuppies, and old-style cowboys. The city carefully nurtures and markets its Old West image. An arch stretches across the main street, proclaiming WELCOME TO GOLDEN—WHERE THE WEST BEGINS. It's just a few days until Thanksgiving, and the town has already begun putting up Christmas decorations.

In the courtroom, thirty-one-year-old Judge Kim Goldberger is facing his first criminal trial. Seven Denver attorneys have agreed to volunteer their time to represent the protesters. The prosecuting attorneys, including two on loan from the DOE, are paid by the government. Several days are spent selecting the six-member jury, three men and three women. Expert witnesses are called; some are local, but many have traveled from out of state or from other countries, at their own expense.

The goal of the Rocky Flats Truth Force is to increase public awareness that Rocky Flats should be closed or converted to non-nuclear work, as recommended by the Lamm-Wirth Report. Members of the

Truth Force freely admit to camping out on the railroad tracks and to attempting to obstruct the activity of the weapons plant. But they plead not guilty to the criminal charge of trespassing. Attorneys for the activists base their defense on a little-known Colorado "choice of evils" law. The statute states that an illegal act is justified if it is done to prevent a greater, imminent evil or crime. For example, the law would allow an automobile driver to exceed the speed limit if the purpose was to save a life or escape immediate danger. The defense states that the activists are working to prevent a catastrophic event as well as make residents aware of potential health effects from environmental contamination. Trespassing is nothing in the face of what's at stake.

But Judge Goldberger begins by ruling that he, and not the jury, has the right to determine if the choice-of-evils defense is applicable, and he decides it is not. In denying the use of it, Judge Goldberger says, "The courts may not be used as political or legislative forums." Gay Guthrie, the deputy district attorney, notes, "People are not to usurp the democratic process. People are not to be so arrogant to take yourself to a point to where you sit yourself down on another man's property and say that what he is doing is so bad that I can infringe on his constitutional rights. This is rule by anarchy."

Although Judge Goldberger agrees to listen to several days of testimony from defense witnesses, he won't allow the jury to hear the testimony. It's filmed, however, in case he later rules it to be relevant. With the jury excused, the defense calls a number of experts to the stand to testify regarding the reality of radiation hazards imposed on local residents by Rocky Flats.

The first witness is Dr. Karl Morgan, a professor of health physics at Georgia Tech, one of the top engineering and physics universities in the country, and an international authority on radiation-induced illness. Dr. Morgan had been hired by the Manhattan Project to be director of health physics at the Oak Ridge National Laboratory, where he spent twenty-nine years determining the radiation limits for workers. His testimony takes up most of the first day. "There is no safe level of radiation exposure," he says. "So the question is not: What is a safe level? The question is: How great is the risk?" There is no such thing, he states, as a "permissible" dose of radiation; the slightest quantity can be enough, in a susceptible human, to cause some form of cancer. Present safety standards are dangerously inadequate. People living near or downwind of Rocky Flats are subject to a greater risk than those in other areas, and the EPA has, in his estimation, failed to consistently enforce even its own inadequate safety measures. Rocky Flats should never have been located close to a large population, he says. The plant should be shut down or relocated in a remote place, "preferably deep inside a mountain."

Two more days of taped testimony follow, still with no jury present. Local scientists unanimously support Morgan's statements. Dr. Edward Martell reports the results of his survey and confirms that high levels of plutonium have been found in soil far beyond the plant boundaries. Dr. John Cobb, professor of preventive medicine at the University of Colorado Medical Center, testifies that the most significant danger comes from the residue from more than five thousand leaking barrels filled with plutonium—plutonium that has been picked up by the wind and blown as far away as Denver. One way to quantify the presence of plutonium is to determine the rate of decay of radiation as measured in disintegrations per minute. In 1973 the Colorado State Health Department proposed an "interim" standard for soil contaminated with plutonium, setting the maximum allowable concentration at 2 disintegrations per minute per gram of soil. The radioactive sand under the barrels measured at 30 million disintegrations per minute, 15 million times the state standard.

All of the nuclear physicists and physicians who testify believe the plant is a public health hazard and must be closed or relocated.

The prosecution then asks whether the potential dangers of Rocky Flats justify the actions of the defendants. Given the fact, Dr. Morgan says, that ordinary political means have failed to produce necessary change, he believes that nonviolent action is probably justified to publicize the problem—even though, he quickly adds, blocking the railroad tracks will not "miraculously" decontaminate the eleven thousand acres already polluted. Dr. John Gofman, from the University of California, Berkeley, states, "Protest is always justified when it is the only means to make a deaf government listen."

After eleven days of testimony, the trial is adjourned until after the Thanksgiving holiday.

Judge Goldberger has not allowed the jury to hear the testimony of expert witnesses, which he's deemed irrelevant, but after the recess he allows some of the defendants to make statements directly to the jury. The prosecution—including attorneys from the DOE—incessantly interrupts. Truth Force member Roy Young, the Boulder geologist, takes the stand. "I was on those tracks not to commit trespass but to prevent random murder on the population of Denver."

"Objection, your honor!"

"Objection sustained."

"And if I thought," Young continues, "that by staying on those tracks . . . I could close the plant tomorrow, I would be willing to stay there for the rest of my life."

"Objection, your honor!"

"Objection sustained."

Skye Kerr, a twenty-three-year-old registered nurse and student at the University of Colorado, speaks of her training at Boston Children's Hospital and the long-term effects of radiation-caused cancer and leukemia. "It [cancer] happens years later," she says. "You can't see or feel or touch radiation, but it's as real as a gun."

"Objection, your honor!"

"Objection sustained."

"I felt the only thing I could do," she continues, "was to bodily put myself on the tracks. I knew that laws much, much higher [than trespassing] were being broken."

"What kind of laws?"

"Laws of human life. You know, violations of rights you have as a human being."

"Objection, your honor!"

"Objection sustained."

Finally Marian Doub's mother, Nancy, takes the stand. She describes the night she was arrested with her daughter. "We went out on the tracks," she explains to the jury, "and walked up the tracks in the dark, with our flashlights, singing 'We Shall Overcome' and 'We Shall Not Be Moved.' It was a very moving experience, standing next to my daughter. It's not the usual thing you imagine for mother-daughter activity. It meant a lot to be standing beside my daughter." She pauses to collect herself. "You know, it shouldn't be just the young people who are worried about this. It's not fair to give them that burden. So I was glad to be there."

In his closing remarks, chief defense attorney Edward Sherman appeals to the jury as "the conscience of the community." On the other hand, prosecuting attorney Steve Cantrell argues that this case is merely "a case of simple trespass. We are not here to change the policy of the U.S. government."

The judge reads the instructions to the jury, reminding them of the fact that the choice-of-evils defense does not apply. The jury is to disregard the emotional appeals of the protesters and consider "only, and nothing but, the formal charges of obstruction of traffic on a public right-of-way and trespass against U.S. government property."

After five hours of deliberation, the jury says it cannot reach a decision. Judge Goldberger excuses them for the evening. The jury returns in the morning and deliberates for another five hours. Their verdict: all defendants are guilty on charges of trespassing, but innocent of obstructing traffic. Each protester faces a maximum fine of five hundred dollars or six months in jail.

The jurors explain that they sympathize with the defendants but, under the instructions of the court, could not acquit them of trespassing. One juror passes a note to the defendants: "My support and prayers are with you all." A reporter tries to interview a juror and the juror stumbles midsentence, leaving the courtroom in tears. Another juror pulls aside defendant Jack Joppa and says, "We all support you and your cause."

By evening, Patrick Malone and his fellow pirates are back in the tepee on the tracks.

☢

CONTAMINATION LIES hidden like land mines not only in the wind-swept landscape of Rocky Flats but in the bodies of the people who live there. In November 1978, Dr. John Cobb of the University of Colorado Medical Center, in conjunction with the EPA, reports that lung and liver tissue taken during autopsies at local hospitals from the bodies of 450 people who lived near Rocky Flats contains plutonium. Further analysis confirms the presence of plutonium-239, the "fingerprint" of weapons-grade plutonium produced at Rocky Flats. Dr. Cobb hypothesizes that plutonium is present in the reproductive organs as well, where it could affect sperm and show up in future generations as cancers and deformities.

The study is halted, however, before Dr. Cobb has a chance to fully analyze his results or begin testing reproductive organs. Begun in 1975, it ends after Reagan takes office in 1981 and James Watt becomes secretary of the interior. Hundreds of frozen sex organs from people who lived near Rocky Flats are sent to freezers at Los Alamos National Laboratory, where they will remain for nearly fifteen years. The presence of plutonium in the bodies is undisputed, but whether it's enough to cause cancers or genetic defects is never determined. The organs are finally sent to Colorado State University in 1994. When data is finally published, the executive summary is rewritten by the government to reflect more favorably on Rocky Flats.

A quiet epidemic shadows the highly touted safety record for workers at Rocky Flats. James Downing, a maintenance machinist who worked in the glove boxes in the plutonium processing buildings, sustained first- and second-degree burns on his hands during a plutonium fire on January 6, 1961. He inhaled an undetermined amount of plutonium during a glove-box fire nine years earlier. The long, lead-lined gloves he wore while working on glove boxes often ripped, and his hands and arms were contaminated. During his time at Rocky Flats, he was injured or exposed at least forty-eight times. On November 28, 1978, he dies of esophageal cancer at age forty-four.

At a court hearing following Downing's death, the former manager of radiation at Rocky Flats admits that forty-eight is not considered a large or unusual number of accidents for an employee at Rocky Flats.

Downing is just one of the first of many employees to pay for their employment with their lives.

Potential homeowners had been asked to sign a waiver for years, but in 1979, the Federal Housing Administration (FHA) establishes a legal requirement that anyone buying a home in the vicinity of Rocky Flats with FHA mortgage insurance or any other HUD assistance must be informed of plutonium contamination in the area. They must sign something called the Rocky Flats Advisory Notice, which says: "This notice is to inform you of certain facts regarding the United States Department of Energy Rocky Flats Plant which is located within ten miles of your prospective residence. You should be aware that there exist within portions of Boulder County and Jefferson County, Colorado, varying levels of plutonium contamination of the soil. However, according to the information supplied by the Department of Energy, the soil contamination in the area in which your prospective residence is located is below the limits of the applicable radiation guidance developed by the Environmental Protection Agency (EPA)."

Many local homebuilders and business owners are upset, and some believe that sales and property values may suffer. A spokesperson for the Rocky Flats Monitoring Council notes, "Very few residents know about Rocky Flats. The problem is, how do you tell them certain things without creating a panic situation?" A Rockwell spokesman tells the press, "We [Rockwell and the DOE] try to stay out of local politics. . . . The public is confused about Rocky Flats, and I have to lay that directly in the laps of the local media."

FOR YEARS I have prepared for my father's death. There are all the DUIs, the fender benders, the nights of waking up late to hear him stumbling in the foyer. The night he comes home with a broken jaw after some fracas with a police officer. The late nights I race down from college—sometimes with Mark—to be with my mother when he pounds on the door. They have separated, more or less, and Dad is living in an apartment in old Arvada. My mother changes the locks on the house and finds solace in long hours at work, and white pills and red wine when she gets home. Kurt is still living at home, and he watches and tries to intervene—sometimes physically—as the situation deteriorates. One evening he and my mother come home late to find that Dad has broken into the house and is lying unconscious on the bearskin rug, sick with alcohol poisoning, his distended abdomen churning. They rush him to the hospital.

"Why," I ask my brother, "does he keep breaking in? Why does he keep coming around?"

Kurt shrugs. "I guess he can't give up the idea of his marriage to Mom," he says. "He wants to be part of the family."

Maybe he wants help, I think. But he'll have nothing to do with re-habilitation programs or therapy. None of us knows what to do. I don't know how I will feel about my father's death when it happens, but it always seems imminent.

Partly through Mark, I have become interested in yoga and meditation. Yoga seems to help my neck, which often pains me, and meditation gives me a sense of calm. I start taking a meditation class at a local church one night a week that is led by a woman with a low, soothing voice, and I feel peaceful just being in the room with her.

One night, just as we are beginning our session, the class is interrupted by a voice in the hall. I hear my name. A young woman peeks in. "Is there a Kris here?"

Everyone exchanges glances. "I guess that's me," I say.

"You have a phone call in the church office."

That's odd, I think. *Who would know to find me here?* I follow the woman into a small, carpeted room with a single desk and a telephone. I pick up the receiver. "Hello?"

"Kris?" It's my mother. Her voice is clipped.

"What's wrong?"

"Something's happened."

Bad news is not unanticipated. "What's happened?"

"I can't tell you."

"Why?"

"You just need to come home."

"Tell me what happened, Mom." I sigh and look at my watch. "I'm in class until nine and then—"

"Get in your car and come home right now, honey."

There's something different about this time.

"Fine." I set the receiver back in its cradle, think briefly about going back to the class to explain, and decide to head straight for the parking lot. This is unusual. My family is always in turmoil with one thing or another, but usually the news can wait.

I pull out of the parking lot and turn onto Wadsworth Boulevard. Few cars are on the road. At the first stoplight I debate whether I should run it. Is this an emergency, or just my mother's dramatic overreaction to something? I pull up to the next light—why am I hitting all the lights?— and I reconsider the tone in her voice. I drive through the red light. By the time I reach our driveway and see every light in the house blazing, I know something is seriously wrong.

The front door is unlocked. I find my mother in the kitchen, alone in the bright light. She sits stiffly at the breakfast counter. Her face is pale.

"What happened?" I ask.

"Kris." She stands.

"It's Dad, isn't it?" I say.

"Honey—"

"Just tell me. Just say it."

"It's Mark."

"What?"

"It's Mark, honey." She speaks rapidly, her voice a dead calm. "It happened at Castle Rock." Mark and I have been to that spot many times; it's where he taught me how to use a carabiner. "He was climbing with his friend Gary, teaching him how to belay, and something went wrong. Gary was up top and Mark was on belay and he fell."

I can't speak. The floor seems to have dropped out from under me.

"Mark fell at Castle Rock?"

My mother nods.

"He's climbed that a million times."

"Something went wrong this time. He fell—"

"He's in Boulder, then? At Boulder Memorial?"

"No—"

"At his mom's?"

"No, honey." Her eyes meet mine. "He died. He died at the scene. They sent an ambulance all the way up the canyon, but it was too late."

The words seem impossible. My body feels like ice.

I hear the front door open and close. It's my father. My mother must have called him. "I can't talk now," I say. I fly up the stairs to Karin's room, the room we once shared as sisters. I shut the door and turn off the light and sit on the edge of the bed in the dark. It feels as if a knife has been shoved in my gut. The air roars around my ears. I hear my father's voice and then my mother's. "That's your daughter," she yells. "The least you can do is go up to her. Go to her! Go to her!"

I hear my father's step on the staircase. "No," I pray. "No, no, no."

There's a long pause as I stifle my sobs. He knocks on the bedroom door. "Kris?"

I can't speak.

"Kris? Are you in there?" He puts his hand on the door handle but doesn't turn the knob.

"Kris," he says again, his voice muffled through the door. "Let me in."

I try to get up. How can I let him in when a thousand times he has cast me out? I pull the dark of the room around me and feel my soles sink into the carpet, a thousand pounds of weight. "No," I whisper. "No."

We bury Mark on a cool fall morning. We kneel on the stiff brown grass at the side of the church Mark grew up in, the church his mother still attends, a few miles from Rocky Flats. The ground is cold but not frozen. Mark's brothers dig a deep, narrow hole with a shovel and we scoop the ashes from the cardboard box and mix them in with the soil, the three of us together: me, Mark's mother, my mother. Our fingers are numb with cold. We pat down the earth and ash and Mark's brothers push a small pine tree into the hole, and we build a protective wall around the roots and slender trunk.

My mother offers me some of her little pills. I look away without taking them. She suggests I talk to the minister. I hear nothing from my father.

I think of Mark's ashes mixed in with the hard soil, rocky and unforgiving. The soil that he believed was contaminated.

The tree that is Mark takes root. I don't think it will survive the winter, but it does.

☢

BY 1978, news of Dr. Carl Johnson's research begins to reach the public, and it is sobering. In his published study, he demonstrates a stark pattern of "excess incidence of all cancer in all age categories" for both genders in areas exposed to Rocky Flats contamination, from cancer of the lung, leukemia, and lymphoma to thyroid and brain cancer in females. In the five years after Rocky Flats was built, leukemia deaths in children rose substantially. By 1962, five years after the 1957 explosion, leukemia deaths in children who had lived near the plant were twice the national average. Before the construction of the plant, they'd been below the national average. Birth defects are higher in Arvada between 1975 and 1977. In Area I, which extends thirteen miles downwind from the plant, Johnson finds that males have a 24 percent higher cancer rate, females a 10 percent higher rate. Lung and bronchial cancer for males is about 33 percent higher than in unexposed areas. In Areas I through III, Johnson makes a "most unexpected discovery": forty cases of testicular cancer, "an unusually high incidence." Ovarian cancer is also higher than expected, at 24 percent.

When Johnson turns his attention to workers at Rocky Flats in 1980, he finds that they have eight times more brain tumors than expected, as well as triple the number of malignant melanomas. In addition to plutonium, other radioactive and toxic materials keep showing up in the environment: cesium, curium, strontium, and carbon tetrachloride. When scientists go out to collect soil samples in residential areas around Rocky Flats, they wear protective face masks and don't stay on the property for very long. "So how can children play in it year-round?" Johnson asks.

In January 1980, for the first time the EPA admits to the press that it believes cancer deaths from Rocky Flats contamination may occur among Denver residents. This conclusion is based on reports prepared by the DOE itself and the fact that air-monitoring stations at Rocky Flats have consistently shown higher levels of plutonium-239 than any others in the western hemisphere.

All this grim news is becoming a problem for the county commissioners, the county board of health, the DOE, and Rockwell. For them, business is the name of the game, and no one wants to hear about contamination. It's a tight group: the county commissioners, who monitor business growth and new home development, appoint the county board of health. The county board of health in turn appoints the health director. Although Rocky Flats is run by Rockwell, it is regulated by the DOE. Rockwell receives bonuses from the DOE based on production. All that counts is the number of triggers produced. There is little transparency or oversight: Rocky Flats makes its own rules. The Atomic Energy Act of 1954 exempts nuclear weapons plants from environmental laws. If there's going to be any environmental monitoring at all, Rocky Flats and DOE officials want to regulate themselves.

The question is how to keep someone like Johnson from releasing more alarming reports to the press. Dr. Otto Bebber's attempts to censure Johnson in 1977 and 1979 were unsuccessful. However, on May 1, 1981, the balance of power shifts when the county board of health appoints the president of the Jefferson County Homebuilders Association to the board. Two weeks later, the board votes 3–2 to demand Johnson's resignation. Otto Bebber calls Carl Johnson and suggests they have lunch. The men agree to meet at a family-owned restaurant in the area.

Carl Johnson is an unlikely renegade. He's a quiet, gray-suited man with black-framed glasses that give him a scientific look, a man conservative in his politics. He has an uneasy feeling about this luncheon, though, and he doesn't have to wait long to see why. The waitress takes their order and Bebber cuts to the chase. He says that the county board of health no longer has confidence in Johnson. He suggests he start looking for another job. He knows Johnson is a family man with three kids. "Why don't you resign?" Bebber asks. "Then you can keep your benefits and not have any publicity or get adverse marks on your record."

Johnson is not surprised. He's been fighting since the day he took office. But he stands by his studies. The lunch ends with little agreement. When the county board of health meets on May 15, the board decides to give Johnson the choice of being fired and losing all his accrued benefits, or of resigning immediately.

Unwillingly, after more than seven years on the job, Johnson resigns.

But some people stand up for him. One county commissioner tells the press that Johnson was dismissed for fueling anger about falling property values and curbing housing development. "The reason they're getting rid of him is because he was looking at what nobody else wanted to look at," she says. Dr. Edward Martell, who is feeling the heat for his own soil studies at Rocky Flats, calls Johnson "the only man in the Denver public health community who is concerned about public health." And a member of the county board, a hydrologist with the U.S. Geological Survey, notes that most of the county commissioners want "a house on every tenth of an acre in the county" and Johnson is standing in their way.

Within days of his termination, Johnson files suit. When the case goes to trial, an attorney for the county board of health states that the director of the Jefferson County Health Department serves "at the pleasure of the Board." Another member testifies that Johnson "could be fired for the color of his tie" if the board didn't like it. The judge rules that because Johnson serves at the pleasure of the board, he cannot be reinstated.

Jim Stone never dreamed he'd get into trouble by simply doing what he was hired to do in the first place. Stone was one of the first engineers employed at Rocky Flats. He was hired in 1952 to work on the original design of the plant, primarily the power and the ventilation systems for two of the plutonium processing buildings. Stone is proud of the ventilation system. Usually buildings are pressurized to exhaust air out of the building. At Rocky Flats, the opposite occurs. Fresh air is brought into the building for workers, pressurized through the equipment and the glove-box line, and then pushed through the filters and back outside again.

For years Stone worked off and on at Rocky Flats. He took other engineering jobs, too, and worked in Alaska, on the missile silos in Idaho and Wyoming, and in Greenland, where he surveyed a pipeline. He loves being an engineer, and he has worked hard to become one. He and his brother and sister were separated at a young age when their parents couldn't afford to care for them during the Depression, and the children were left in an orphanage in St. Louis. Stone was educated in Catholic schools and eventually attended Washington University. He hopes someday to make enough money to bring his siblings to Colorado, where they can be united after decades apart.

In November 1979, managers at Rocky Flats persuade Stone to come back full-time. They need him, they say, as they intend to "build in-house capability," relying less on independent contractors. Stone hesitates. The plant is thirty years old, run-down, and largely obsolete, and he's disillusioned with policies and procedures there. Maintenance has been poor. *But if they're committed to fixing things up, well,* he thinks, *that might be interesting.*

Things don't go as planned. The plant is not redesigned. But that's okay. Based on his long years of experience with so many different aspects of the facility, Stone's role evolves into in-house troubleshooter. His job is to prepare weekly reports that identify key problems and then recommend how they should be fixed. He loves it. He likes the variety of things that come his way, and he likes the idea that he can try to remedy some of the things that have bothered him for years, particularly air and water contamination both on- and off-site.

Stone takes issue with a number of things. There's far too much radioactive waste stored at the plant, and Rockwell has a hard time figuring out what to do with it all. Drums and containers full of plutonium are piled up, and plutonium-laced liquid from contaminated pond water is sprayed through an irrigation pump onto fields of grass, even in winter, when there is so much ice and snow on the ground that the contaminated water flows into ditches and waterways running off the Rocky Flats site. There is inadequate monitoring of water wells. Plutonium is clogged in the ductwork of buildings, especially Building 771, and the amount of material unaccounted for (MUF—an odd acronym by anyone's standard) is growing at an alarming rate. The company is burning plutonium-contaminated waste in an incinerator that's not properly licensed or filtered, and those contaminants are floating up into the Colorado sky.

But the problem that worries Stone most, in the short term, is pondcrete.

Pondcrete is the result of yet another desperate effort to deal with the overflowing radioactive waste at the plant. Since the 1950s, Rocky Flats has been storing liquid hazardous waste in five shallow man-made ponds, similar to small swimming pools. The liquid—low-level radioactive waste and sewage sludge—is poured into the ponds, where it is heated by the sun to evaporate moisture and reduce its overall weight. The DOE wants the ponds phased out. Rocky Flats officials intend to mix the toxic sludge from the ponds with concrete in order to immobilize the hazardous waste, and then pour the toxic pudding into plastic-lined cardboard boxes the size of small refrigerators. They plan to ship the boxes to the Nevada Test Site for burial.

In October 1982, Stone sends a memo to Rockwell's management. Pondcrete is not going to work, he writes. The cement will not harden and the plutonium will not stabilize. His co-workers, who have warned him in the past that he is going too far, caution against the memo. "Oh, you're going to get it now!" they tease him. "That last one was a doozie!" But Stone is adamant. His job is to tell Rockwell what they need to know, whether they like it or not.

Morale at Rocky Flats, Stone knows, is low. Some of the managers he works with are discouraged, and the feeling trickles down to the workers. Some managers, Stone hears, were booted from the aerospace industry in California and feel they've been transferred to the dirtiest plant in the country. They don't give a damn, he thinks privately. The policy is to protect your job and keep things quiet because it's the best money in town. "Someday," he says to his co-workers, "someone will come in here and find a lot of skeletons in the closet, find out about violations of environmental laws, and just plain dumb engineering things, and wasting money and jeopardizing the workers and the community. And there's going to be hell to pay."

Despite the admonitions of his friends, Stone isn't worried about his job. He's doing what he was hired to do. He knows where the skeletons are buried, and the company needs to know it, too. He knows he'll never rise in the ranks at Rocky Flats, but he also knows that his job is absolutely secure.

☢

MY FAMILY never talks about feelings, and we certainly never talk about plutonium. It's hard to take something seriously if you can't see it, smell it, touch it, or feel it. Plutonium is a cosmic trick. The invisible enemy, the merry prankster. Can it hurt you or not? None of us know.

Plutonium is the darling and the demon of the nuclear age.

The story of plutonium began with radium, an element discovered in 1899 by Marie Curie. She called an element "radioactive" if its nucleus was unstable and it decayed and released particulate radiation. A radioactive atom gives off radioactivity because the nucleus has too many particles, too much energy, or too much mass to be stable. The nucleus of the atom disintegrates in an attempt to reach a stable or nonradioactive state. As the nucleus disintegrates, energy is released in the form of radiation. Different terms are used for measuring radiation, depending upon what is being assessed. The amount of radiation emitted by a radioactive material is measured in *curies*, named after Marie Curie. The radiation dose absorbed by a person—that is, the amount of energy deposited in body tissue—is measured in *rads*. A person's biological risk of health effects due to exposure to radiation is measured in *rems* or *sieverts* (the sievert, which equals 100 rem, is the international standard). Radiation workers wear film badges called dosimeters, which measure exposure in rems or sieverts.

At the beginning of the twentieth century, though, radium was not viewed as a health risk. On the contrary, it was hailed as a cure for all kinds of ailments. As early as 1906, radium salts were used to try to shrink or eliminate cancerous tumors. Radium-laced water, facial creams, and baths were also popular. But the excitement about radium was short-lived; it wasn't long before people began to experience health effects like bone fractures, bone cancer, and jawbone infections.

Perhaps the worst cautionary tale was that of the "Radium Girls." In the 1920s, about seventy young women worked long days at the U.S. Radium factory in Orange, New Jersey, employed by a defense contractor that supplied glow-in-the-dark watches to the military. Each girl painted

250 watch dials a day with radium paint, earning about a penny and a half per dial. Chemists at the plant wore masks and used lead screens and tongs, but the female workers were told the paint was harmless. They licked the tips of their paintbrushes to sharpen the points, and some girls even painted their fingernails and faces with the glowing substance. The girls suffered from anemia, bone fractures, and necrosis of the jaw, and some died. An ensuing lawsuit, settled in the fall of 1928, created a media sensation and helped establish a new labor law protecting employees from hazards in the workplace.

In response to concern over the dangers of radium, the National Bureau of Standards established an occupational standard for radium, just two months before the discovery of plutonium. Six years later a group of scientists at the University of California, Berkeley, led by Glenn Seaborg and Edwin McMillan, synthetically produced the solid, silvery-gray element using a five-foot-long cyclotron.

McMillan called the element plutonium after the recently discovered planet Pluto, which had been named by an eleven-year-old schoolgirl in Oxford, England, who won five pounds for her efforts. She took the name from the Roman god Pluto, god of the underworld, god of the dead, the Destroyer. It seemed an appropriate name for a cold, dark planet made of rock and ice.

Seaborg suggested the abbreviation *Pu* as a joke. No one seemed to get the gag, and *Pu* passed into the periodic table without comment.

A paper describing the new element was prepared for publication in March 1941, but was abruptly withdrawn after the discovery that plutonium was capable of sustaining a nuclear chain reaction, and thus might be useful in an atomic bomb.

In 1939, amid concern that Germany might be in the early stages of developing a nuclear bomb, physicists Edward Teller and Eugene Wigner sent a letter to President Franklin D. Roosevelt, suggesting that the United States should begin its own research into a bomb. The letter was also signed by Albert Einstein. (Einstein, as well as other prominent physicists, later regretted the letter, as it led to the development and use of the atomic bomb against civilians. The physicist J. Robert Oppenheimer himself, known as the "father of the atomic bomb," was later opposed to nuclear weapons.) Niels Bohr, who had won the Nobel Prize for Physics in 1922, was prescient. When asked if enough uranium-235 and -238 could be separated to produce a bomb, he said, "It can never be done unless you turn the United States into one huge factory." Years later he repeated this to his colleague Edward Teller. "I told you it couldn't be done without turning the whole country into a factory," he said. "You have done just that."

Following the Japanese attack on Pearl Harbor on December 7, 1941, the U.S. government initiated the Plutonium Project at the University of Chicago, with the goal of creating a nuclear chain reaction for plutonium-239 and developing an atomic bomb. In a makeshift lab under Chicago's Stagg Field in a project euphemistically called the Metallurgical Laboratory (Met Lab), a team of scientists led by Enrico Fermi achieved a sustained nuclear reaction in the world's first nuclear reactor. When the project was taken over by the army in the summer of 1942, it became the Manhattan Project. The Manhattan Project would bring together some of the greatest physicists in the world to try to build a workable nuclear bomb from scratch in three years.

General Leslie Grove was chosen to lead the project, and J. Robert Oppenheimer managed the day-to-day operations for the research and design of an atomic bomb. The project moved to Los Alamos, New Mexico, near Santa Fe, to a piece of land on a beautiful high mesa that previously had been the site of a ranch school for boys. A hidden city with a deadly secret blossomed in the high desert. There were two other research and production sites associated with the Manhattan Project: the plutonium-production facility at the Hanford site in Washington State and the uranium enrichment facilities at Oak Ridge, Tennessee.

Upon the death of President Roosevelt, Harry S. Truman assumed the presidency on April 12, 1945, and learned of the secret wartime project. When Germany surrendered on May 8, 1945, the Manhattan Project was just months away from producing a workable nuclear weapon. Two types of atomic bombs were developed. The first, a gun-type fission bomb, was made from uranium-235. This weapon design ultimately proved inefficient, and Los Alamos developed an implosion bomb in which a fissile mass of uranium-235, plutonium-239, or both was surrounded by high explosives that compressed the mass, resulting in nuclear fission. The plutonium triggers eventually produced at Rocky Flats were fission "pits" that when detonated trigger the more powerful fusion explosion of a thermonuclear or hydrogen bomb.

No one used the word *bomb*; they called it a "gadget."

The first test of a nuclear bomb was in New Mexico on July 16, 1945. Oppenheimer named the test bomb Trinity, reportedly in reference to lines from the poet John Donne: "Batter my heart, three-person'd God" and "As West and East / In all flat Maps—and I am one—are one / So death doth touch the Resurrection." Twenty miles from the blast, Edward Teller put suntan lotion on his hands and face, even though the early-morning sky was still black. Enrico Fermi stood beside him. For many onlookers, the tremendous explosion evoked powerful, ambivalent feelings: pride and joy at the extraordinary—and swift—scientific achievement, and horror at the deadly weapon that had now been unleashed. The Trinity bomb had a plutonium-239 core.

Three weeks later, on August 6, 1945, the crew of the Boeing B-29 Superfortress bomber *Enola Gay*—named after the pilot's mother—dropped an atomic bomb on the Japanese city of Hiroshima. "Little Boy" was a gun-type bomb. Ninety thousand to 160,000 people died within the first four months. "Fat Man," the atomic bomb dropped on Nagasaki, Japan, on August 9, 1945, three days later, killed 100,000 people within the first two to four months, and tens of thousands more were injured in both blasts. "Fat Man" was a plutonium-core weapon.

The existence of plutonium was made public only after the bombs were dropped. From 1945 to 1989 the United States produced tens of thousands of nuclear warheads in its arms race with the Soviet Union. Mutual assured destruction—known as MAD—was the governing philosophy. The MAD program was intended to act as a deterrent in that if one country attacked another with nuclear weapons, the attacked country would immediately retaliate and both countries would be destroyed.

After the Second World War, to formulate safety standards for the nuclear industry, American scientists carried out studies of the effects of plutonium on humans, including tests in which researchers exposed or injected people with plutonium without their informed consent. These studies determined that for humans, even 1 microgram—that is, one-millionth of a gram—should be considered a potentially lethal dose. Experiments with animals demonstrated that within the body, plutonium was distributed differently and more dangerously than radium.

The term *body burden* was used to describe the amount of radioactive material present in a human body, which acts as an internal and ongoing source of radiation. The DOE established a permissible "full body burden" for lifetime accumulation of radiation within the body on the assumption that a worker whose exposure did not exceed this level would not suffer ill effects. Although some workers whose body burden was near the limit did not experience any adverse health effects, others exposed at levels far less than the permitted full body burden developed various types of cancers. Exposure to plutonium was linked to cancers, brain tumors, and reproductive disorders, but plutonium was determined to be most dangerous when taken into the lungs. Particles of plutonium weighing 10 micrograms or less can easily be inhaled.

Robert Stone, head of the Plutonium Project Health Division at the Met Lab in Chicago, made the first estimate of a permissible plutonium body burden. He set the limit at 5 micrograms. In July 1945, scientists at Los Alamos reduced that standard by a factor of five, to 1 microgram. In 1949, in the wake of the disturbing new results from animal testing, representatives from the United States, the United Kingdom, and Canada at the Tripartite Permissible Dose Conference set an even stricter standard: they agreed that the maximum body burden for plutonium should be 0.1 microgram.

Officials from the AEC were not pleased. Workers at Los Alamos were already operating with a limit ten times higher than that, and they pointed out that this "extremely conservative" standard would add millions of dollars to the construction of buildings at Los Alamos. They held the level at 0.5 microgram. In 1977 the International Commission on Radiation Protection (ICRP) established a guideline for the maximum occupational plutonium dose for workers of 5 rem annually from both internal and external radiation. In 1991 the ICRP proposed that workplace exposure be lowered from 5 to 2 rem per year, but the United States has not accepted this change. The level of 5 rem per year, set in 1958, is still the U.S. standard.

Plutonium was supposed to be a savior, to save us from the enemy. It wasn't supposed to leak and burn and blow away, seep down into the water table and fly up into the sky. It was supposed to pay attention to borders and fences and property lines. It was supposed to know the good guys from the bad guys.

We don't worry about it too much, though. The government will let us know if there is any real danger.

DESPITE JIM Stone's stern memos, Rockwell decides to go ahead and manufacture pondcrete. Workers call the production of pondcrete the Jelly Factory. But the plutonium pudding, a huge concoction of pond sludge, radioactive substances, hazardous chemicals, and concrete, takes longer than expected to gel. Managers pack the gooey mess into cardboard-and-tarp boxes the size of small refrigerators anyway, resulting in 12,000 one-ton blocks that stand out in the open. Unprotected from sun, wind, and snow, many of the blocks of pondcrete are part liquid, and the boxes are piled on top of one another like huge, soggy, sagging Lego blocks. In less than a year the blocks start falling apart. As they disintegrate, liquid containing nitrates, cadmium, and low-level radioactive waste leaks and leaches into the ground, where it runs downhill toward Walnut Creek and Woman Creek. Workers test the thickness and consistency of the sludge by sticking their thumbs into it. When they report the problem to management, one foreman tells them to "cap" the soft pondcrete blocks with fresh cement over the spot where inspectors usually stick their measuring instruments.

Then there's another problem. By the end of 1986, approximately two thousand blocks of pondcrete have been shipped and buried at the Nevada Test Site. Suddenly shipments are halted. Inspectors there have discovered that the pondcrete blocks contain radioactive material as well as hazardous chemicals, which means they are classified as mixed waste. Regulated under the new Resource Conservation and Recovery Act (RCRA), which governs the disposal of solid and hazardous waste, the Nevada facility doesn't have a permit for mixed waste.

Waste regulation at nuclear weapons sites is problematic, even with the passage of RCRA and the Clean Water Act, enacted under a Republican administration. Either DOE bomb plants are exempted from the new laws, or politicians and government officials dodge the issue with vague language and deadlines that are rarely met. DOE managers vigorously resist new environmental laws. At Rocky Flats, it's common for managers to blindfold EPA investigators before taking them through the plant.

What no one talks about is the fact that from the moment environmental laws like the RCRA were passed, or when those laws began to apply to DOE facilities, no one—including Rockwell, the EPA, or other defense contractors around the country—has much choice when it comes to allowing the storage of unpermitted mixed hazardous wastes. Facilities like Rocky Flats have to break the law to continue operating. Production cannot stop.

With much fanfare, that same year the DOE, the Colorado Department of Health, and the EPA sign a joint compliance agreement to address environmental problems at Rocky Flats. But a confidential internal DOE memo acknowledges that the plant is "in poor condition generally in terms of environmental compliance. . . . Much of the good press we have gotten from the agreement in principle has taken attention away from just how bad the site really is."

The DOE may admit internally that the plant is a mess, but they want to keep that information to themselves and put on a good face. Jim Stone continues to write his reports and grows more adamant, despite the fact that Rockwell is becoming increasingly annoyed and tells him to withhold his reports from the DOE. There are layoffs that year. Layoffs are not uncommon given the ever-changing budget, and sometimes people are even hired back or continue to work for Rocky Flats as contractors. Jim Stone isn't worried. The layoff deadline passes.

One day he's in a meeting when there's a knock on the door. He opens it to a manager and a security guard. "You have one hour to leave," he's told.

He's stunned.

"I'm in the middle of a meeting!"

"Doesn't matter," the manager says. "You're out of here."

Under the watch of the armed guard, Stone returns to his office and, as ordered, gathers up as many materials as he can in the time allowed. He stuffs the boxes into his car and turns in his security badge.

He will never work again in the nuclear weapons industry.

Eventually, 16,500 one-ton pondcrete blocks will be stacked on the asphalt pad, exposed to the elements, just upwind of Arvada, Broomfield, and the city of Denver.

I GO back to my classes and my job and my apartment. I don't tell people what happened. I can't even say Mark's name. Sometimes it's hard to speak at all. Heavy dreams invade my sleep. A clutch in my chest and ringing in my head make me feel as if I were the one who fell from the cliff.

As a kid I kept secret note pads and spiral-bound journals with curled-up edges. I wrote in the margins of my Big Chief wide-ruled pads. I collected pens, and only the right kind would do: a very fine-tipped black ink that didn't smear. Things were happening all around me and I had to write them down, fast. In third grade I read *Harriet the Spy*. Harriet was like me. She snuck around with a secret notebook and took notes on everyone. I stuck ragged little notebooks in the back pockets of my jeans, carried them in my book bag, stashed them under my mattress, and stacked them under my bed. I wrote about everyone: my friends, my enemies, my family, my teachers, my horse, the lake. As I grew older I was always running out of space and wrote on anything that was handy: receipts, napkins, bookmarks, the backs of movie tickets. I couldn't keep track of it all.

When Mark dies, I stop writing.

My year and a half with Mark fills three journals: one thick spiral notebook; a dog-eared leather book with unlined paper; and a tall, narrow-lined book with a beautiful red cloth cover, dotted with coffee stains. I put them in a box under my bed.

I come home on occasional weekends. My family tries to keep up appearances. My mother has her hair done in an elaborate bouffant helmet at the Arvada Beauty Parlor, just as always. Karma and Karin take turns driving Dad to his office and to his court dates when he loses his driver's license over another DUI. Bills go unpaid and the house slips into decline. Even my grandfather can't save the law practice from sinking into a financial abyss. My father grows rail-thin except for his distended abdomen. "I don't understand it," my mother says. "He never eats." He comes home only sporadically and we never really know where he is. Around us, he's silent and sullen. His white shirt is crumpled and stained, his pants loose and baggy. He rarely changes clothes. My mother wonders if he's living on the street, or in his car, or with his parents. Does he spend nights on a barroom floor? No one knows what's true anymore. We're all waiting, but for what?

The months stretch on. My mother refuses to file for divorce. "And how," she asks, "can I support four kids?" She thinks about selling the house, but in the meantime she sells her diamond engagement ring to pay for groceries.

The Rocky Flats Truth Force has been occupying the railroad tracks for a full year, through the summer's heat and the winter's snow and cold. When anyone is arrested or removed, another person takes his or her place. The vigil unites groups opposed to Rocky Flats. In the spring of 1979, Patrick Malone—the intrepid pirate of tepee fame—along with other Truth Force members and two chanting Buddhist monks, engage in a 242-mile walk around the area affected by Rocky Flats, going from Boulder to Golden, to the state capital, and back through the cities of Arvada, Broomfield, and Lafayette. They circle the plant three times over a period of three weeks, showing that Rocky Flats is within walking distance of nearly everyone in the Denver metropolitan area.

They arrive back at the plant just in time for a two-day rally beginning on Saturday, April 28. Police anticipate two to three thousand people. Fifteen thousand show up. The crowd includes people from more than twenty states and a twenty-two-member delegation from Japan. Hundreds of protesters, led by a Buddhist monk, walk the twelve miles from Boulder to Rocky Flats, singing and chanting along the highway. Rockwell security guards, along with nearly fifty police officers from the Jefferson County Sheriff's Department, stand on alert. Sister Pam Solo opens the rally by calling for the end of nuclear weapons. "We are determined to put an end to this technology before it puts an end to us," she says. "We demand the conversion of Rocky Flats to an industry that is environmentally safe and socially productive."

People roar their approval. Kites flutter in the wind and Frisbees sail at the back of the crowd—a sea of blue and green mountain parkas, orange Hare Krishna robes, and yellow rain slickers. T-shirts are emblazoned with "No Nukes" and "Rocky Flats Sparkling Water—Don't Think, Just Drink." The highway is lined with parked cars, and nearly seven hundred bicycles stand along the barbed-wire fence. Signs wave above everyone's heads: "Hell No, We Won't Glow" and "Better Active Today Than Radioactive Tomorrow." Albert Einstein gazes placidly from a placard onstage, flanked by two tall speaker towers. A group of Native Americans who've walked eleven miles offer a spiritual blessing.

Jackson Browne and Bonnie Raitt perform onstage. Speakers include Daniel Ellsberg, Helen Caldicott, and George Wald. One month earlier, in Pennsylvania, there was a serious accident at the Three-Mile Island nuclear reactor involving the partial meltdown of a nuclear core. Now Ellsberg calls Rocky Flats "Denver's own Three-Mile Island." George

Wald, a retired Harvard biology professor and winner of the Nobel Prize, notes that current stockpiles of nuclear weapons in the United States and Russia are the equivalent of sixteen billion tons of TNT. "That's four tons of TNT for every man, woman, and child on this planet," he says. "Is there any greater madness than the concept of a *limited* nuclear war?" He's older than most of the crowd. "I'm here to represent the generation gap," he says, white hair ruffling in the wind. "I've had my life, but it's highly questionable whether you will have yours, my children." Dr. Helen Caldicott is even more direct. A doctor serving as president of Physicians for Social Responsibility, she calls Rocky Flats a "death factory." "We're killing ourselves," she says, "to make bombs better to kill ourselves better."

The rally is peaceful, and at noon brightly colored balloons are released to the sky, with written warnings of how far radiation from Rocky Flats can travel attached to their strings. Some reach as far as Canada. Between speakers and music sets, or whenever there is a pause in activity, the crowd begins a low chant that grows louder with each repetition, fifteen thousand people in unison: *Close Rocky Flats. Close Rocky Flats. Close Rocky Flats.*

On Sunday, 286 men and women are arrested and charged with trespassing in violation of the Atomic Energy Act of 1954. They offer no resistance. As they're loaded into school buses to be taken for booking, they toss flowers out the windows at the security guards.

When the first three protesters appear before the judge, he calls them "arrogant," finds them guilty of trespassing, and, in a lengthy statement before sentencing, quotes U.S. Supreme Court justice John Paul Stevens: "If a religious, moral, or political purpose may exculpate illegal behavior, one might commit bigamy to avoid eternal damnation, steal from the rich to give alms to the poor, burn and destroy, not merely public records or perhaps buildings, but even public servants as well, to implement a utopian design." He then issues formal instructions to all Denver federal judges that "the morality or immorality of nuclear weapons or nuclear power is not something to be tried." He rejects the trio's request to do community service rather than pay their $1,000 fines and, he adds, if they don't pay their fines on time, they "can stay in jail forever."

Most of the defendants are convicted.

BY 1979, nearly 3,500 people work at Rocky Flats. The plant bills itself as one of the safest government facilities in the country, with the number of "continuous safe hours"—work hours during which no reported accidents occurred—posted proudly on a sign near the plant entrance.

But signs can lie.

Fresh out of high school, in 1970 Don Gabel was working as a fry cook when he got the call that he'd passed the security check and could begin working as a janitor at Rocky Flats. A year later he transferred to Building 771, the notorious plutonium processing building, where he could make an additional fifteen cents an hour in "hot pay." Like most employees, he was told that the plant was the safest place he'd ever work.

Don learned to operate a furnace that melted plutonium, and a good part of his workday was spent with his head a few inches below a furnace pipe with a sign more appropriate for a shopping mall: DO NOT LOITER. Monitors showed the pipe to be highly radioactive, but Gabel was told that he was not in danger and that levels of radioactivity were well within government standards. One morning, just a few weeks after his transfer, he was instructed to tear construction tape from a contaminated tank. "That tape was really hot," he recalled later, and sure enough, his hands, face, and hair measured 2,000 counts of alpha radiation per minute. He had to wait at the plant nearly an hour for help, which was not unusual, he later testified, as "incidents" in the hot areas were common. When Gabel was tested again in January 1976, after working at Rocky Flats for six years, radiation levels on his body measured 1 million counts per minute. Chromosomes in his blood and brain cells had been altered by radiation. He developed a tumor on the side of his head the size of a grapefruit. On September 6, 1980, Don Gabel died, leaving behind a wife and three young children.

On his last day of work he was making $8.60 an hour.

Within hours of his death, Don's wife received a surprise call from the DOE. They wanted to examine her husband's brain. She gave permission for them to analyze the brain and there was an autopsy. The DOE took the brain, and she heard nothing. Finally the DOE admitted they'd lost the brain for three months. When it was rediscovered, it had deteriorated to the point where it could not be tested for the presence of plutonium.

Only a few of Don's friends at the plant know the circumstances of his death, and many employees continue to stick to the company line regarding safety. Rocky Flats engineer Larry McGrew is one of them. He makes public presentations on the safety and desirability of nuclear energy and the products that are manufactured at the plant. "It's ludicrous to call it a trigger," he tells a local Optimist Club. "We're not making triggers, folks. We're making fuel cans." Those fuel cans have to be shipped somewhere else to become nuclear weapons, he says. The term *fuel can* sounds innocuous enough. As to what's in those cans, he says they're nothing more than "two or more subcritical masses of fissionable material processed into a classified configuration that would allow them to be triggered into a nuclear explosion by being violently shoved together to form a critical mass." Few people understand the jargon.

In response to the negative publicity surrounding the April demonstration, some Rocky Flats workers and retirees form a group called Citizens for Energy and Freedom, and they decide to hold their own rallies. Kathy Erickson, a spokesperson for the group, tells the press that she and other workers and supporters want to dispel fears about plutonium production. "We aren't dying of cancer and our children aren't deformed," she says. "And we're not murderers." Although not as large as the anti–Rocky Flats rally, the counterrally has a strong turnout and speakers include Peter Brennan, labor secretary during the Nixon administration. The crowd is served popcorn, soda, and sandwiches along with miniature American flags. T-shirts are sold with "Power for the People" printed below an image of an atom. Bumper stickers read "Save Rocky Flats—Move Denver," "Pro-Nuke and Proud," and "Let the Bastards Freeze in the Dark."

To this group, the nuclear industry means patriotism, freedom, and equality.

As families spread out picnic blankets and rub on sunscreen, a rock band ironically blasts out the 1960s song "Wipe Out." One organizer says, "We want the world to know that the pro-nuclear supporters are fun-loving people."

Most employees feel that Rocky Flats is a good and safe place to work, and if it weren't, they say, they obviously wouldn't have bought houses and raised their families in the communities surrounding the plant. In commemoration of its thirtieth anniversary, Rocky Flats hosts a Production for Freedom celebration, and thousands of employees and retirees attend with their families.

ON THE cold morning of September 26, 1979, when the sky is still pitch-black, seven people "experienced in pacifist civil disobedience," as the press later referred to them, use a pair of pliers purchased at a local hardware store to cut through the barbed-wire fence along Indiana Avenue, the boundary of Rocky Flats. Carrying lighted candles, they walk until they're about a mile from the secured area surrounding the plutonium processing buildings, where they are confronted by guards and arrested just as the sun hits the mountains.

"It's a good thing we were nonviolent," says one of the arrestees later. "I could see the plant quite clearly. I realized that any terrorists could have had their way with that plant."

At least one Rockwell official feels the activists got off too lightly. "I'm pretty liberal," he says, "but if I'd been guarding out there, I'd have shot them. And if I were a terrorist trying to get into the place, I'd dress like a hippie pacifist. I don't know why the guards didn't shoot those people. They were lucky."

One Sunday morning, as I drive down from the college to meet my mother for brunch, I pass by the prayer group that meets just outside the gate. I slow down to take a look. I cautiously pull my Volkswagen bug over to the shoulder of the road until the tires crunch on the gravel.

The protesters smile and wave their signs. "Hey!" a woman yells. "Come join us!"

I'm curious, and I stop and get out of the car. Who are these people, really? What are they saying? The air is fresh and cool and the sky cobalt blue. I feel the breeze ruffle my shirt and the wind lifts my hair. There is a boy standing there who could be Mark. The same color hair, the same jeans, the same boots.

But no. He turns and I see it's someone else.

I get back in the car. I can't join those people. I lack courage or conviction. Or maybe I need convincing. I don't know if those people are crazy or heroic. If we're contaminated, if I'm contaminated, maybe it's better not to know.

I just want a glimpse of Mark. I would give anything to see him again.

KARMA HAS no such qualms about her feelings concerning Rocky Flats. She begins to attend anti–Rocky Flats meetings and attends a screening of a film about Rocky Flats, *Dark Circle*, in the basement of a church with a group of activists. Even though the film has won the Sundance Grand Jury Prize and an Emmy Award, it's not shown on television or widely distributed because it's considered sensational and too antinuclear. My friends and I have heard of it but no one's seen it. It's a big secret that everyone whispers about, like an X-rated movie.

One sunny weekend in October 1983, Karma drives out to the plant with Karin, Kurt, and her friend Laurie. It takes them awhile to find a place to park. Thousands of cars—as well as bikes, motorcycles, and scooters—line the highway.

It's the day of the Rocky Flats Encirclement. Protesters from all over the country are planning to link hands around the plant's perimeter. Organizers estimate they need at least twenty thousand people to reach all the way around the seventeen-mile border of the plant. More than six hundred people have been arrested at the plant in the past five years, but no arrests are anticipated today. Organizers want "a legal protest with no civil disobedience . . . a good day with good spirits." The encirclement is planned so as not to interfere with workers coming in for the 3:30 p.m. shift change.

Karma believes in closing the plant and saving the environment. Laurie's father still works at Rocky Flats, and, even though he disapproves, she wants to protest on behalf of workers and residents. Karin and Kurt are mostly curious and think it might be fun, although Kurt is disappointed that his best friend, Shawn, refuses to join them. Shawn's father, sister, and mother all work at the plant.

They get out of the car and put on their jackets—the day has turned cool. They're greeted by volunteers in purple bandannas who wear shirts saying "Link Arms to End the Arms Race." The volunteers help them get into line and prepare to link hands. Police are everywhere. State Trooper Dave Harper tells a reporter, "Bombs have been around before I was born and they'll be around after I'm dead—that's the way I see it. This," he says, waving a hand toward the growing circle, "isn't going to do anything." Security guards from Rocky Flats patrol the road with loudspeakers. "You won't make it," a guard shouts as Karma, Karin, Kurt, and Laurie walk along the road. "You don't have enough people."

They join the circle along with thousands of other people: high school students, a political science professor, a small child in a gray jogging suit with his hood up to shield him from the wind. Some people wear gas masks and dust filters, or scarves tied around their noses. A tall man in black clothes and a Grim Reaper mask walks along the line.

"Listen for the trumpets," the peacekeeper says, and hands them a flyer. At precisely 1:55 p.m., trumpet players stationed at intervals around the circle play "Taps," meant to signal the end of nuclear proliferation. Karma reaches out for the hands of those standing next to her. She looks across the landscape. The human chain runs for miles, snaking up and down hills, people lined up against the barbed-wire fence. Blankets, ropes, jackets, backpacks, baby strollers, and bandannas are used to fill in the gaps. American flags wave in a rainbow sea of balloons, and signs and slogans are everywhere: "Freeze or Burn," "No Nukes Is Good Nukes," "It's Hard to Hug a Child with Nuclear Arms." Across the road, signs held by counterdemonstrators read "Nuke the Liberals" and "When Nukes Are Outlawed, Only Outlaws Will Have Nukes." The spinning rotors of four news media helicopters are nearly deafening.

The trumpeters play the first two measures of "We Shall Overcome," and then the entire crowd sings, in unison, the verses printed on their flyers. After the third verse, brightly colored balloons are released from the four corners of the site.

Near the gate, a group of about a hundred counterdemonstrators from the Colorado Conservative Union chant, sing, and burn miniature Soviet flags. "Where is your circle?" a man taunts. "You don't have enough people!" Many employees share their sentiment. Jack Weaver, a plutonium production manager, notes, "Well, the peaceniks are back. . . . Don't you have something better to do in life than to just stand out here and hold hands and chant around the plant site?" As he drives past them on his way to work, he thinks, "I'm doing something that I think is valuable to the country. And oh, by the way, the reason you're out here able to protest is because I'm doing what I'm doing."

The circle does, in fact, fall short on the southeast corner, despite people stretching the line with jackets, sweaters, and anything else they can come up with. But few people seem willing to stand in the southeast area anyway—the area that leads to our neighborhood—which is directly downwind from the plant, and where soil contamination is reputedly the most severe.

Eventually, both Shawn and Laurie lose their fathers to cancer linked to their work at Rocky Flats, although nothing can ever be proven for sure. Rocky Flats seems to touch the lives of nearly everyone we know, in one way or another.

Not long after the encirclement, Dr. LeRoy Moore, an adjunct professor at the University of Colorado and a protester who had been arrested at Rocky Flats in 1979, joins five other protesters to create the Boulder Peace Center, later known as the Rocky Mountain Peace and Justice Center, in rented space at a local church.

Karma and Laurie feel emboldened after the encirclement. One summer afternoon, Karma takes the keys for our father's copper-orange Ford

Maverick, the two-door beater with a cracked windshield and overflow-ing ashtray. Feeling adventurous, the girls decide to drive out to Rocky Flats and see if they can get past the security guards. It's not the first time they've had the idea. Every kid in Bridledale has looked over that barbed-wire fence and wondered what's over the hill. And security is rumored to be surprisingly lax, even though guards are supposedly under orders to shoot to kill. Karma drives steadily. They see the guard behind the window at the gate and slow down, but he merely looks up and waves them on.

"Oh my God," Karma mutters. "I can't believe that just happened!" She drives through the gate, heart pounding. "What now?"

"Keep driving," Laurie says. "Look straight ahead."

The girls drive. They have no idea where they're going, and they begin to feel giddy. Finally they pull over next to a building and get out. No one takes notice. Karma walks to the back of the car and opens the trunk. She looks around to see if anyone sees them.

"We could have a bomb in this car," she says loudly. "Right here in the trunk!"

"Yeah," Laurie agrees. "We could cause some serious damage!"

They stand next to the trunk and wait for someone to notice, for security to show up.

Nothing happens. Finally the girls close the trunk, pull the car back onto the road, and drive out. No one comes after them.

I take it as further proof that maybe Rocky Flats isn't really as dan-gerous as people say it is.

PAT MCCORMICK of the Sisters of Loretto travels to Seattle for train-ing in nonviolent ways to oppose the buildup of the arms race. When she returns, the prayer group that meets every Sunday at the west gate of Rocky Flats has grown to almost eighty people. They arrive with signs and songs and thermoses filled with hot coffee, and often chat with the guards. There is a strong sense of camaraderie.

It won't be long, though, before those same guards will be facing them in court. Two friends of Pat's, Sister Pat Mahoney of the Blessed

Virgin Mary and Sister Marie Nord of the Order of St. Francis, decide to create fake security badges and try to get inside the plant to stage a protest. They arrive with the morning shift on the east side of the plant, wave their faux badges, and cruise through the security checkpoint. The nuns drive toward what they believe is the plutonium production area, park, and hang a sign on a fence comparing the plant to the internment camps of World War II.

The women are quickly arrested and later convicted of a felony for falsification of documents. During their trial, the sisters repeatedly refer to Rocky Flats as "a nuclear bomb factory." No more euphemisms, they say.

They spend a year in federal prison.

Throughout the year that Sisters Pat Mahoney and Marie Nord are incarcerated, Pat McCormick writes and visits. Pat Mahoney will eventually serve two terms in federal prisons for her protests at Rocky Flats.

Pat McCormick begins to think that she, too, might want to protest inside the gates of Rocky Flats. She prays with her friend Mary Sprunger-Froese, a member of the Mennonite community, and together they come up with a plan. They meet a group of friends at the Catholic Worker soup kitchen and each person gives blood, enough to fill two baby bottles. The blood in the bottles symbolizes the children who will be born in a world threatened by nuclear weapons. Then they construct two foot-long wooden crosses and cover them with pictures of people from all over the world.

As a Catholic sign of repentance for the first day of Lent and an act of prayer that Rocky Flats should be permanently closed, the women decide to enter the plant on Ash Wednesday. They leave well before sunrise. Their car is old and clunky, a donation from a parishioner, and it refuses to go into reverse. But it chugs up the hills leading to the plant just fine. Keeping in mind that their two friends ended up in prison for falsification of documents, they carry no IDs and are prepared to be stopped at the gate and stage their demonstration there.

But they aren't stopped. When they drive in with the 6:30 a.m. shift—a long stream of headlights in the fading dark—the guard waves them right through. "Wow!" Pat exclaims. "They're taking us in just as if we're going to a Broncos game!"

Stunned, the women drive down the main road, toward what they suspect is Building 771, where the triggers are made. There is a parking place right next to the fence. They get out of the car, lean their crosses against the chain-link fence, and splash a little blood. Then, hearts beating in fear, they kneel, begin praying, and wait to be arrested.

And wait. It's cold and still fairly dark, and even though the women are dressed warmly—nuns no longer have to wear habits—they start shivering. Both women are middle-aged. Their knees hurt.

Twenty minutes pass, and finally a group of security guards shows up with trucks and lights. The security manager is furious. The women are handcuffed and put in the back of a security van while the guards ransack their ramshackle car. "Where are your security badges? How did you get in here?" they demand.

"We just drove in," Mary says.

The guards don't believe her. "Where are your fake IDs?" they repeat.

"We don't have any," Pat responds. "We came in without them."

"That can't be true," a guard says.

"This is a nightmare to them," Mary says quietly.

One of the guards walks over to the van and recognizes Pat from the early-morning prayer meetings at the west gate. He's shocked to see her in handcuffs. "What are you doing here?" he asks.

"Well, it's Ash Wednesday," Pat says. "This is my act of repentance and prayer."

"Oh my God," he says, anger temporarily forgotten. "I forgot to go to Mass!"

Pat laughs. "This can be your mass, okay?" she says.

The nuns are arrested and booked at the Denver County Jail. During their trial, they—like the two nuns before them—insist upon calling Rocky Flats a bomb factory. "We need to tell the truth about it," Pat says. "It's only through resistance that Rocky Flats will become visible."

The nuns spend two months in jail for trespassing. When Pat returns to her job, the other nurses have covered her shifts and held her position for her. She returns to her weekly prayer meetings at Rocky Flats, a commitment she keeps for twelve years. She is saddened when she hears that the security manager, a man she knew and liked, has contracted cancer and died.

NOT ALL protesters are hippies or nuns. On August 9, 1987, Ann White joins her friend Allen Ginsberg as well as Daniel Ellsberg and thousands of others in a protest to commemorate the August 9, 1945, bombing of Nagasaki. The event is a planned civil disobedience: protesters will sit down in the middle of the road so cars can't enter the plant.

She leaves her home in Cherry Creek, an exclusive Denver neighborhood, at 5:00 a.m. to meet her elderly friend Ruth and drive out to Rocky Flats. It's still dark. As she says good-bye to her husband, she warns him that she is planning to get arrested. He reminds her of their son's big engagement party at a fancy restaurant in Denver that evening. "Don't worry," she says. "I'll be back long before then."

Ann and Ruth drive to the plant, park their car along the highway, and walk up to join the other protesters. They form a large circle of linked hands and stand together as the sun rises and the press arrives. A line of police in riot gear stands ready to act. The protesters begin to move toward the line of police.

Rocky Flats guard Debby Clark stands in that line. She's been told to hold her ground, and that the protesters will merely approach the line of guards, stop, and sit down before coming close to the plant's boundary. Debby doesn't like protesters. Stories circulate about protesters who spit on guards, who rub baby oil on their skin so the guards can't pick them up. Nothing like that has happened to her, but the stories rankle. She hopes she doesn't have to use her club.

Ann, on the other hand, has heard stories of activists being kicked and beaten by guards and police officers. She hopes her gray hair and air of propriety buy her a little respect.

Debby watches as the protesters walk calmly toward the line of guards. Then, to her surprise, they walk right through the ranks. Debby looks over at her supervisor. "They're not supposed to do that!" she mouths. He gestures for Debby and the other guards to run ahead of the protesters and take out their billy clubs. "You can sit down or get knocked down," the supervisor yells to the protesters, "but you cannot enter the plant site." Debby moves up, takes her billy club in both hands, and stands firm. It looks like she might have to hit someone after all.

The protesters sit. Some protesters sprawl on the pavement like mock corpses.

Ann and Ruth plunk down on the pavement. An officer approaches them and leans down. "Would you like to be arrested or not?" he asks, almost gently. Ann and Ruth exchange glances. He's giving them a chance to leave. Ann pauses. "I am here to be arrested," she says. Ruth nods in agreement. The women help each other up and walk on their own rather than be dragged like some of the other activists. The officer escorts them to waiting school buses, where the women join Ginsberg and Ellsberg and more than three hundred others, all waiting to be taken for booking. This is Ginsberg's second arrest and Ellsberg's fourth. Of the throngs of protesters arrested that day, Ann and Ruth are among the oldest.

The protesters are bused to a large chain-link enclosure where they're held while waiting to be fingerprinted, booked, and ticketed for obstructing a roadway and disobeying a police officer. They're kept in the enclosure without water or access to a phone until the paperwork is completed, which takes hours. Ann slips her phone number to a fellow protester released ahead of her, and he contacts her husband. When she's finally allowed to leave, she finds her husband waiting in the parking lot with a dress and shoes stashed in the backseat of the car. They're late to their son's party, but she tells her astonished guests that it was worth it. And though she's convicted of trespassing, her parole officer lets her work off her sentence at the Denver Art Museum.

PLUTONIUM IS a radioactive imp. It flares and burns unpredictably. Like a lethal bee flying from flower to flower, plutonium taints everything it touches. What becomes contaminated with plutonium becomes contaminating itself.

A boxcar full of plutonium is a busy hive.

Like the barge of garbage from New York that roamed up and down the Atlantic Coast seeking a dump site, a boxcar from Rocky Flats is making a grand tour of the western states.

In the single-minded drive to win the Cold War, thousands of tons of radioactive and toxic waste have accumulated around the country. Every site in the DOE complex is contaminated, and there is no central disposal site for nuclear waste. An agreement between the DOE and the state of Colorado stipulates that the amount of radioactive plutonium waste held at Rocky Flats cannot exceed 1,601 cubic yards as of May 1, 1988—roughly the equivalent of a 5,000-square-foot single-story house filled floor to ceiling with plutonium. Rocky Flats has a full house.

For decades, Rocky Flats has been shipping waste to the Idaho National Engineering Laboratory near Idaho Falls, a "temporary" storage facility. In the high desert, millions of cubic feet of plutonium-contaminated waste sit in rusting barrels, drums, and boxes, waiting for the federal government to find a permanent dump site. This dump site is not safe. Sitting nearly six hundred feet above the Snake River Plain aquifer, the site is in an earthquake zone and a floodplain. The aquifer below it supplies water for Idaho farmers. The Idaho site currently holds 3.5 million cubic feet of plutonium-contaminated waste that is not expected to stabilize for ten half-lives, or 240,000 years.

Eighty-five percent of the radioactive waste comes from Rocky Flats, which began shipping waste to Idaho in 1969, following the Mother's Day fire. In the early years, drums and barrels were dumped off the trucks like normal household trash. Now, twice a month, rail cars leave Rocky Flats filled with fifty-five-gallon steel drums and fiberglass-coated plywood boxes. When they reach the Idaho site, the drums and boxes are stacked on an asphalt pad and then covered with plywood, plastic, and three feet of soil.

Federal officials promised that the Idaho site would be used only until 1970, when the waste would be moved to a new site. The trouble is, there is no new site. The $700 million facility that was intended to provide permanent storage, the Waste Isolation Pilot Plant (WIPP) near Carlsbad, New Mexico, is plagued with "technical, structural and management deficiencies." Located twenty-six miles east of Carlsbad and 2,150 feet beneath the desert, WIPP has yet to prove that brine seeping into the waste repositories won't harm the barrels holding waste, and that potential leaks from aquifers both above and below the site won't threaten local water supplies.

Idaho governor Cecil Andrus is fed up. In October 1988, he bans any further shipments from Rocky Flats.

Just as the governor issues the ban, a single, steel-lined red boxcar covered with the familiar warning signs of radioactive material crosses into Idaho. Filled with 140 waste drums at fifty-five gallons of radioactive waste per drum, it would put Rocky Flats well over its limit. The boxcar comes to rest temporarily in Blackfoot, a small town that bills itself as the Potato Capital of the World. It sits on a spur of the Union Pacific railroad behind an old granary, guarded by a small team of last-ditch sentries: a detective from the Union Pacific, a safety specialist from the Idaho National Lab, and several Idaho state police officers.

Governor Andrus says Colorado must take the boxcar back. "If you can't store it, don't make it," he says. Colorado won't take it back. The DOE wants to expand storage for radioactive waste at Rocky Flats, but Colorado's governor, Roy Romer, who took office in 1987, refuses. And he stands his ground. A spokesman tells the press, "Governor Romer doesn't want [the boxcar] back and doesn't want any long-term storage at Rocky Flats." Romer chastises officials from the DOE. "If you can't dispose of the waste," he says, "you can't produce the material. If the limit is reached, Rocky Flats should not operate." He threatens to close Rocky Flats if a solution can't be found.

Andrus is equally adamant, despite the threat of legal action. His state has been dumped on for decades. "The legal grounds are not near as important as the moral and political grounds," he says. "And I can use the courts till you can step on my beard." The Reagan administration begins looking for alternatives, including potential sites at military bases and in New Mexico, Colorado, Washington, Idaho, Nevada, South Carolina, and Tennessee. "No sale," says Washington governor Booth Gardner. All of the governors say no. The administration suggests a site in Pueblo, a city of 102,000 south of Denver. Citizen opposition is swift. "We're totally against putting that stuff here," says Pueblo's city manager. "They cannot expect our city to take their problems. Nobody here wants it." Governor Romer proposes a short-term solution of moving boxcars of plutonium waste to Piñon Canyon just east of Trinidad, a town of nine thousand residents two hundred miles south of Denver. Piñon Canyon is owned by the Army and often used as a training ground by Fort Carson. The idea is squashed.

The red boxcar filled with radioactive material sits for days in Blackfoot, Idaho, until, in a deadly game of high-level atomic NIMBY ("not in my backyard"), Governor Andrus sends the boxcar back to Colorado. For weeks it sits on a railroad sidetrack—along with six other armored boxcars awaiting a destination—on the Rocky Flats site while officials try to figure out what to do with this "orphan waste." Governor Romer asks the DOE not to unload the car. Finally, in December 1988, top DOE officials meet with Romer and the governors of New Mexico and Idaho. They come to a tentative agreement. They will pressure Congress to open WIPP more quickly, and the DOE will provide financial assistance to Colorado, New Mexico, and Idaho while they search for a temporary, interim storage site. Romer is reluctantly persuaded that it is important to keep Rocky Flats open as a production facility. The closed boxcar sits at Rocky Flats for months before finally being moved to WIPP.

THE INCINERATOR at our little old house in Arvada is long gone and turned into a garden planter by the new owners. We don't have an incinerator at the house in Bridledale; the burning of household trash is now banned and a noisy truck rattles down our street once a week to pick up the bags of trash we lug to the end of the driveway.

Rocky Flats follows different rules. A $2 million, three-thousand-square-foot "trash compactor," or fluidized bed incinerator, was completed in July 1979 and housed in Building 776, the same building in which the 1969 Mother's Day fire occurred. Designed and built by Rocky Flats workers to burn plutonium-contaminated waste, the "plutonium recovery" incinerator is the only one of its kind in the United States and perhaps in the world. Intended to operate around the clock, the incinerator makes its first continuous 108-hour run in 1979. But no one outside the plant learns of the active incinerator until 1986—seven years after it began operating. Even then, no one knows exactly what the incinerator burns or if it adequately filters waste. Dr. Edward Martell and a group of scientists submit a paper to the government, the health department, and the EPA opposing the incinerator, as the incinerator releases plutonium and other contaminants into the air and creates a health and environmental hazard. Rocky Flats and DOE officials deny any danger and continue to operate the incinerator. Following a meeting of several hundred local residents at a high school gymnasium, Gene Towne, a spokesman for Rockwell at Rocky Flats, says "the majority of the material was office paper and paint thinner," although Rockwell's director of plutonium operations later admits to the press that the incinerator had been used to burn eighteen tons of radioactive material.

The problem with Rocky Flats is not just a smoking chimney or a hole in the dike. The weapons plant is like a bag filled with ultrafine sand—a bag filled with millions of glittering, radioactive specks too tiny to see—and the bag has been pricked with pins.

WHO WILL bless this seemingly blighted land?

Charles Church McKay, whose family homesteaded on Rocky Flats land, has been waiting a long time to get things settled with the government. Nine years have passed since litigation began in the case in which McKay and two other landowners sought $23 million in damages from the DOE and its codefendants, claiming their contaminated land not only had diminished market value but was unfit for human habitation. According to the DOE, McKay's land surrounding Rocky Flats is safe, but the state of Colorado and Jefferson County refuse to issue building permits or allow any commercial or residential development.

When the case finally goes to trial, officials from the Colorado Department of Health, Rockwell, and the EPA testify. Ousted county health director Carl Johnson presents his studies, stating that radioactive emissions from Rocky Flats have caused an excess of cancer and there should be no new housing within ten miles of the plant because of ongoing contamination.

DOE officials disagree. Stanley Ferguson, epidemiologist at the Colorado State Health Department, testifies against Johnson. He says there is no proven connection between the activities of Rocky Flats and an increase in cancer, and no reason to consider closing or moving the plant. He concedes that the health department's stance "has been and always will be that Rocky Flats was built on the wrong place and still is on the wrong place," but he asks, "What is the case for moving it at considerable expense to the taxpayers?"

U.S. district judge Richard P. Matsch admits that land near Rocky Flats is contaminated in varying degrees by radioactive plutonium and americium. Nonetheless he believes that areas near the plant "are suitable" for future development and use, and that Johnson's standard for plutonium in surface dust is unnecessarily conservative. The judge agrees with Ferguson that "no measurable increases in cancer incidence resulting from operations at Rocky Flats have been demonstrated by any appropriate scientific method" and "there should be no human health effects different from those resulting from background radiation."

Ultimately the suit is settled out of court for about $9 million in favor of McKay and the other landowners for about two thousand acres of contaminated land. Other details of the settlement are not released. Under the terms of the settlement, all confidential documents received during discovery are returned to the DOE. This effectively seals off information about contamination from use by journalists, scientists, or concerned citizens. Hundreds of contested acres will remain "open space and recreation," where residents can hike or bike. In other areas, residential development will continue as long as homebuilders take remedial measures such as plowing contaminated soil below the surface level before laying the foundation for a house—an act that in itself can redistribute plutonium into the air and does not take into consideration movement of soil by weather, burrowing animals, or other forces and conditions.

Under the settlement agreement, the state of Colorado agrees to provide the McKay family with a certificate attesting that levels of plutonium and americium in their land are at or below the state standard of two disintegrations per minute per gram of dry soil. Charles McKay, the nephew of Marcus Church, is pleased with the result. The settlement allows development to continue on land with levels of contamination five times higher than what Johnson and others consider safe, and it also eliminates further controversy for the DOE by sealing the records. "The purpose for the lawsuit was to get our land back," McKay says. "Our basic intention was to have the federal judge bless the land."

Even without the pulpit of his job, Carl Johnson doesn't let up. "Radiation is sneaky," he says to the press. Cancer and leukemia can take years to show up. In fact, he reminds the press, in 1984 the DOE prepared a point-by-point response that confirmed most of the data in the Howard Holme and Stephen Chinn report, demonstrating a likely connection between contamination and illness. Further, Johnson notes, the court system is flawed. "The burden of proof is on the victim, not the defendants. I think the nuclear industry has traded on that fact. Officials have permitted excessive plutonium exposures knowing that they will be through with their careers and retired before the evidence is apparent."

THE CHORUS of people calling for relocation of the plant grows. Representative Pat Schroeder (D-Colorado) is very vocal about moving Rocky Flats away from Denver for public health and civil defense reasons. Other politicians agree. The Colorado Medical Society unanimously calls for the removal of plutonium operations at Rocky Flats. The group Physicians for Social Responsibility calls Rocky Flats a "creeping Chernobyl." The DOE, on the other hand, is pressing for expansion of the plant, based on production quotas and waste storage requirements, but even the governor—who seems to have been reluctantly convinced of the necessity of weapons production—is opposed to that.

Under political pressure, the DOE agrees to do a study on the future of Rocky Flats, focusing primarily on worker safety and upgrading facilities. Pat Schroeder insists the study be conducted by independent experts. "I have yet to see an agency study itself and turn itself in," she says.

Indeed, rather than focus on risk and potential health effects, the results of the DOE study emphasize the economic loss Denver would suffer if Rocky Flats were relocated. Moving plutonium operations would result in a loss of 3,500 jobs and a payroll of more than $40 million, and moving the entire plant would mean a loss to the general Denver metro area of six thousand jobs and millions of dollars. Colorado fears the bomb, but can't live without it.

If Denver wants to keep Rocky Flats for economic reasons, then perhaps the best thing is to give people a good emergency warning system. But the Radiological Emergency Response Plan continues to be locked in debate. Officials can't agree on what type of accident could actually occur and, if something does happen, what they should tell people to do. Should evacuation centers be established? If so, officials argue, should people be able to bring their pets? The proposed plan begins by stating that "the actual danger posed by the plant is very small," but in the event of a radioactive emergency, residents will be instructed to remain indoors, close all windows and doors, and turn off circulation systems. People who have been outdoors will be told to take a shower and change clothes. There are no instructions for full-scale evacuation.

Dr. Carl Johnson, who has filed an appeal in his lawsuit against the county board of health, feels that the response plan is completely inadequate. A radioactive plume, he says, is similar to a dust storm, and plutonium is likely to lodge in windowsills, bricks, and cracks in wood. "It's not just a matter of a sort of gaseous cloud passing overhead that will soon dissipate," he says. "We're talking about particles that drop to the ground and remain dangerously radioactive. . . . Even getting in a car and rolling up the windows may be safer than staying indoors." He also believes the government's levels for acceptable plutonium dose exposures are far too high and would result in a rise in radiation-related cancer and spontaneous abortions.

Felix Owen, director of information services for Rockwell, says there is no reason for anyone living around Rocky Flats to fear for their health. Not only is an accident very unlikely, he says, but small amounts of plutonium have only "negligible" effects on human health. "I personally feel low-level radiation is overemphasized," he says. "We live in a sea of radiation. Man has built an immunity to it. The human body is already developing defense mechanisms against it."

However, Johnson states, "Very little of Colorado's normal background radiation comes from alpha-emitting plutonium. People in this state are just not usually exposed to inhaling alpha-radiating particles unless they live around a facility like Rocky Flats. Low-level radiation from internal sources is particularly dangerous; we certainly have no natural immunity to it." The emergency plan, he feels, should be based on the premise that all of Denver might have to be evacuated should a radiological accident occur.

The plan never moves beyond the drafting stage and is never fully tested, due largely to lack of support from Rocky Flats. Niels Schonbeck, the biochemistry professor and member of the Rocky Flats Environmental Monitoring Council, later notes, "My impression is that the public is without a clue as to what they would do in the face of an accident at Rocky Flats. . . . [To implement an effective emergency response plan] would alarm an otherwise uninformed public." The government stance seems to be that educating the public about what to do in the event of an actual radioactive emergency would only result in panic and confusion.

Rocky Flats continues to grow in order to meet production quotas. Back in March 1980, officials announced the completion of Building 371, a $215 million plutonium processing building intended to replace Building 771, the site of the 1957 fire. Designed to last twenty-five years, Building 771 is now thirty years old and, in the words of General Edward Giller of the AEC just after the fire in 1969, is an "old, outmoded, and increasingly hazardous operation" that must be replaced. "The present facility," he said, "is deteriorating due to the severely corrosive atmosphere inside the glove box lines. . . . The corrosion causes equipment and glove boxes to fail, resulting in spills or leaks of contaminated materials into the working areas." Nearly a decade has passed since then, with 771 still in full swing.

Manager Jack Weaver is assigned to oversee the new Building 371. There are problems, though. In Building 771, everything is on one floor. Building 371 is three floors and contains seventy-seven miles of pipes for processing plutonium, from the ground floor to the basement and subbasement. During construction, Weaver and his crew find cracks in the concrete walls and other problems.

Building 371, despite the best efforts of Jack Weaver's crew, never quite works right. Equipment and filters malfunction. Plutonium is lost in the system. In Weaver's first inventory of the facility in April 1983, he reports that 25 percent of the plutonium in the facility is snagged in the piping or ducts. Building 771, on the other hand, is unable to account for only 2 to 3 percent of its plutonium, according to its twice-yearly inventories.

Two years after Building 371 begins operations, Rockwell shuts it down for safety and security reasons. It never reopens. Full production resumes in Building 771, now referred to by the media as "the most dangerous building in America."

In March 1982, Bruce Shepard, a Colorado Springs developer and administrator at the Department of Housing and Urban Development (HUD), recommends abolishing the Rocky Flats Advisory Notice, which has been in effect since 1979. The notice required that homeowners within a ten-mile radius of the plant be informed of plutonium contamination and emergency response procedures before buying a home with FHA mortgage insurance or any other HUD assistance. Other local developers support his efforts, believing that it has had a negative effect on property values and home sales.

HUD lifts the requirement. Someone buying a home in Bridledale will no longer have to sign a piece of paper as my parents did. The developers involved in rescinding the Advisory Notice have invested in at least two housing developments near Rocky Flats.

MY MOTHER reluctantly puts our house on the market. A real estate agent in bright lipstick and pantyhose appears and puts a sign at the end of the driveway. "I don't know where we'll go," my mother says, "but at least I won't have this house hanging around my neck anymore." She stays inside and rarely answers the phone. She sits at the dining room table, lighting one cigarette after the other, gazing out the front picture window.

But the house doesn't sell. Month after month it sits on the market and no one comes to look. We paint the fence and have the curtains cleaned. Even the animals are on their best behavior. No one calls.

"It's Rocky Flats," says the real estate agent. "People are nervous to buy in this neighborhood."

"Nonsense," my mother snorts. It's just family luck, she's sure. Or the blundering real estate agent. "Your dad always said not to trust women agents," my mother muses as the agent click-clacks in her high heels down the driveway to her car. "Maybe I should have believed him."

The neighborhood is changing. Everyone's gone: the packs of kids who swam in the lake and irrigation ditches, who ice-skated on the pond and galloped their horses around the perimeter of Bridledale and Meadowgate, who roared their minibikes up and down the street—all are off to college or vo-tech school or working at the supermarket or waitressing. Randy Sullivan is working as a bouncer in a bar. Even the turkey farm is gone, the land turned into open space as part of a legal settlement.

Karma knows she will leave home. She packs a duffel bag and asks Mom to drive her and her friend Laurie out to the freeway. Our mother slides behind the wheel of our old green station wagon with her usual fortitude and grace. She doesn't have a second thought about allowing her youngest daughter to hit the road. Just as she put us out the back door and didn't allow us to come back until dusk when we were small, she expects each of us to stand on our own.

The girls throw their bags in the back and count the dollars between them. When they reach I-70, Mom pulls over to the side of the road. The girls get out, swing their duffel bags over their shoulders, and Karma reaches into her pocket for a quarter. "Tails, east, heads, west," she says. She flips the coin into the air and catches it in her palm.

"Heads," Laurie says. "West."

"West it is," Karma agrees.

Mom turns the car around and waves as the girls trudge up to the shoulder of the highway to catch a passing truck. In an instant, Karma is gone.

My life feels rudderless. I'm grasping at straws, looking for an invisible god. I meet a boy named Andrew at the university and he stops by the English department, gets my class schedule, and waits for me after my Chaucer class. Six months later, while we're driving down a Boulder street in my red VW bug, he proposes. I choose a wedding dress off the rack at the mall and there's hardly enough time to send out invitations. The ceremony takes place at a small community church in Boulder, and the reception is held at our tiny duplex with our two cats locked in the bathroom. My new husband and I are barely old enough to drink the champagne at our own wedding.

In December 1983, our house finally sells at a bargain-basement price. The house my parents designed and built, filled with Scandinavian knickknacks and secondhand Italian furniture, leaky pipes, and a flea-bitten bear rug, the house weathered by a steady torrent of found cats, wayward dogs, mean Shetland ponies, fast-reproducing hamsters, hardy mice, parakeets always making a break for it, and a goat just passing through, is gone. The horses are sold or given away. My mother leases a small apartment in Arvada and packs up as many boxes as she can fit into one of the bedrooms. Most things she has to sell or give to the thrift store. She and my brother take one pet with them, a grumpy gray cat named Colby. The apartment feels cramped and dark, with orange carpet and an avocado kitchen. My mother works full-time and takes night classes to become a hospice nurse while Kurt finishes high school. Two weeks before graduation, Kurt is suspended from school for a senior prank that earns the respect of all his friends. He and his buddies paint brightly colored polka dots on the "temporaries"—the long wooden trailers parked behind the school building to accommodate overflow students, trailers that became permanent long ago. In a final salute, the boys ring the flagpole in front of the school with radial tires. When the security guard catches them, Kurt's friends run, leaving him suspended on top of the pole.

The principal threatens to deny graduation and hold him back. Our mother vigorously defends him. It's decided that Kurt will graduate, but can't attend the graduation ceremony.

That's all right with Kurt. On graduation day he crawls up into the heating duct above the gym and scoots across the ceiling on his belly. When the ceremony begins, he lights small firecrackers and drops them on his unsuspecting compatriots in crime as they walk across the stage.

He's caught again. The principal withholds his diploma for weeks. "That Kurt," my mother says with a tinge of pride. "I just don't know what to do about him."

Karin leaves to take classes in music at the state college and dreams of being in a rock band. Karma is still hitchhiking. No one's heard from her; no one knows where she is. My father is no longer in our orbit.

THE DOE isn't unaware of problems at Rocky Flats. An earlier study of plutonium operations details, among other things, "poor quality instruments, improper use of radiation monitoring equipment and faulty record keeping, concerns over antiquated fire detection and alarm systems, waste shipping and storage problems, and poor communication of safety and health matters up and down the organizational structure." There is "little indication" that work was performed according to federal requirements. But little has changed. And the government will continue to press Rocky Flats to make bombs for at least another decade.

In response to the discovery of leaking pondcrete, two officials from the Colorado Department of Health, accompanied by two engineers from the EPA, try to inspect Rocky Flats for violations of the Resource Conservation and Recovery Act (RCRA), passed in 1976, which governs disposal of hazardous waste. So far the EPA and the Colorado Department of Health haven't been very successful in strong-arming Rocky Flats. They can't even get data about what's being released into the environment. Rockwell is not happy to see them. The men are told they're endangering national security by trying to conduct an inspection and are refused entry to certain areas.

In the fall of 1988, however, the stakes change when a DOE official himself gets contaminated. On September 29, three people—a special inspector from the DOE and two Rocky Flats workers—are inspecting a room in Building 771. They walk into a room with several bins filled with contaminated tools and clothing next to an air vent. A sign telling the workers to put on respirators has been covered up by a cabinet. The alarm suddenly goes off, and a lab test confirms they've all been exposed to plutonium.

Rocky Flats workers are accustomed to occasional misplaced or missing signs, dosimeter badges that don't work or have been zeroed out, little slips and mishaps that make for a strong sense of camaraderie and no small amount of gallows humor. But the inspector from the DOE is more than a little annoyed. He files a report detailing "very serious" violations at the plant that leave "no margins for safety," and criticizes "slow and unskilled" work by radiation monitors. He writes that radiological monitoring equipment is "unchecked, out of service, or out of calibration," and that there is "haphazard" posting of warning signs. Among his laundry list of other problems is the note that "attitudes are complacent."

On October 7, 1988, the DOE closes Building 771, at least temporarily. Production is scheduled to resume within weeks. Incensed, the Rocky Flats Environmental Monitoring Council, the citizens' watchdog group appointed by Governor Roy Romer and Representative David Skaggs, calls a public meeting in Westminster, a community adjacent to Arvada, on October 25. Officials from Rocky Flats and the DOE agree to show up and answer questions. More than four hundred people jam the room.

Earl Whiteman, a slender, soft-voiced man who serves as the DOE's manager of Rocky Flats, begins with an informational slide presentation. He briefly describes the contamination incident that led to the closure of Building 771.

"Shut it down!" someone yells from the back. Another voice chimes in. "Shut it down!"

The building *is* shut down, Whiteman explains calmly. It was shut down in response to the September 29 incident. Beginning November 30, the incinerator must be shut down as well, the result of a court order following a successful lawsuit by the Sierra Club, and remain closed until February 28, 1989.

The crowd boos. "Shut it down for good!" someone yells.

"Tell the truth!" another person calls. The phrase is picked up by others in the crowd. "Tell the truth! Tell the truth!" The room is filled with shouts.

Jim Wilson, chairman of the Rocky Flats Environmental Monitoring Council, calls for everyone to settle down. We don't need a fistfight, he entreats, because "there are babies in the room."

Hilda Sperandeo, a sixty-one-year-old schoolteacher from Arvada, stands up to speak. Her husband died of cancer several years earlier. They never knew what was going on at Rocky Flats. She's not sure if her husband's illness was associated with Rocky Flats or not. "We just thought it was a factory," she says. "All they do is lie to you. They don't care about anything but making bombs."

Whiteman tries to reassure the crowd that the danger from the plant is minimal, and the amount of plutonium that contaminated the three people on September 29 was minimal as well.

"How much plutonium was involved?" someone asks. "How much exactly?"

Whiteman concedes that the exposure was ten times the amount normally found in the plutonium processing areas at Rocky Flats, and because the bins were near an air vent, radioactivity might have spread throughout the entire building.

Questions turn to the radioactive boxcar refused by the governor of Idaho. "What's in that boxcar?" someone asks.

Like other boxcars, Whiteman explains calmly, the railcar holds 140 drums of waste. Each fifty-five-gallon drum is permitted by DOE regulations to contain up to 200 grams of plutonium. Production at Rocky Flats generates one boxcar per week.

The crowd erupts. It's a well-established fact that a millionth of a gram of plutonium in the body can cause cancer. Further, biochemist Niels Schonbeck stands and declares that 200 grams is twice the amount capable of causing a serious nuclear accident, based on information that Rocky Flats itself has released to the public. Harvey Nichols, the University of Colorado biology professor who studied the pollen and snow around Rocky Flats, agrees. If a small amount of moisture gets inside one of those barrels, he says, an explosion could be triggered.

Just like the rallies at Rocky Flats, a low murmur moves through the crowd. *Close Rocky Flats. Close Rocky Flats. Close Rocky Flats.*

Not everyone chimes in. Workers from the plant are in the crowd as well. Just before the meeting ends, a man stands and says he is one of the three people who were contaminated on September 29. He refuses to give his name. "I think the whole incident has been blown way out of proportion. I think it is a safe place to work." There is sporadic applause.

But the DOE is finding itself in deeper and deeper trouble. Shortly thereafter, a comprehensive DOE study of 160 contaminated sites at sixteen nuclear weapons facilities around the country is released to the public. Rocky Flats is ranked number one—the most dangerous site in the United States—primarily due to hazardous waste in the groundwater and the large population directly downwind and downstream. Groundwater is of particular concern; in addition to plutonium contamination of soils and sediments, the solar evaporation ponds have contributed to nitrate contamination in water supplies. Two buildings at Rocky Flats make the list of the ten most contaminated buildings in America. Number one is Building 771.

Other DOE sites are in trouble, too. At the 570 square-mile Hanford site near Richland, Washington, liquid radioactive and toxic wastes, dumped into trenches for four decades, have contaminated large underground reservoirs used for drinking water and irrigation. Large amounts of radioactive materials have been released into the Columbia

River and into the air, including approximately 740,000 curies of iodine-131, and the result is a high occurrence of thyroid cancer and other thyroid conditions. The Nevada Test Site, the Pantex facility, the Feed Materials Production Center, and the Savannah River Plant are all deeply contaminated and have affected surrounding areas.

Six weeks later, Building 771 reopens for business, with limited use of the incinerator. The DOE says it has commissioned exhaustive studies to support its contention that Rocky Flats poses no threat to residents. But even some government officials are beginning to disagree. "I don't believe that it's possible to reverse the harm that has been done at Rocky Flats," says Bob Alvarez, an investigator for the U.S. Senate's Governmental Affairs Committee, which is responsible for monitoring nuclear weapons complexes. The groundwater and soil at Rocky Flats are so full of radioactive materials and toxic chemicals that Alvarez and other experts expect Rocky Flats to likely become a "national sacrifice zone," an area that will remain toxic for so long that no living creature will be able to enter without endangering its health.

NUNS AND hippies, housewives and physicists, attorneys and Buddhist monks. History makes for odd alliances.

In 1987 two men from separate government agencies form an unlikely team. Jon Lipsky of the FBI's new Environmental Crimes Division and William Smith of the EPA's National Environmental Investigation Unit quietly begin to look into alleged abuses at Rocky Flats. They seem straight out of a Rocky Flats version of *Butch Cassidy and the Sundance Kid*—but with a happier ending.

Progress is slow. Lipsky, a former Las Vegas street cop who's not afraid of plutonium, the government, or hard work, has a casual demeanor, but he's been the lead investigator in thirteen environmental cases. He's persistent. Years of secrecy and threats have made workers and activists alike very nervous around government types. No one wants to talk. Finally, after months of trying to build trust and convince people of his sincerity, Lipsky gets lucky. An activist gives him the name of Jim Stone.

Stone has been waiting months—no, decades—for a call like this. He loads up all his boxes and takes them down to the FBI's Denver office. A polite, slightly stout man wearing a fedora, Stone doesn't look much like a rabble-rouser to Lipsky and Smith. But he has a story and he can't wait to tell it.

Lipsky has few confirmed facts about contamination at Rocky Flats, but one thing he knows for sure is that the incinerator in Building 771 is supposed to be shut down. Yet he's heard numerous rumors that it's being operated illegally. "Are they still burning plutonium out there?" he asks Stone.

"Oh yeah," Stone replies. "They have so much waste out there that they have to fire up that incinerator. I told them there are better ways, that you don't have to do it that way. That incinerator is not protected with suitable filters. It's not even designed to burn common trash properly without causing air pollution. But they said no, this is the most expedient, we're going to do it this way."

Rockwell and the DOE have always contended that the 771 incinerator is exempt from RCRA regulation because Rocky Flats is a "plutonium recovery" facility and thus granted an exclusion.

"How do you know these things?" Lipsky asks. Both Lipsky and Smith are having a hard time keeping up with Stone. Stone has waited a long time to talk.

"Well, I worked in that building all the time," Stone says. "The new incinerator, the fluidized bed incinerator, never did work. They've tried it a few times but could never get it certified. There's a limit on how much hazardous and radioactive waste they can store, and they have no room for it. So they burn it. They burn waste contaminated with plutonium, low-level and medium-level waste."

"What else do you know?" Smith asks.

"I can tell you about a lot of things," Stone replies. "Standley Lake, for example. Not only is there plutonium and americium and uranium and you-name-it, but I know by the stratus in the lake sediment when that contamination occurred."

Lipsky looks over at Smith. They have their first source.

"A lot of contamination goes up the stack [the Building 771 incinerator smokestack] and into the environment," Stone continues, "because the filters leak like a sieve. The wind prevails from the west. It's the same thing with the groundwater, with Great Western and Standley Lake just downhill, right on down to the Platte River. Denver is sitting at the gravity base of all this pollution coming down from Rocky Flats. And it has to be stopped at the source." He pauses. "That's always an engineer's primary objective: determine the cause of the problem, get at the source, and correct it there."

Stone tells Lipsky and Smith about his long history at Rocky Flats, about how he helped build the plant and knows the facility inside and out. Workers have inadequate protection, he says. But he also talks about how workers mess with or remove the filters, because filters slow down production. He talks about how productivity trumps safety or environmental laws. There's a lot of plutonium missing, Stone adds, some in the ventilation ducts and piping, some blowing around outside.

"They blackballed me. The industry is spooky about whistle-blowers," he says. "But I don't see myself as a whistle-blower. I see myself as a good engineer."

With this ammunition in hand, Lipsky and Smith contact Ken Fimberg, assistant U.S. attorney for the state of Colorado and a Harvard-trained lawyer with a strong track record in environmental issues. If Rocky Flats is burning contaminated waste in that incinerator, the men say, the technology exists to detect it. Lipsky explains that he wants to do a flyover of the plant and do infrared photography. Contaminated waste gives off heat. Forward-looking infrared (FLIR) will reveal whether there's anything thermally hot.

There's a hitch, though. Rocky Flats has guards under orders to shoot if necessary. "I want a letter of immunity for me to take pictures of Rocky Flats," Lipsky says. "Your office prosecutes activists all the time for trespassing over there. And it's under the exact Atomic Energy Act section of trespass that you're not allowed to take pictures. So I'll be violating the law, and I want a letter of immunity."

At first, Fimberg looks at Lipsky and Smith as if they're out of their minds. But he knows something about Rocky Flats: the fires, the leaking barrels, the lawsuits. Fimberg served on the board of the Colorado Wildlife Federation and clerked for the Environmental Defense Fund. He decides to support their investigation. Lipsky gets the letter.

Lipsky, Smith, and Fimberg then bring their case before U.S. attorney Mike Norton. They're not sure how he'll respond. Norton has run twice as Colorado's Republican candidate for Congress. Both times were unsuccessful. He's thinking about running for governor. Despite a lack of criminal experience, Norton's just been appointed by President George H. W. Bush as U.S. attorney for Colorado, with the Department of Justice. He's still waiting for his Senate confirmation when the three men visit his office to discuss Rocky Flats.

The meeting seems to go well. If they're burning plutonium out there, Smith emphasizes, we can catch them.

Somewhat to the men's surprise, Norton gives them a green light. But he cautions that the investigation must proceed carefully. He has to seek approval from the Justice Department in Washington. The EPA is an independent agency, but the FBI falls under the Department of Justice.

The Justice Department approves.

The first flyover occurs in October when Building 771 is supposed to be closed. Then, on the cold nights of December 9, 10, and 15, 1988, an FBI plane armed with an infrared heat-sensing camera flies directly over the plant. Jon Lipsky, Ken Fimberg, and another EPA agent are on board. Lipsky hates to fly—especially in puddle-jumpers—as he gets motion sickness. But he's not thinking about his stomach. They take photos of the Building 771 incinerator—shut down by court order until February 28—and other areas in and around the plant. "Look at that," the EPA agent says, pointing to the monitor. The men can see white plumes rising from a smokestack and long white ribbons spreading out from the plant in lines, shapes, and swirls, as well as occasional white spots. White indicates thermal activity.

The photos are sent to an EPA laboratory for analysis. The results are dramatic. The photographs indicate that, contrary to statements by Rockwell and the DOE, the 771 incinerator is thermally active and likely in operation, burning radioactive waste. Further, the plant appears to be illegally discharging radioactive liquid waste into Woman Creek. Streaks of light splay out from the "spray fields" where contaminated waste is sprayed. Narrow white rays that stretch across Indiana Street and toward Great Western Reservoir seem to indicate the movement of radioactive material beyond plant boundaries.

Lipsky is shocked by the results. The heat signatures show the runoff fanning out just like a spiderweb. Capillaries spread down to Woman Creek and then to Standley Lake, which provides drinking water for nearby cities. Even on the coldest night of their flights, when it was just 7 degrees Fahrenheit, Rocky Flats was still spray-irrigating with radioactive waste.

Never before have two government agencies—the FBI and the EPA—planned to raid a third government agency, particularly one as powerful as the DOE.

Based on the videotape and other evidence they've accumulated, Lipsky and Smith begin to prepare a 116-page affidavit that will lead to a search warrant. The affidavit states that the DOE and Rockwell tried to prevent the public from learning "just how bad the site really is. . . . There is probable cause to believe that Rockwell and the Energy Department officials have knowingly and falsely stated Rocky Flats' compliance with environmental laws and regulations and concealed Rocky Flats' serious contamination."

Lipsky and Smith decide to call the raid Operation Desert Glow.

Fimberg flies to Washington to brief supervisors at the Justice Department and gain their approval. On January 10, the head of the Environment and Natural Resources Divisions approves. In March, Attorney General Dick Thornburgh signs off, and in June, James Watkins, the secretary of the DOE, who has brought a new spirit of openness to the agency, signs a memo of understanding about what is to happen.

☢

CARL JOHNSON doesn't live to see the raid on Rocky Flats. In Colorado politics he was a pariah, but beyond state boundaries he became an internationally renowned expert on the effects of radiation. His work included a significant study of people in Utah, the "downwinders," who were subject to the fallout from nuclear bomb testing in the 1950s and 1960s.

Five months after he was terminated from his job as Jefferson County Health Director in May 1981, Johnson's "Rocky Flats Area Cancer Incidence Study" was published in the highly respected journal *Ambio*. In 1983 his case against the county board of health and the county commissioners—who, according to the press, had appointed a new member to the board for the express purpose of firing Johnson—was appealed to the state supreme court. One week after the Church lawsuit—initiated by owners of the land on which Johnson had tried to prevent residential development—was settled for about $9 million, Jefferson County commissioners offered to settle with Johnson for $150,000. He accepted. It was less than ideal—what he really wanted was to have his job back—but at least he felt somewhat vindicated.

On December 18, 1988, less than two weeks before his unexpected death and six months before the still-secret raid, Johnson published an article in the *New York Times* titled "Rocky Flats: Death, Inc." "The actual number of people who have been injured or died because of the operations of Rocky Flats and other such plants can never be fully known," he wrote. "Thus, communities near nuclear weapons and nuclear power facilities must insist on detailed investigations of all activities and emissions. I was a whistle-blower. . . . I was forced out of office. If the nation is to be properly protected, all studies of nuclear contamination and associated health effects should be conducted primarily by independent scientists who are insulated from cynical retaliation from the nuclear establishment as well as advocates of urban development."

Eleven days later, the man who overcame tuberculosis as a child died of complications from heart surgery at the age of fifty-nine. He was buried with military honors at Fort Logan National Cemetery. "It was probably the most respect Carl J. Johnson ever got from the government," wrote a reporter from the *Boston Globe*.

In *An Enemy of the People,* Henrik Ibsen—the Norwegian playwright whom my mother insisted we read when we were young—tells the story of a doctor who discovers that the baths at a popular vacation spot are contaminated by toxins from a local tannery, and citizens are becoming ill. He expects to be commended for saving the townspeople from disease, as well as the many tourists who visit the town. Instead, he is denounced, driven away, and declared an "enemy of the people."

The strongest man in the world, Ibsen wrote, is the man who stands most alone.

ON THE morning of June 6, 1989, at 8:00 a.m. sharp, a U.S. magistrate in Denver issues the search warrant. More than seventy-five FBI and EPA investigators are waiting outside the gates of the Rocky Flats Nuclear Weapons Plant as FBI agent Jon Lipsky, search warrant in hand, drives from the Denver courthouse to the foothills, green with spring, to meet them.

A RAID AND A RUNAWAY GRAND JURY
1989

Marriage is like a boat; I hope to find steadiness on a rocky sea. Instead the boat seems less stable than the waters, and I find myself constantly peering over the bow. I try to follow my mother's advice—you have made your bed, now you must lie in it—although I sometimes worry that my hasty marriage is playing out unhappy patterns from my parents' situation. I work as a secretary to help my husband, Andrew, through his last year of college, and a year after he graduates we move to Germany for his engineering job. I begin writing again and find work as a freelance journalist. The money is slim, but I relish the chance to travel.

On April 26, 1986, a large radioactive cloud travels across the sky near our home in Germany and continues its rounds across Europe. No one seems to know exactly what happened with Reactor Number 4 at the Chernobyl nuclear power plant, but my neighbor is afraid to eat the blueberries in her garden and everyone stays inside for a few days. It probably doesn't matter, I joke with my friends. Whatever there is to get, I probably already got from Rocky Flats.

Our son Sean is born in June 1989 in a hospital in Frankfurt. The birth is traumatic but swift; an emergency C-section saves Sean's life and he greets the world with round pink cheeks and indigo eyes. I lie awake all night and hold him, watching as the full moon travels across the black sky.

As I lie in a hospital bed in Germany with a newborn in my arms, the raid at Rocky Flats unfolds.

JUST BEFORE 9:00 a.m., on June 6, 1989, FBI agent Jon Lipsky drives through the gate of the Rocky Flats Nuclear Weapons Plant with a search warrant in his pocket, a .357 by his side, and the knowledge that just outside the gate stands a line of vans and vehicles filled with more than seventy FBI and EPA agents waiting for the signal to enter. Rockwell has no advance warning of what's about to happen, but in a memorandum of understanding between the Department of Justice, the DOE, and the EPA, the DOE has agreed that security guards at Rocky Flats will not prevent the raiding officials from entering the plant. Lipsky pulls into the parking lot, aware that he's being watched.

The administration is housed in a plain gray two-story building with a no-frills factory feel. Some doors have coded locks, others don't. Secretaries look up as Lipsky walks past. He raps on the door of the conference room once, then again, and enters.

The conference table is ringed with men: FBI agents in dark suits on one side and Rocky Flats officials on the other. A badge hangs from every neck. Plates of coffee cake and mugs of coffee are half submerged under documents, flow charts, and maps of the facility. The room is calm and cheerful. The FBI has told Rocky Flats officials that this meeting is to be a briefing on a potential "ecoterrorist" threat from Earth First!, the radical environmental advocacy group.

Lipsky takes a seat at the table. He nods briefly to the special agent in charge, Tom Coyle, to let him know that the search warrant is signed. The men wait, sipping coffee, until the Rockwell manager of Rocky Flats, Dominick Sanchini, arrives. Sanchini seems rushed. He's eager to get the meeting started.

But the FBI is calling the shots. Suddenly an agent standing near the door nods to Coyle. That's the sign. All is ready.

Coyle informs the men at the table that ninety FBI and EPA agents are invading the plant to begin an official investigation of Rockwell and the DOE for environmental crimes at Rocky Flats. Everyone is to remain seated. Guards, managers, and workers will be instructed to be cooperative.

"You can't be serious," sputters Sanchini.

"We are serious," says Coyle.

Dominic Sanchini, or "Dom," has been manager at Rocky Flats for three years. He has an impressive background. He has a degree in mechanical engineering from Lehigh University and a law degree from the University of Southern California. In the past, his work with Rockwell included helping to develop the main engines for the Space Shuttle orbiters. At sixty-two, he's beginning to think about retirement.

The vehicles pass through the gate. Agents emerge from the vans, some in protective gear and carrying face masks. They enter buildings around the facility, one after another, confiscating documents and examining equipment. They set up a command post inside an administrative building with their own telephones and portable computers. Someone alerts the press, and a TV crew arrives.

Workers are stunned. Many are scared or angry.

Back in the conference room, Lipsky and Coyle are busy playing defense. Rockwell's first response is to say that the agents don't have appropriate equipment—masks, respirators, or hazmat suits—to protect them from radioactivity and other dangers in the more hazardous parts of the site. The FBI, Coyle says, has taken that into account. The agents have proper gear.

A Rockwell official protests that the agents can't go into high-security areas anyway because those operations are classified. He's informed that the FBI has full permission to access all areas of the plant.

Another Rockwell official points out that all visitors need to go through a security check and obtain a proper security badge from the

DOE before entering Rocky Flats or any other nuclear weapons plants. Security checks could take days or weeks. The FBI and the EPA agents are, Coyle says, perhaps with a touch of irony, definitely not visitors.

Sanchini is reluctant to let agents into his office to go through his files. He tells them that he's seen notices of noncompliance from regulatory agencies, but they were minor and quickly corrected. Problems got solved, he says, if the DOE wanted to pay for them.

It's a long day, but at the end of it, Lipsky is pleased. The raid will require more long days of investigation, sampling, and interviews with employees—eighteen days, as a matter of fact. But the hard part, Lipsky thinks, will be finding someone who's willing to open up about what's actually going on behind the scenes at Rocky Flats without fear of retaliation. Already it's become apparent that few employees are willing to talk to the FBI. There is a strong sense of collegiality among workers and they've observed a code of silence for years.

JACQUE BREVER started working in the cafeteria in 1982, when she was divorced with a kid and putting herself through school. Like Debby Clark and many other Rocky Flats employees, she'd heard that Rocky Flats paid well. She'd been taking art classes, but soon after she started in the cafeteria, Jacque added a few math and physics classes so she could qualify to become a chemical operator and work with plutonium, where the real money was made.

When she qualified, she was immediately assigned to Building 771.

Jacque didn't know much about the facility, and she knew very little about nuclear weapons. All she really cared about was that she had a daughter to raise on her own, and she was about to make more money than she'd ever thought possible. Production at Rocky Flats was at an all-time high.

Her first day in Building 771 was a shock. She'd expected a tidy lab with flasks and vials and workers in lab smocks, but the building was forbidding, the equipment looked old, and the room where she was to work was crowded with glove boxes and piping. It felt like a factory. "Oh, my goodness, what have I gotten myself into?" she thought.

There were other challenges too. She was one of the first women hired to work in Building 771—or any of the hot areas—and some of the men made it tough for her. She learned to cuss and talk back and hold her ground. But there was also a strong sense of camaraderie and bravado that she liked. Practical jokes were common. Other workers talked about Flats employees as being one big happy family, but she considered it to be one big family "working through denial." The work was dangerous and everyone knew it, but no one talked about it. You couldn't work in that kind of environment if you thought too much about the potential consequences.

Most of all, the pay was great. She joined the Steelworkers Union. In addition to her salary, Jacque earned extra money by working in the hot areas, and she could work all the overtime she wanted. Some nights she slept in the locker room between shifts to get in the extra hours. Eventually she was promoted to crew leader and began training other chemical operators. Her own training had been spotty. She bought a little black notebook and began keeping a daily work journal to help with training and to keep track of what she did and what she learned.

Some things made her nervous. Managers seemed to look the other way or cover things up in order to meet production quotas. Sometimes there were accidents—spills or leaks or procedures gone awry. Everyone called them "incidents." One day, at the beginning of her shift, she was called to an incident that could have been fatal for her and her crew. A problem in the plutonium line was very near the criticality stage. Jacque shut down the line just in time. Her journal took on new significance. "It eventually evolved," she would later testify, "[to the point where] I was writing down things we shouldn't have been doing, like safety concern stuff." Working conditions, she felt, were intolerable. She complained to Rockwell and filed grievances with the union, and got herself labeled a troublemaker.

On the morning of the raid, Jacque has just arrived at work when FBI agents abruptly enter the processing area, wearing coveralls and carrying radiation-detection equipment. She and her crew watch as the agents order the managers to leave their offices. They begin going through desk drawers and filing cabinets, seizing documents. They order everyone to leave the building while they assess the amount of stored radioactive material and count the number of drums filled with radioactive waste.

Jacque can't help but feel a little amused at the managers, usually so calm, scrambling to deal with the situation. *They look like a bunch of chickens with their heads cut off,* she thinks. *At last the cavalry has come. Finally they'll see what an unsafe place this is, and maybe somebody will do something about it.*

But she has no intention of talking to anybody. Especially the FBI.

The next morning the raid is splashed across the front pages of the Denver newspapers. Readers learn that all production at Rocky Flats has been halted. Processing lines with plutonium in various stages have stopped in the middle of production.

It's the first time that most Coloradoans hear about problems at the plant. Even the governor of Colorado is shocked. Although he had been attempting to work with Rocky Flats, he didn't know about the investigation leading to the raid, the raid itself, or the extent of the potential environmental violations. "I have been victimized," Governor Roy Romer tells the press. "I am outraged—absolutely outraged. . . . I have been trying to say to people that this [Rocky Flats] operation is an operation that we're monitoring closely and it's not endangering your health. Today I have to say to people, 'Wait a minute. I don't know yet.'" He threatens to close the facility if there is any real threat to public health or safety, but it's unclear if he has the authority to do that. Romer flies to Washington to meet with Secretary of Energy James D. Watkins and DOE officials to ask them to begin cleanup immediately at Rocky Flats, rather than wait for the FBI and EPA to complete their criminal investigation of alleged cover-ups.

A special telephone line is established by federal investigators to take tips from whistle-blowers and angry and concerned citizens. A local radio station begins playing the Rocky Flats jingle, sung to the tune of "Rocky Top," which includes the lines: "I was born next door to Rocky

Flats / That's why I have two heads" and "Rocky Flats, you'll always be / radioactive to me / Good Old Rocky Flats / Rocky Flats you lied to me!"

Suddenly everyone—not just the protesters—is talking openly about Rocky Flats. A series of billboards appears along Highway 93, near the West Gate of Rocky Flats: CLOSE ROCKY FLATS, and THINK GLOBALLY, ACT LOCALLY. Now, for the first time, the controversial 1983 film *Dark Circle* is shown on public television. The film alleges many of the same crimes now being investigated by the FBI.

Dark Circle includes interviews with sick Rocky Flats workers and their families, including Don Gabel, the worker who died of a brain tumor from working under radioactive piping, and Rex Haag, the builder of our house in Bridledale, who talks about the death of his eleven-year-old daughter, Kristen. "The plutonium that went out with that fire must've carried right into her sandbox," he says. "It just tears me up to think about it now. We were right downwind."

The rumors, the film, the findings of Dr. Carl Johnson and Dr. Edward Martell, even the protesters who made headlines in the newspapers, had never been enough to provoke anyone to take action in my neighborhood. There were no protests or petitions. Anyone who criticized Rocky Flats—or even spoke of it—was ridiculed or ignored. Suddenly the atmosphere has changed.

Many families who bought houses around Rocky Flats, homes with "a million-dollar view," begin to doubt the government's assurances that Rocky Flats isn't a threat. Property values plummet. People stop gardening and stay indoors. Some people attend community meetings. But not everyone is galvanized by the news. Billy Chisolm built his home near Rocky Flats seventeen years ago. When asked about the plant, he just smiles and shrugs. David Weatherspoon bought his house two months before the raid and can see the lights of Rocky Flats from his driveway. "We've kind of looked the other way, because we didn't want to face it," he tells a reporter.

There's a new surge of activism. The director of the group Colorado Freeze Voter (part of the Nuclear Freeze movement) plans a second encirclement of the plant to take place on August 6, 1989. LeRoy Moore, one of the founders of the Rocky Mountain Peace & Justice Center, begins a thirty-five-day water-only fast in front of Denver's capital.

On the second day of the raid, however, FBI agent Jon Lipsky feels like the rug has been pulled from under his feet. U.S. attorney general Dick Thornburgh—who had been governor of Pennsylvania at the time of the Three-Mile Island accident—instructs the Justice Department to unseal the affidavit and application for the search warrant for the raid. The affidavit outlines Lipsky's game plan. It lays out almost the entire investigation, as well as the evidence the FBI has already collected. It describes a pattern of illegal dumping, burning, and polluting by Rocky Flats as well as cover-ups and instances of false statements and concealment by Rockwell International and the DOE. It reveals that the flyover established probable cause that the Building 771 incinerator was in fact operating in December 1988, and that a supposedly closed solar evaporation pond was "thermally active." The affidavit notes the millions of dollars Rockwell has received in bonuses for high production and safety over the previous fourteen years. Only two years before the raid, in May 1987, Rockwell had been awarded a performance bonus of $8.6 million for its excellent management of Rocky Flats.

Only one plant employee is named in the affidavit: Rockwell's top manager at Rocky Flats, Dominic Sanchini, who is accused of lying about Building 771. Suspicions about Sanchini seem to be playing out. On the fourth day of the raid, FBI agent Edward Sutcliff looks into a cabinet in Sanchini's office and discovers a large box of steno pads. Sanchini says they're just notes from when he worked at NASA, and he intends to use them to write a book. Fine, the agent thinks. No problem. But he digs around a little more and finds more steno pads. This time they appear to be Sanchini's diary of his experiences at Rocky Flats. Some of these entries will end up as evidence and show up in the newspapers. "Environment becoming a big deal," says one entry from July 1, 1986. "The EPA can destroy us." On May 6, 1987, Sanchini had written, "DOE doesn't follow the law."

By the time Lipsky knows the affidavit has been unsealed, it's already in the hands of Rockwell attorneys and high-level officials at Rocky Flats. Attorney General Thornburgh explains that he authorized disclosure of the affidavit to reassure the public that "this investigation does not signal any major new environmental or safety concerns."

Lipsky's been worried all along that high-level officials would be given advance notice of the raid. With enough warning, it would be possible to hide things. Lipsky suspects that the Criminal Division at Justice Department headquarters in Washington, aware of the investigation, informed the DOE of the details of the planned raid, even though high-level DOE officials are suspected of criminal acts. Two months before the raid, the highest-level DOE official on-site at Rocky Flats was transferred to Washington, D.C.

Lipsky presses on. The investigation continues until 5:00 p.m. on June 23 and involves more than ninety FBI agents and EPA investigators, who seize thousands of documents and hundreds of samples of waste. Evidence indicates that for more than thirty years, spills, leaks, and waste disposal practices have contaminated dozens of sites around the plant. The most significant public health issue is groundwater pollution. Groundwater at Rocky Flats is relatively shallow, just twenty-five feet below the surface. The most seriously contaminated sites include the 881 Hillside, the Mound area, and the East Trenches. One of the most shocking discoveries is that there is a great deal of material unaccounted for (MUF): more than 2,640 pounds of plutonium is missing. And despite the fact that Rocky Flats officials have always insisted that there has never been a criticality at Rocky Flats or an incident of uncontrolled fission, a memo reports an average of two "nuclear criticality infractions" each month.

Lipsky's partner, William Smith, handles the pondcrete investigation. He's astonished that no one has really taken much notice of the huge blocks that have been sitting out in the open for years. "It was on the main road. . . . People had been driving by this forever, even EPA people, and never knew it was something that wasn't legal," he tells the press. "I think it was incompetence, to be honest. How could you not get permits for pondcrete? You couldn't hide it."

The FBI and EPA's final allegations against Rocky Flats include concealment of environmental contamination, false certification of federal environmental reports, improper storage and disposal of hazardous and radioactive waste, and illegal discharge of pollutants into creeks that flow to drinking water supplies. They cite an internal memo written in 1987 by a DOE manager working out of the Albuquerque office that instructed a DOE official in Washington to send "a message to EPA that DOE and its management contractors are willing to 'go to the mat' in opposing enforcement activities at DOE facilities."

Rockwell officials have no comment regarding specific federal allegations at Rocky Flats. On the day of the raid, Albuquerque is removed from the chain of command and Rocky Flats will now report directly to Washington.

FOR THE first two weeks of the raid, Jacque keeps a low profile. She gets most of her information from the newspapers, because the managers at Rocky Flats don't tell the workers anything specific. One day she and a few of her co-workers decide to find things out for themselves. A worker tiptoes into a manager's office and discovers a copy of the affidavit. Much of it is written in legalese and is hard to decipher. But some things are clear, like the section about the flyover, in which the FBI alleges that there was "illegal midnight incineration" going on.

Oh my gosh, Jacque thinks. *That was one of my overtime days in Building 771. And oh, I think we did that.* She's scared, and worries that she and her crew might end up in jail.

Jacque looks at her work journal but she's still not sure of the exact days she worked the incinerator. She talks to her manager. "I think we did that," she says, "and I'm concerned." The response isn't what she expects. Management holds a meeting with all of the chemical operators. "The FBI might be wanting to come to talk to some of you," they say. A top-level manager is more direct. "Whistle-blowers," he says, "will be dealt with severely and completely."

Supervisors ask Jacque to produce information from her notes and journals about when she worked the incinerator. Only later does it occur to her that providing such material may, in the long run, make it easier for Rockwell to hide information and structure their case. She thinks about how talking to the FBI might affect her job, her life, and the life of her daughter. But after she confers with Karen Pitts, another woman who works in Building 771, both women decide to talk to the FBI.

Rockwell is cornered, and there's a lot of money at stake. Although the annual fee that Rockwell receives to run Rocky Flats is not publicly disclosed, experts believe it to be in the range of $10 million, excluding bonuses. In 1987 alone, Rockwell received a bonus of $8.7 million from the DOE for management and safety excellence.

Rockwell decides to sue the Department of Justice, the DOE, and the EPA, claiming that they can't meet government contract requirements if they have to also conform to environmental standards. The company also takes out full-page ads in Denver's two daily newspapers, the *Denver Post* and the *Rocky Mountain News,* claiming that they have been the victim of "turf battles, political disputes, and unfair news coverage," and that they have managed Rocky Flats with "proper concern for public health, safety, and the environment." They deny any criminal wrongdoing.

The day after Rockwell argues in court that it can't fulfill its DOE contract without violating environmental law, Energy Secretary Watkins terminates Rockwell's contract with Rocky Flats, effective December 31, 1989. The defense contractor EG&G will take over. On September 28, 1989, the EPA adds Rocky Flats to its Superfund cleanup list.

IN JULY 1989 Wes McKinley, a forty-five-year-old rancher and math teacher in the southeastern corner of Colorado, receives a notice in the mail.

The grandson of homesteaders, Wes spent his first eight years of formal education in a one-room sod schoolhouse with no electricity or running water. He married his high school sweetheart, also a teacher, and they live with their four kids on the land his family has owned for generations. Wes spends a lot of time on the land and in the saddle. When he wears his dusty white cowboy hat and red bandanna, he isn't acting.

The morning he reaches into his mailbox and pulls out the notice, at first he thinks it's one of those computer-generated ads that say you've won a million dollars. Call with your credit card in hand. He almost throws it away. He tucks it into his pocket instead, and when he goes into town, he shows it to a friend. "Do you know what this is?"

"Sure," the friend replies. "You've been summoned to serve on a grand jury."

Wes isn't sure what that means or what it involves. But he makes the long trip to Denver, where he learns that he's been called to serve on Colorado Special Grand Jury 89-2, a federal jury specifically empaneled to determine whether there is enough evidence to go forward with a trial against the operators of Rocky Flats. As with any other grand jury, all of the proceedings will be kept secret. The names of all those summoned for jury duty are put in a drum and then drawn out one by one. Wes thinks to himself that when his name is called, he'll ask to be excused. He has a ranch to run, a ranch that's at the tail end of the state, over three hundred miles away.

But he's intrigued. He's never even heard of Rocky Flats. How could all this be going on, and he's never even heard of it? When they call his name, he decides to say yes. He looks around the room. The other jurors include a bus driver, a hairstylist, a swim coach, a letter carrier, a repairman, a lawyer, and a retired police officer.

On August 1, 1989, the case begins, and after another juror resigns, Wes McKinley is named foreman for Special Grand Jury 89-2. He has no idea how complex and time-consuming—and even dangerous—the case will become. U.S. district judge Sherman G. Finesilver takes a full hour to read the jury their instructions. "It is every person's duty to conform his acts to the laws enacted by Congress," he says. "All are equal under the law, and no one is above the law. . . . [I]f the members of a grand jury after deliberation believe that an indictment is warranted, then you will request the United States attorney to prepare a formal written indictment."

For the next two and a half years, the jurors meet in Denver one week out of every month to hear testimony from one hundred witnesses and sift through 760 boxes of evidence.

Wes leaves his old pickup at home and takes the bus to Denver each month. Some jurors don't last and have to be replaced. Wes is always there, and always on time. "Attorneys must study how to be late," he jokes. They're supposed to start at 9:00 a.m., and things don't seem to get going until 9:15. Wes is a rancher. He'd start things at 6:00 a.m. if he could. It's just starting to dawn on him how important this case really is.

The first person to testify is FBI agent Jon Lipsky. He describes the raid, then focuses on the illegal operation of the incinerator that burns radioactive waste.

Jim Stone is eager to take the stand. Stone talks about the pondcrete, the incinerator, and the spray fields. He talks about the plutonium in the air duct pipes and how there is enough "lost" radioactive material in those pipe ducts to make several bombs. There's so much sandy material in the pipes, he says, that it looks like a windblown desert.

Dr. Edward Martell testifies, and later repeats his words to the press. "It takes minuscule amounts of plutonium to cause cancer or promote cancer. We know there is an awful lot of plutonium out there. The soil-borne contamination has been progressively redistributed by wind in the direction of the heart of Denver. Plutonium-induced cancers in people may take twenty or thirty years to develop. In effect, everybody living within eight or ten miles east and southeast of Rocky Flats may be guinea pigs."

John Cobb, a professor of preventive medicine at the University of Colorado Medical School, presents to the jury the data from his EPA-sponsored study of plutonium in tissue from autopsies of people who lived around Rocky Flats. He discusses the leaking barrels and the radioactive contamination that has seeped into the soil and water, and escaped into the air.

One fact that emerges in the testimony is how Rockwell manager Dominic Sanchini responded to pondcrete problems. When DOE inspector Joseph Krupar warned Sanchini that the pondcrete blocks were splitting and leaking, Sanchini "defined his access" at the plant and surrounded the blocks with a barbed-wire fence and a sign that read UNAUTHORIZED PERSONNEL KEEP OUT.

Approximately fifty Rockwell employees receive grand-jury subpoenas and they negotiate immunity agreements with the prosecutors. The first Rockwell employee to testify before the grand jury—right after Jon Lipsky—is Jacque Brever. She talks in detail about her work and the near-constant operation of the incinerator. Work she has done herself.

Employees at Rocky Flats are well aware that Brever is going to testify. Her co-workers are already talking about how she's going to shut down the plant and make them all lose their jobs. Officials send out a plant-wide memo announcing the day that grand jury testimony will begin. Shortly before the court date, Brever discovers that someone has poked a hole in one of the lead-lined gloves she wears when she puts her hands into the glove box. She's exposed to plutonium and americium. Several workers confront her afterwards. "That's what you get for making waves," they say.

She testifies anyway, despite continuing threats at work and at home. Another employee, a manager named Ron Avery, also testifies that he operated the 771 incinerator when it was supposed to be shut down. Still, the threats continue. Brever tries to keep her job at Rocky Flats and reluctantly resigns in April 1991. With co-worker Karen Pitts, who has had a similar experience, Jacque files a lawsuit saying she was threatened, harassed, and forced out of her job after talking to the FBI about problems at Rocky Flats. The case is dismissed when the judge decides it is "not detailed enough" to continue.

Shortly thereafter, Jacque and her daughter go into hiding. She is diagnosed with thyroid cancer, one of the more treatable types of cancer, and reactive airway disease. Years later she returns to Colorado to pursue a master's degree in environmental policy and management. "The best thing I can do," she tells an interviewer, "is what my conscience tells me to do while I'm here. I'm not afraid of dying, but at least I can do something to clean up the mess we made. I'm really ashamed that we're leaving this mess for people like my daughter and her generation."

☢

TESTIMONY BEFORE the grand jury continues. The 771 incinerator charge is fiercely contested by Rockwell and DOE general counsels. Yes, they admit, they've been storing and burning hazardous waste in the incinerator for years without a permit. But is it a type of waste that is subject to RCRA and EPA jurisdiction? Rockwell argues that emerging environmental law is filled with ambiguity, and that the Atomic Energy Act exempts nuclear weapons facilities from laws such as RCRA and Superfund.

Other charges are more difficult to defend. Waste in a series of holding ponds continued to contaminate groundwater even after regulators had closed the ponds. The "spray irrigation" was done to reduce levels of waste and to allegedly avoid the scrutiny of regulatory agencies and the public.

But is it fair to single out Rocky Flats? Plenty of other DOE facilities, including Hanford in Washington State, Oak Ridge in Tennessee, the Savannah River site, and the Fernald plant in Ohio have severe problems with radioactive and toxic waste and storage. Some are worse than Rocky Flats.

The issue of "permits" is particularly troublesome. RCRA and Superfund gave the EPA responsibility for regulating hazardous waste "from cradle to grave," and the EPA has at times issued permits that allowed for some pollution. But the 771 incinerator does not have an EPA permit at all, which is why the Sierra Club was able to sue and get it temporarily shut down in 1989. The year after the shutdown, the Sierra Club scored another victory when the lawsuit was decided in their favor. The ruling directed Rocky Flats to manage plutonium residues as hazardous waste and said the residues were subject to RCRA regulation.

Jim Stone is right about more than pondcrete. In the fall of 1989, following the raid, an independent study finds that there is, indeed, enough "lost" plutonium in the exhaust ducts at Rocky Flats to create the possibility of an accidental nuclear reaction. The experts who conduct the study estimate that approximately 12 kilograms of plutonium could be caught in the piping. Energy Department officials justify it by saying that the plutonium is not harmful to workers within the buildings, and that filters prevent the plutonium from getting into the air outside the buildings. Nonetheless, the DOE agrees to conduct its own study.

Six months later, in a meeting with plant officials, Melinda Kassen, a lawyer with the Environmental Defense Fund in Boulder and a member of the Rocky Flats Monitoring Council, asks point-blank if the study promised by the DOE the previous October has been completed.

Kassen is told that the study is done, but results are being "withheld."

Under pressure, the DOE reveals to Kassen the next day that the quantity of plutonium in the air ducts is 28 kilograms, or about 62 pounds—more than twice what they had expected to find, and enough for seven nuclear bombs. Spread over six thousand feet of pipe, the plutonium got stuck in the ducts partly because, as the filters became clogged and automatically closed down operations, frustrated workers punched holes in the filters so that air could pass through the system. Never mind that the air was contaminated with plutonium.

Leo Duffy, director of Waste Management and Environmental Restoration at the DOE in Washington, defends the government. He notes that the plutonium in the ducts at Rocky Flats is the residue of thirty-eight years of operation and is only a "very small fraction" of the amount of plutonium that has been handled by the facility. Nonetheless, "none of this," he admits, "is a satisfactory way of running an operation."

Meanwhile, in an appearance before a House armed services subcommittee on nuclear weapons production, Energy Secretary James D. Watkins strongly implies that the problems at Rocky Flats could delay the deployment of Trident II ballistic missiles. The weapons production complex needs to get back to full operation. "I can guarantee if we don't move aggressively," he says, "there will be severe ramifications."

THE GRAND jury investigation drags on. The lives of the jurors are affected in unexpected ways by two years of week-long sessions and absences from work and family. Relationships suffer. The court sessions are long and often involve detailed, dull testimony. One juror serves for nearly eighteen months before she truly grasps what's at stake. At a restaurant in Brighton, a small town near Denver surrounded by cattle ranches, the waiter brings to the table water that tastes "like cow waste." "A light bulb came on," she recalls. "I knew why I was on that grand jury."

After twenty-one months of work, on May 18, 1991, the jurors, led by cowboy Wes McKinley, vote to indict Rockwell, five of its employees, and three others working for the DOE. But in a surprising turn of events, U.S. attorney and Department of Justice prosecutor Mike Norton refuses to sign the indictments. Not a single Rockwell or DOE official is indicted, despite the fact that more than four hundred environmental violations occurred for decades.

Instead, Norton negotiates a plea bargain with Rockwell. In addition to no indictments, dumping and incineration charges are dropped. No individuals will be charged, and the settlement guarantees their immunity. Rockwell agrees to plead guilty to criminal violations of the federal hazardous waste law and the Clean Water Act, admitting to five felonies and five misdemeanors. The criminal conduct includes "possible exposure of workers and local citizens to radioactive and hazardous waste that was sprayed into open pools and even stored in ventilation vents." Rockwell is required to pay an $18.5 million fine—at that time the second largest in U.S. history for an environmental crime, following the 1989 *Exxon Valdez* oil spill in Alaska—yet still a smaller amount than what the corporation has collected in bonuses for running Rocky Flats and meeting production quotas for that one year, representing just one-sixth of 1 percent of Rockwell's yearly sales. Rockwell routinely received millions of dollars in performance bonuses despite the DOE's own ranking of Rocky Flats as "the most dangerously contaminated site" in the nation's nuclear weapons complex. Rockwell is also allowed to file for reimbursement of $7.9 million from taxpayers for case-related fees and costs. Contractors hired by the federal government to operate Rocky Flats, or any nuclear weapons facilities, are indemnified from any damages. The more shocking charges—midnight burning of waste in incinerators that were supposed to be shut down, and secret dumping of poisons into waste ponds that were also supposedly closed—are not prosecuted.

Perhaps most devastating to local communities, however, is that the plea agreement indemnifies Rockwell from any further claims and closes the door on all future prosecution, whether criminal or civil. All court records will remain secret. The agreement stipulates that Jim Stone's charges cannot be mentioned in the plea agreement, and trial-related documentation regarding past and ongoing contamination will be permanently sealed. The agreement also allows Rockwell to receive future contracts with the government. Judge Finesilver approves the plea bargain and tells the grand jurors that their work is finished and they can all now go home.

The jurors are stunned. They can't believe what just happened. They refer again to the instructions of Judge Finesilver: "The federal grand jury . . . is independent of the United States attorney . . . it is not an arm of the United States attorney's office. Please keep in mind, you would perform a disservice if you did not indict where the evidence warranted an indictment. . . . The government attorneys cannot dominate or command your actions."

But they are bound by an oath of secrecy about the case, even after the case is closed. They had two questions: what was done, and who ordered or allowed it? The Justice Department, as a matter of policy, declined to sue its sister agency, the DOE. And Rockwell says that everything it did was at the behest of the DOE, and the company insists that no individuals be charged. It's not uncommon for the Justice Department to allow corporate officials involved in environmental crimes to avoid individual indictments in exchange for plea bargains and fines. In most cases this is reached before a grand jury can see the evidence of the crime. However, the Rocky Flats case is different. The Rocky Flats grand jury endured almost three years of damning testimony. They can name names. They want indictments, and they want the public to learn the truth of past and ongoing contamination practices at Rocky Flats.

"We were studying a million pages of documents," McKinley tells the press. "A million. Memos, internal information, balance sheets, and other information. We had a little disagreement as to whom to indict. If the case had gone as the government planned, we would have indicted various workers at the plant. The people who went out there, earned a day's wage and followed orders. Those were not the responsible people. [To indict average workers] would have targeted the throwaway people, and things would have gone on as usual. They would have replaced one valve-turner with another. We wanted to get the people responsible so that we could actually enforce environmental laws. These were the people in the DOE and the executives in the higher echelon of Rockwell International who, we felt, actually committed the crimes."

Rather than go home, the jurors consider their oath, consult the Constitution, look over the evidence again, and write their own grand jury report—without the help of prosecutors. They ask Judge Finesilver to read it and release it to the public. He refuses.

On September 25, 1992, Finesilver rules that the grand jury report will remain sealed. Three years earlier, the same judge had ordered the jurors to find the truth. Now he asks them to keep the truth under wraps.

But somehow, a few days later, much of the report prepared by the grand jurors is leaked to the press, and a detailed account of the grand jury report and its accusations against Rocky Flats—with large sections still missing—is printed in a Denver alternative newspaper. *Harper's* magazine publishes excerpts in its December 1992 issue. Finesilver calls for the Justice Department to investigate whether grand jurors have breached their oath of secrecy and should face criminal charges. Publicly revealing secret information from a grand jury can result in high fines and a jail sentence. The DOE wants to discover who leaked the report and prosecute the grand jurors. The jurors become known as "the Rocky Flats 23," the first grand jury in history to risk personal incarceration for revealing information critical of the Justice Department. Twelve of the jurors write a personal letter to President-elect Bill Clinton, asking him to appoint a special prosecutor to investigate the government's plea bargain with Rockwell. Foreman Wes McKinley and juror Kenneth Peck, a lawyer in private practice, call the press and read the letter from the steps of the Denver courthouse. There is no response from Clinton. In January, seven of the jurors appear on *Dateline NBC* to discuss the trial and why they feel it is a miscarriage of justice.

Both Mike Norton and Ken Fimberg defend their decision not to prosecute corporate or government individuals. Many people were simply carrying out orders, they say. The people who might have been prosecuted, Fimberg tells the press, were basically mid-level people who had not established the policy, but acted consistent with it. Further, they've both begun to realize how the whole ballgame involves not just a contaminating weapons plant, but forty years of public policy. "I'm not going to prosecute conduct well known to regulators," Norton says. There is no question of groundwater contamination and toxic runoff, but health effects in nearby communities are hard to prove. And the DOE's policy of indemnifying weapons plant contractors means that the Energy Department—thus the taxpayer—will have to pay any fine levied against Rockwell. Only if they settled could they make Rockwell pay its fine.

The plea bargain with Rockwell puts all of Colorado up in arms. In Washington, an oversight subcommittee of the Committee on Space, Science and Technology, chaired by Representative Howard Wolpe (D-Michigan), issues subpoenas to a number of players in the grand jury investigation, including Fimberg, Norton, and Lipsky, for eight days of private hearings in September 1992. Prosecutors emphasize that they've been the first to initiate criminal prosecution of environmental crimes at a DOE plant, but what they found was an "institutional culture unchallenged by Congress or regulatory agencies." Is it fair to indict and prosecute individuals, Fimberg asks, for acting consistently within that culture? Wolpe responds by asking whether a situation has been created where there is literally no accountability that can be imposed on government agencies. "Are you telling us," he asks, "[that] the culture of an agency, even if it violates a law that has been passed by Congress, represents a kind of defense? . . . Isn't the purpose of law to change behavior?"

In January 1993, the Wolpe committee issues a report revealing evidence of high-level intervention by Justice Department officials to reduce charges and fines against Rockwell. "The most important thing that federal prosecutors bargained away in negotiations with Rockwell was the truth," Wolpe says. Jonathan Turley, a Washington lawyer representing the grand jurors, is more blunt. "The Justice Department," he says, "is in complete denial."

One person who does not live to see the controversy surrounding the grand jury investigation is the previous Rockwell manager of Rocky Flats, Dom Sanchini, who is ill and has stepped down from his job. On November 17, 1990—seventeen months after the raid begins—he dies at age sixty-three after a struggle with cancer. He was fifty-eight when he began working at Rocky Flats, and he was one of the individuals the grand jurors had wanted to indict. Sanchini worked for Rockwell for a total of thirty-seven years. He leaves behind a wife, two daughters, and three grandchildren.

SOME WORKERS and their families begin taking legal action. Workers' compensation claims are filed on behalf of fourteen workers who were involved in fabricating plutonium and sustained burns, exposures, and puncture wounds as a result. Represented by attorney Bruce DeBoskey, thirteen of them died of cancer and the surviving worker has bladder cancer. Many of these cases date back to the 1970s, and the "Widows of Rocky Flats," as the press calls them, have been fighting the government for years. In the compensation hearings, government scientists claim that the cancers and other illnesses suffered by workers could not be caused by conditions at Rocky Flats because their exposure to radiation was within the level that the government considers safe and that deaths were caused by smoking or other causes. Experts testifying for the plaintiffs argue that exposure to radiation at Rocky Flats caused the cancers, and the levels for permissible exposure established by the government are much too high. In a number of studies, the DOE has confirmed higher incidences of cancer and unsafe working conditions at Rocky Flats, yet at the same time, a report by Rockwell in 1988 puts the plant's rate of worker injury at 3.2 injuries per 200,000 hours of work, a rate less than half the national average for industrial plants across the country. "This is probably the safest place they'll ever work," says Clayton Lagerquist, manager of radioactive protection at the plant.

DeBoskey disputes the government's position. "If society insists on spending its resources to produce nuclear weapons," he tells the press, "it ought to do it understanding that there is enormous human cost. That hasn't been factored into the equation yet."

In September 1990, the case of Rocky Flats worker James R. Downing, who died in 1978 at the age of forty-four, becomes the first in U.S. history in which a judge determines that occupational exposure to radioactive elements was solely responsible for a worker's death from cancer. Downing's exposure to radioactive elements was within the limits considered acceptable by the DOE.

LOCAL RESIDENTS aren't sure what to believe. Property values have gone down. Is it due to nothing more than bad publicity? While residents wonder, the General Accounting Office (GAO)—the audit, evaluation, and investigative arm of the U.S. Congress—reports that the latest inventory records show that 2,900 kilograms of plutonium as well as 97,000 kilograms of solid residue and 14,000 liters of liquid residue contaminated with plutonium are stored at the plant. The DOE can't move the nuclear materials in their present form, as they do not meet shipping and disposal requirements.

Mark Silverman takes over in 1993. Silverman is a West Point–trained engineer who did a combat tour in the U.S. Army and worked at another bomb factory in South Carolina. Fourteen tons of plutonium are spread in various forms all over the facility. Plutonium operations ended in 1989 with the FBI raid. Because the stoppage was supposed to last only a few weeks, nuclear materials have been left in tanks, pipes, and containers designed only for temporary storage. Plans to restart weapons production were stalled again and again. Deployment of the W88 warheads on the Trident II missiles—the most sophisticated strategic thermonuclear weapon in the U.S. arsenal—was slowed in 1989 by the raid on Rocky Flats, and plans for production at Rocky Flats were finally scrapped in 1992 when the W88 weapons program ended. Silverman finds himself facing not only extraordinary environmental problems, but a plant that has "no production, no clear mission, no real deadlines, and regulatory and political turf battles."

No one's sure what will happen with Rocky Flats. Will it be closed or moved? How much of it can be cleaned up? The plant has more employees than ever before, but no one's sure exactly what is being accomplished.

Silverman is a little different from the managers Rocky Flats has had in the past. The most daunting part of his job is to overcome the distrust, anger, and hostility between Rocky Flats and the public, including the regulators, citizens, news media, and elected officials at all levels. In an interview with Silverman, *Time* magazine reports that "aging buildings are tainted by plutonium spills from leaking pipes, valves and containers, and from compartments known as 'infinity rooms' because their level of radioactivity is so high. Barrels of radioactive waste are stacked fifteen feet high. Fields contaminated with radioactive oil are covered by only one layer of asphalt. Now that suburbs have crept within three miles of the plant's perimeter, the plutonium that has periodically leaked into the air and nearby streams poses new dangers."

I KNOW nothing of what's happening at Rocky Flats. My sisters are in California, and my brother in Arizona, all of us building lives of our own. My mother and father, both still in Arvada, pay no attention to the drama at the bomb factory. It's old news to them. I haven't been watching the news in Colorado; my life is filled with more immediate concerns. In the fall of 1989, Andrew and I return to Denver and soon I'm pregnant with our second son. Nathan arrives in the world like a happy seal, smiling and kicking up his toes. The boys are so close in age that as they grow they often pass as twins, although Sean has dark hair and Nathan's is lighter, like mine.

But even the addition of another blithe spirit can't save us. Our marriage dissolves in clouds of sad and bitter emotion, and I leave with the children and as much furniture as I can fit into the back of a pickup truck. I move into a small duplex in Arvada, not far from where I grew up. I'm not sure what lies ahead.

Before we left West Germany, I had waited anxiously for the Berlin Wall to come down. Surely, I thought, this dark, ominous symbol of war and violence would soon fall, and when it did, it would be a joyous occasion. I wanted to be there. But the months dragged on, and we returned to the States. When I finally see the Germans celebrate the fall of the Berlin Wall on November 9, 1989, it is on television in my living room in Colorado. I feel overcome with emotion, and I know my friends in Germany feel the same.

On November 21, 1990, President George H. W. Bush declares the end of the Cold War.

In Arvada, Colorado, the Cold War is far from over. My life feels war-torn as well. Everything is turned upside down. I decide to go back to school. But this time I'm a single parent with a baby and a toddler, a silky black cat named Jasmine, and a pile of bills.

I need a job.

The Rocky Flats Nuclear Weapons Plant is about to get a new face: a change in management and, four years down the road, a new name altogether.

On January 1, 1990, EG&G, a company originally founded by Manhattan Project scientists who helped develop the first atomic bomb, replaces Rockwell International as the new contractor at Rocky Flats. EG&G manufactures everything from valves and meters to airport security systems, but the fastest-growing aspect of its business is managing government nuclear facilities, including the Idaho National Engineering Lab and the Nevada Test Site. EG&G's four-year contract is estimated to be worth approximately $500,000 a year in operating and management costs, with an additional $10 million in profit that will be paid as performance bonuses.

EG&G is not crazy about the public reputation of Rocky Flats.

If the plant's reputation has been tarnished, its production record remains unsurpassed. For almost forty years, the entire U.S. nuclear arsenal has been dependent on Rocky Flats. It is a key part of the federal government's nationwide nuclear weapons complex, which depends on private industry and corporations. There are seven plants around the country, each with its own function. Like Rocky Flats, each of the other six is run by a major U.S. industrial corporation: Union Carbide, DuPont, Bendix, Monsanto, General Electric, and Mason & Hanger. Much of the work is done in secret. At the height of its production period, the Oak Ridge, Tennessee, site, which produces and purifies uranium, employs tens of thousands of employees, many of whom don't fully know or understand what they're working on.

Rocky Flats is the primary facility to have produced plutonium triggers. From the day the first plutonium trigger rolled off the assembly line to the day Jon Lipsky entered the plant with a warrant in his pocket, Rocky Flats has produced approximately seventy thousand plutonium cores, each about the size of a slightly flattened softball, each costing about $4 million. If all the triggers produced at Rocky Flats were stacked end on end, the height would be greater than eighteen Empire State Buildings.

When EG&G assumes management of Rocky Flats, the company plans to resume weapons production. It's a lucrative business. This begins to change in January 1992 when President George H. W. Bush declares in his State of the Union address that the United States has won the Cold War and he is canceling the W88 Trident warhead program. Now EG&G has to start thinking about possible cleanup, closure, and transferring or terminating workers.

But the public image of Rocky Flats is what requires the most immediate attention. The plant needs a new name that reflects the plant's new focus. Officials hold a name competition, open to plant employees and the general public. Response is swift, and more than six hundred names are submitted, including a few from representatives of groups such as Citizens Against Nuclear Disinformation in Denver (CANDID). Suggestions include "Radiation Acres," "Rad Rocks," "Doom with a View," "Never Dark Park," "Hazy Heights," "Glowing Waters," and "Toxic Town." One person notes that "Death Valley" is already taken.

It is not until 1994 that Rocky Flats officials settle on a name they feel "best represents the changed mission of the site": the Rocky Flats Environmental Technology Site (RFETS).

To his credit, DOE manager Mark Silverman calmly continues to face the press. He knows he's sitting on a time bomb. There's enough radioactive waste on the plant premises to cover a football field to a depth of twenty feet. Plutonium lies in various stages of disarray and there is no safe, permanent storage site for radioactive waste, although some material is being shipped on a mostly temporary basis to the Nevada Test Site and the Envirocare Company in Utah.

The DOE projects that it will take at least until 2065 and cost American taxpayers more than $40 billion to marginally clean up the nuclear waste at Rocky Flats. A DOE official comments to the Senate Armed Services Committee that some weapons plants, like Rocky Flats, may never be cleaned up because the technology to do so at a reasonable cost doesn't exist.

I DECIDE to go see my father.

My life in the duplex with the boys is cramped but good. My brother has moved back to Colorado, and he and his wife live next door—they have two small daughters, and I have my two boys, so we decide to join forces. Kurt is always ready to show up on the doorstep with a beer in hand. My mother stops in when she gets off her shift at work. The boys and I have a cat, Jasmine, and a pair of honey-white doves who sing us awake each morning. I divide my time among taking classes, driving the boys to school, and trying to pick up enough freelance writing to keep the bills paid. But they won't stay paid. I lie awake at night, worrying about the utilities and the groceries and the rent.

I adopt the time-tested family strategy for getting through tough times: add another pet to the household. A friend's dog has puppies, and suddenly the boys and I are owners of a goofy, floppy-eared, pink-bellied basset hound with big paws and wrinkly bags under deep-set eyes. We name him Heathcliff because of his general good looks.

Some nights when I can't sleep I pull Heathcliff—who seems to double in size every week—up on the bed with me. The slow rhythm of his puppy breath is a comfort.

One evening my mother informs me that my father is in the hospital. It's the same thing as always—health problems related to his drinking— but this time, she says, it's worse. Critical, in fact.

Sean and Nathan have never met their grandfather.

The next day I drive to the hospital, bundle the boys into their double stroller, push it into the main lobby, and ask for my dad's room number. At the nurses' station, the broad-shouldered nurse is solicitous but circumspect and I have the feeling, as I have often had in the past, that others manage to connect with my father, to find him funny and smart and endearing, in ways that I can't.

I peer into the room, where he is lying under a white blanket. He looks up and I see his face is gaunt.

"Hi, Kris," he says in a guttural voice. His speech is thick and clotted.

"Hi," I say. "How are you?" The question seems formal and ridiculous.

"I'm all right," he says. "They haven't killed me off yet." He has an IV taped to his hand and his skin looks yellow. He waves his other hand in the air. "So who have you got there?"

"This is Sean." I start to lift Sean out of the stroller and then change my mind when I see the nurse standing in the doorway, hands on her hips, as if my time were already up. "And this is Nathan." I brush the top of Nathan's head. Both boys gaze at him.

"Well, hello, guys," Dad says. He sits up slightly.

"This is your grandpa," I say solemnly. Nathan is too young to fully understand. Sean just smiles. Dad looks at them briefly and then settles back into his pillow, as if he's too tired to do anything more. He closes his eyes. The nurse steps forward with a stern look.

"I hope you feel better soon," I say. Inept. Absurd.

"Thanks, Kris." He opens his eyes. "Can you find your way out of here?"

"No problem, Dad." I feel like touching his hand, his fingertips, and then decide against it. I adjust the boys in their stroller and turn to go.

Outside, I whirl around to face the nurse, who's a half-step behind me. "What's his condition? What's wrong with him?"

She stands silent.

"I'm his daughter."

She purses her lips. "Only certain people who are on a list are supposed to know about his condition."

I feel chagrined.

"It's his heart," she concedes. There is no compassion for me in her voice, just a statement of fact. She wants me to leave.

"And his blood alcohol level?"

"Is high." She takes my elbow and walks us down the corridor toward the elevator.

"Will you let me know if he gets worse?"

She avoids my eyes. "We'll take good care of him," she says.

EACH MORNING over coffee I scour the want ads. I need a job with flexible hours that pays well, and I've had enough of waitressing. And then, there it is, a large ad: administrative skills, flexible hours, $12.92 an hour. The Rocky Flats Environmental Technology Site is hiring. *Start Immediately*, it says.

The Rocky Flats Environmental Technology Site? They must have fixed things up out there, I think. And it's just down the road. I debate whether or not it's safe to work there. The money is sorely tempting, and I reason to myself that if there is contamination out there, I've already been exposed to it. After all, I grew up with it. A little more won't hurt, will it? Most of all, though, I want to see what it's like inside. My childhood has been shadowed by two enormous fears: my father's alcoholism and Rocky Flats. Maybe I can demystify one of them.

I call the number and make an appointment with what turns out to be the Sunnyside Temp Agency. "I'm calling about the Rocky Flats job," I say.

"Yes, we've been getting a lot of response to that," the receptionist replies. "We're not associated directly with the Department of Energy or EG&G," she adds. "We just provide contract workers. Secretaries, file clerks, that sort of thing. You'll need to come to our office for testing and an interview."

At nine the next morning I show up at their office near downtown Denver, in a skirt and pantyhose. The pantyhose are a big concession; since middle school I've hated that tight Spandex feel. They're hot and uncomfortable. Men don't have to stuff themselves into sausage casings.

The temp agency's office is in a high-tech building, all glass and steel, and it's nice: plush carpet, piped-in Muzak, dark wood furniture. Several women of various ages, all looking somewhat resigned, sit in the waiting room. I fill out pages of paperwork: employment history, education, what software programs I'm familiar with. I don't need to go through a background check, as I won't have a Q, or high-security, clearance. "Government grunt work," the woman next to me murmurs as she fills out the same paperwork. "We're going to be peons."

"Why is Rocky Flats hiring?" I whisper.

"Big layoffs. They need temp workers to take their place."

I hand my clipboard to the receptionist and mention casually that I'm a Ph.D. candidate at the university.

"Oh, we don't care about that," she chirps. "We just need to know how fast you can type."

It's been years since I've taken a typing test, and I'm suddenly as nervous as a cat. "You can have a few minutes to warm up," the receptionist says. "And then I'll turn on the timer." She leaves me in front of a keyboard and a blank computer screen. For lack of a better idea, I type my name over and over again. *Kristen Kristen Kristen Kristen Kristen.*

"Ready?" She smiles. Her makeup is flawless, eyes and lips outlined against a porcelain mask. "Here you go." She sets a paper next to the screen. "You have ten minutes. Type as fast as you can. Each mistake counts."

I begin typing, a row of numbers and then the same sentence over and over. *The quick brown fox jumps over the lazy dog. The quick brown fox jumps over the lazy dog. The quick brown fox jumps over the lazy dog.*

The timer rings and she tallies up the result. "Oh, my dear," she whispers, leaning close to my ear. "Would you like to try that again?"

"Sure." My cheeks burn.

The second time around goes a little better. "I think you'll be fine," the receptionist purrs as she escorts me to the office of the employment consultant, a woman in corporate attire with a stack of applications on her desk. She looks over my file briefly and has only one question. "When can you start?"

AFTER YEARS of allowing the marriage to drift and dissipate, my mother finally gets an envelope in the mail confirming that her divorce from my father is final. She seems surprised, disappointed, and relieved all at once. "In Bridledale we had a beautiful home," she says when I stop by with Sean and Nathan to tell her I've found a good job. "Then your father went right down the tubes. As a matter of fact, the whole thing went down the tubes."

She pauses to take a drag from her cigarette. "I should quit," she says. But she never will. She's smoked since she was a kid on the farm in Iowa. She tries to hide it from Sean and Nathan, sneaking off to the bathroom or the back porch to light up.

"Grandma," Sean announces solemnly one day, "we know you smoke. You don't have to hide it from us. You don't have to hide anything from us." Nathan nods in agreement.

"Well, all right," my mother says. She grins impishly, like a naughty schoolgirl. "But I want you to understand that it's a very bad habit, and you must never start."

"Okay," the boys chime.

Now she smokes in front of Sean and Nathan with the same ceremony and savoir faire that I remember as a kid. The slim lighter, the long cigarette between her graceful fingers, the touch of the flame, the welcome drag. She guards her ashes with an upturned palm and never leaves a cigarette butt in plain view. "There are so few joys in life," she sighs. "This is one of them."

In a wicker basket next to her coffee table are photo albums, big and small, worn and new. She saves every family photo, and has scrapbooks in the basement that contain pictures of our Norwegian ancestors and her own trip to Norway to trace the family tree. She's a proud member of Sons of Norway, a club that promotes and preserves Norwegian culture, where she's taking a class in the traditional Norwegian folk art of rose-maling, which involves painting traditional images of flowers.

But she wants to talk about the more immediate past.

"There were good times, too," she says. She shows me a black-and-white photo of herself and my father in the kitchen of our first house in Arvada, not long after I was born. With a firm jaw and a white T-shirt on, he looks like a slim Marlon Brando with dark hair. Her hair is in a twisted scarf and she's wearing dark eyeliner and pedal pushers. He's leaning toward her with a smile as wide as Nebraska, and she's nearly doubled over in laughter.

"We were in love then," my mother muses, and takes a drag.

Sean and Nathan look a little embarrassed.

She turns to me. "He's driving a cab now, did you know that? The law practice is gone."

I shake my head. I didn't know.

"Life can be so strange," she says. "You never know what's going to happen." She sighs and closes the photo album. "At least, Kris, you found a good job."

I'M NOT sure the job is worth the risk. Mark's warnings about Rocky Flats float back into my thoughts and dreams.

Still, after school the next day I strap the boys into their seats in the backseat of our tired little Toyota and we go for a drive. Highway 93 between Golden and Boulder is a thin, two-lane snake of road that dips and rises amid fields of low grass spotted with cattle and horses. A handful of crayons has melted into a puddle of hot magenta and forest green on the dashboard, and the passenger seat is covered with Heathcliff's hair. With a jawline nearly as handsome as his namesake in *Wuthering Heights*, Heathcliff likes to ride in the front seat with his head lolling out the window, sporting a bright Hawaiian-style sun visor we bought for him in a drugstore.

I roll down the window and look out at the foothills, rolling golden hills that rise abruptly into dark mounds and slabs of blue rock. A railroad track emerges from one of the canyons and curves out toward the highway into a long, low valley. I pull over to the side of the highway at the gravel road that leads up to the west side of the plant. I can hear meadowlarks. There's no obvious sign of the plant. From here it all looks innocuous.

Can I really do this? I ask myself. "Should I really do this?" I turn and ask the boys.

Nathan waves his fingers in the air. Sean studies his graham cracker. The answer remains unclear.

ON THE morning of September 14, 1994, I wake early to get the boys ready before I drive out to the plant. Sean and Nathan wear matching Thomas the Tank Engine sweaters—it will be another couple of years before they want to stop dressing alike—and I comb their hair and tie their shoelaces. They'll have an hour at day care before Sean goes to kindergarten and Nathan goes to his preschool class.

The morning is crisp and clear and the leaves have just begun to turn. I drive out to the foothills and enter the plant from the west gate, off Highway 93. I've been instructed to pick up a temporary badge at the guard shack, the first building on the right. The guard is an older guy, gray hair, with a belt a little too short for his belly. He cheerfully checks my driver's license and crosses my name off a list. In addition to my temporary ID, he hands me a photocopied map of the plant—there are more than one hundred buildings, all numbered—and a sheet titled "Plant Visitor Information." He points out the Special Nuclear Materials Area, a protected area surrounded on the map by double black lines. "You don't want to go there," he says. "You'd get in trouble. You need special clearance for that."

I mention the Rocky Flats Lounge I passed just across the road from the west gate, a tiny place with a big parking lot and plenty of blinking neon beer signs. "Is that an official Department of Energy building?" I joke.

"No, no," he says with a smile. "But that building does date back to the 1950s. It was the first payroll building. They've got cold Coors and great catfish. You can buy yourself a T-shirt that says 'Get Nuclear Wasted at the Rocky Flats Lounge.' Great place to watch football."

I nod. I don't watch football, which is as close to a sin as a person can get in a city like Denver.

"Good luck!" he adds. "You'll like it here."

My heart thuds as I get behind the wheel of my car. At last I get to see the belly of the beast, the bull's-eye target, the glowing light beyond my childhood bedroom window.

It's late morning as I drive through the gate. The plant spreads before me like a toy village, raw in the garish light of the sun. But there is nothing charming about this village. Cars and trucks fill the main street. There's an air of busyness. But it has a strangely anonymous feel. The buildings are plain, gray, square, mostly concrete block. It reminds me of what little I've seen of East Germany.

I disobey the guard's instructions and drive around before reporting for work. I take the big loop around the plant, past a large empty lot partitioned off with yellow police tape, past rows of stacked wooden crates and cardboard containers, past endless nondescript buildings, and then down the hill past the area surrounded with chain-link fences and razor wire. I see the guard towers. I wonder if anyone will notice a beat-up red Toyota tooling aimlessly around the plant. *I'll say I'm lost*, I think. I'm new. I am lost.

I'm fascinated. And disappointed. It all seems terribly mundane.

I drive to Building 117 to have my photo taken and get my permanent badge. My badge is number 26453. I am now one of 6,232 employees at the Rocky Flats Environmental Technology Site. I'm instructed to wear my badge above the waist and keep it in plain sight at all times.

Official now, I get back in my car and go as instructed to Building 130, the administration building on the east side of the plant, where I'm to report to project management. The parking lot is small and graveled. I notice a number of bumper stickers: I ♥ TOXIC WASTE and I WILL BUILD NUCLEAR WEAPONS FOR FOOD. The wind is blowing so strongly I have to hold down my skirt.

Compared to what I've seen in other areas of the plant, this isn't so bad. There is a square, two-story gray office building, a warehouse, a cafeteria, and an engineering building with a small, bare-bones courtyard. Every door in the building has a punch-key lock.

A secretary sitting behind a tall desk curtly asks my name, then checks it off a list and gives me a code for the punch combination lock that leads to the project management offices. "Don't forget it," she snaps.

I notice a photo of a young girl on her desk—same dark curls and a similar, slightly indignant look. "Is that your daughter?"

"Yes," she says, and lets the question pass. "You're scheduled to work out in the trailers." She nods toward the door. I've noticed rows and rows of small construction trailers lined up next to the building, their corners set on concrete blocks. They're similar to the trailers behind the junior high school, where I had my French class. "If you're here long enough, you'll get to work in the regular building," she says. "But right now we're on overflow and people who are just starting out have to work in the trailers. There are ten trailers, with letters from A to J. You're in Trailer 130F. You can't miss it."

I sling my purse on my shoulder and turn to leave. "Watch the purse!" she barks. "Keep it simple. Keys. Lipstick. You can be searched at any time. The guards will check your purse if and when they feel like it. No briefcases or anything like that."

"Okay," I say.

$12.92 an hour. I don't think I could make that much waitressing, but I'm starting to wonder exactly what this job might involve.

The interior of Trailer 130F is a maze of cubicles with gray fabric partitions. The carpet is worn, also gray with an intermittent pattern of coffee stains, and there is a scent of scalded coffee in the air. A woman sitting at a government-issue desk greets me with a smile and looks over my paperwork. "You'll need to read this," she says, and hands me a sheet with the words "Radiological Health and Safety" across the top.

"The DOE has established guidelines regarding radiation exposure," I read. "DOE Manual N 5480.6 and DOE Order 5480.11 expressly state requirements for radiation protection of occupational workers, unborn children, minors, students, and on-site members of the public. In addition to 'maximum dose' values, the DOE also requires that exposure to radiation be kept as far below limiting values as reasonably achievable." There is an acronym for this: ALARA. As low as reasonably achievable.

Everything at Rocky Flats has a number and an acronym.

Before the secretary can speak again, a voice sputters over the loudspeaker. "Attention. Attention." There is a brief pause. "The plant is currently experiencing winds of fifty-five miles per hour or more. Those in tents should secure the area and move into permanent buildings. Those in permanent buildings should remain inside and not leave the building. Thank you for your attention. Have a nice day."

The trailer shudders slightly as if to emphasize the situation.

"Oh dear," laughs the secretary, noting the expression on my face. "Don't pay any attention to that. The wind is always blowing out here. Every once in a while a car windshield gets blown out. Watch it when you open your car door—be sure to hang on to the handle."

"Okay," I say.

I'm directed to a small cubicle with a tiny desk, a phone, and a computer screen. I sit directly across from a distraught-looking woman with long dark hair. Once the woman at the front desk is out of sight, she leans over and whispers conspiratorially. "Don't worry," she says. "I'll help you. It's very confusing at first, but it gets better." She tells me she's worked at the plant for fifteen years. "I was here during the raid," she whispers. "I had to empty my file cabinet at gunpoint!"

"What raid?"

"You don't know about the raid?"

"No."

"Shh!" she says. "We're not supposed to talk. Have you had lunch?"

I tell her I haven't and she suggests we eat together when I get my break.

But we don't get breaks at the same time. A manager brings me a sheaf of documents. "You don't need to worry about trying to understand any of this," he says. "Just type it." Three hours later I'm given permission to head down to the cafeteria, a small lunchroom with a crew of three

Hispanic women making sandwiches and ground beef burritos. The tables are mostly empty. I order a burrito and walk out to the courtyard, which is sunny and provides a little shelter from the wind. I sit next to a much older man with a brown paper sack.

"Afternoon," he says. "You new?"

"Yes." I nod, and take a bite. "Today's my first day."

"Eleven years, here. Name's George." He takes a sandwich out of the brown sack and unwraps it. "How's the burrito?" There's a slim ring of grease on my plate.

"It's all right."

"I never eat in the cafeteria."

"Really?"

"Nope." He pulls up the top piece of bread to show me. "Turkey, white bread, no mayo. Every day for eleven years."

"You're kidding."

"Nope. I like to know what I'm getting." He laughs. "Besides, there's plutonium in that cafeteria."

I blanch. "Seriously?"

"No," George says. "It's a joke." He smiles. "I'm just joking."

"Pretty funny." I take another bite.

He finishes his sandwich and stands. "Maybe I'll see you tomorrow."

I often see him after that, with his sandwich and brown paper bag, and sometimes we eat lunch together in a comfortable silence.

After a while I start bringing a lunch, too.

Later—much later—I hear through the grapevine that there actually had been a problem with plutonium in the cafeteria. I also hear that it wasn't uncommon for plutonium to be carried in on people's clothes, even though we weren't near the hot areas of the plant.

I QUICKLY learn where I can and can't go on the plant site. Each morning I wait in a long line of cars at the east gate and watch the old shift drive out as the new shift drives in. I flash my badge at the guard who never returns my smile, drive over a gentle rise, and then into the low basin where Rocky Flats spreads out like a little metropolis. More than six thousand people work here. Buildings are numbered according to the type of work done there. Plutonium and other radioactive and dangerous materials are handled in the 300 and 700 buildings, machining in the 400 series, and administration is in the 100 area. Only those with a government Q clearance are allowed in the 300 and 700 buildings. There are several cafeterias, a medical center, and a firehouse.

Once again Randy Sullivan and I pass within a stone's throw of each other.

Randy's dad had been a captain for Continental Airlines, and like his dad, Randy enjoyed being around planes. After high school he moved around a bit and then returned to Denver to work as a mechanic for a small airline. He married and had kids, and began to think about a more ambitious, permanent career. Someone mentioned to him that Rocky Flats had a fire department, and they were hiring. The pay was good.

He'd always dreamed of being a firefighter. And Randy was familiar with Rocky Flats. One of his best friends growing up had a father who worked at the plant. No one knew what he did, and no one knew what Rocky Flats did, but it was a good way to support a family.

Randy filled out an application. A few weeks later they asked him to come out to the plant for a physical agility test and an interview in which he was asked about why he wanted to work at Rocky Flats.

Driving into the plant for the first time was a little intimidating. *Finally,* he thought to himself, *I get to look into Pandora's box.* He stopped at the guard gate for his temporary clearance, and he was given a map of the facility. The plant was larger than he expected, and he got a little lost trying to find the fire station. He was acutely aware of the guards with guns. *I hope I find it quick,* he thought. *If they see me just driving around, I might get into trouble.*

But he eventually found the fire station. He passed all the tests. He was tall, physically strong, and although he had been relieved to escape the Vietnam draft, he was pleased to serve his country at Rocky Flats. On July 1, 1991—the same year that Russia and the United States agreed to dismantle approximately thirty thousand nuclear warheads between them—Randy Sullivan officially became an EG&G employee. He was thirty-three years old. It would be his last job, he thought. It was a real career, and it would make his family proud.

Randy wasn't completely unaware that Rocky Flats was involved in nuclear activities. They'd told him he would be a nuclear firefighter rather than a regular firefighter. He thought that was kind of cool. Maybe, he thought, he would actually get to see some plutonium.

By the time I go to work at Rocky Flats, Randy's an old hand. He's been there for years.

SEPTEMBER AND October can be cold, windy months, and on bitter days we wear sweaters and knit gloves with the fingertips cut off in Trailer 130F. I endure a couple of weeks in the trailer before I'm promoted, thanks to my quick typing speed, to a more permanent position in the administrative building. The main building has heat and a little more status. I earn a few icy stares when I graduate from Trailer 130F.

My new digs are essentially the same, although the carpet is less worn. I am in a cubicle in a sea of cubicles. Only the managers—all male—have offices with windows on the perimeter. The two top managers from EG&G and the DOE walk around in crisp shirts with buttons pinned to their lapels: IT'S THE PLUTONIUM, STUPID, a play on President Bill Clinton's successful presidential campaign slogan, "It's the Economy, Stupid." The DOE, with Hazel O'Leary serving as secretary of energy, claims a new honesty and openness. Even so, the truth about Rocky Flats—what's stored there as well as the leaks, fires, accidents, and contamination problems—is classified and most people know little about it.

My immediate chain of command involves a team of two women who have been around for years and worked their way up the ranks. Debra and Diane are both senior administrative assistants who work directly under the managers, and they wield their power together, ruthlessly and with rigor. Everyone from project engineers to file clerks is more than a little terrified of Debra and Diane.

On my second day in the administration building, Debra, slightly younger and a little less authoritative than Diane, takes in my appearance and offers me some advice. "I know where you can get your nails done right," she says. "If you're going to work here, you should get your nails done. You don't wear much makeup, do you?" Both Diane and Debra retouch their nails and page through fashion magazines when the managers are out of the office. They flirt with the project engineers. They're famous for getting new secretaries transferred out immediately if they don't like them.

Diane is downright dictatorial. "You can't leave your desk without telling one of us," she says. "Not even to go to the bathroom. We need to know where you are all the time."

"I need permission to go to the bathroom?" I ask, incredulous. I'm not sure if this is Cold War security or just gratuitous hazing.

"Permission is needed for everything here," she barks.

I slink back to my cubicle and privately make a note on a yellow stickie: *Permission is needed for everything here.* And *Diane wears enough perfume to gag a goat.* I put the stickie in my purse.

Even though production of plutonium triggers has ceased, stockpiled triggers are still being shipped to laboratories in California and New Mexico for analysis to determine their lifespan. They're transported in specially designed high-security trucks that travel on roads and highways escorted but unannounced. And as in the past, Rocky Flats is involved in other work related to plutonium recovery and defense and weapons production.

No one talks about it, and managers keep their doors closed.

Some of the secretaries have husbands or boyfriends who are guards, or who work in the hot areas. Debra dates a guard. Guards have a reputation among the secretaries for being buff. Working out at the company gym for several hours every day is part of their job. There are strict divisions between employees at Rocky Flats. Blue collar versus white collar, DOE versus EG&G, managers versus hourly workers, guards versus firefighters, men versus women. But we're all Cold War warriors, or at least that's what people like to say.

I feel like an outsider, a rebel in hiding in more ways than one. But the mundane schedule, one day the same as the next, has a kind of mind-numbing comfort to it. Four hours each morning, half an hour for lunch, four hours in the afternoon, with two evenings and one day a week for my classes at the university. I type memos and letters and meeting minutes, and with my promotion-of-sorts I'm now tasked to type the weekly "Hot List," a list of "incidents" or problems, milestones, and events that is sent to the higher-ups at the DOE in Washington at the end of each week. Everything is expressed in acronyms and euphemisms. An ROD is a "record of decision." An OU is an "operable unit," an "environmental restoration unit." (On a Superfund site, areas to be cleaned up are divided up into OUs.) An IHSS is an "individual hazardous substance site." I learn that MUF is "material unaccounted for," that is, missing plutonium. I write about solvent spills and steam leaks and problem solar ponds, the 881 Hillside and its secret long-buried waste, and the West Spray Fields, where contaminated waste is sprayed out onto open fields: contaminated groundwater and carbon tetrachloride, radionuclides, and bacterial waste; sewage sludge and plutonium- and uranium-contaminated waste and plutonium-bearing nitric acid solution.

I hate being a secretary. Word processors are standard, but I miss the old Selectric at my dad's office, where I could bang the keys and get a satisfying whir of the type ball and a firm clunk of the letter on the page. Some memos and letters I type have my initials, lowercase, at the bottom, a tiny emblem of a young would-be writer whose initials fly off into the world under the signature of someone more important.

I don't know what all the acronyms mean, and no one is eager to explain them to me. Frankly, I'm not sure I want to know. The Kafkaesque language has an anesthetizing sameness to it that's both frightening and comforting. And I'm a little ashamed to admit to myself that perhaps I don't particularly care. It's a good paycheck, with decent hours, and I need the money.

There is a sense of bravado among the employees and I feel that, too. I don't talk about Rocky Flats with anyone outside of the plant—not my family or friends, especially not my friends at the university. I say nothing to Sean and Nathan. I just know that I'm spending my days working next to some crazy amount of plutonium. It hasn't killed me yet, I joke with the other secretaries and administrative assistants. We're tough.

Yet part of me is petrified of the place, and always has been. Still, I want to see. I want to understand. I want to get on the inside and figure it all out. So I do what I always do: I take notes. At first it's on envelopes and napkins and note pads and Post-its that I cram in my purse at the end of the day. Gradually I get a little bolder. Rarely is my purse searched; I can flirt with the security guys just as well as anyone else. I've been getting daily lessons from my immediate supervisors. I buy a small notebook and start keeping a daily journal. I am the post–Cold War Harriet the Spy, reporting from the front lines. Except that the Cold War isn't really over. Here, just three miles from my childhood home, it's alive and well.

Debra is full of advice on everything from fashion to dating. "If you ever have to go into one of the hot areas," she advises, "take off your bra before you go. Those guys set the checkpoint so high that a bra with an underwire will set it off, and they like to make you take it off. They're real bored down there. Watch out for the guards." I intend to ask her if she follows her own advice, but she doesn't give me the chance. "And watch out for the activists at the east gate when you come in. They're always there. Boulder crazies. They'll wave a sign at anything. Have you seen those kids with petitions? They don't understand the issues. They're just making a buck. They get fifty cents per name. It's nothing but kids, hippies, and housewives."

I hear the echo of my father in her words.

"Is it true what they're saying about contamination?" I ask. "Is it really polluted out here?"

A look of anger crosses her face. "You'd have to ask a scientist. I don't know. Who am I? Some of these guys really know what's going on, but they don't talk."

I nod. One thing I do understand is silence.

"I've worked here for a long time," she says. "Sure, there's pollution all over the place. But I know someone who's worked down in the Zone for thirty years. And there's nothing wrong with him. Not a thing."

THREE WEEKS later, on October 8, there is a serious "incident." It takes place in Building 771 and involves the unauthorized draining of a process line containing plutonium-bearing nitric acid. Six days pass before the accident is reported to senior management. All plutonium operations immediately come to a halt. Three employees are terminated for failing to adhere to prescribed procedures, violating safety procedures, and other violations of the plant standards of conduct.

I wouldn't have paid much attention if one of the managers hadn't offered to take me to lunch. And the lunch, it turns out, has an impact on my social status.

Hourly employees are required to stay on the plant site during their entire shift. Managers are excepted from this policy. One day Mr. K, a manager who seems very nice and profoundly unsuited for his job, asks me to lunch. He likes to stand around and chat, philosophizing about everything from politics to books to trying to guess how many people the DOE will lay off from one week to the next. "They always hire them back," he says. "It's never very long. Weeks. Days. Hours. Budget up, budget down." He smiles. "I'll drive you to Boulder," he says. "I know a nice French place."

No one looks up as I follow Mr. K past the row of desks and partitions. On the way out he shows me stacks and stacks of empty wooden containers, piled in rows behind the parking lot. "Those are from the pondcrete containers. Do you know about pondcrete?"

I shake my head.

We drive without comment to the restaurant and order before Mr. K begins to talk. I'm a little suspicious of his motives—the office is always buzzing with rumors of who might be having an affair. I guess it's a distraction from wondering who might be next on the layoff list or who might be bringing in a little plutonium on the soles of their shoes. One manager, a short, self-assured man with a solid paunch, spends a good deal of time in the elevator with one of my cohorts, a prim secretary with mincing steps, tinted orange hair, and a ruffled blouse that's a little too revealing. It's a two-story building and the elevator occasionally seems to get stuck between the two floors when they're inside. People talk.

But Mr. K has no such motive. He wants to talk about his job.

"I don't belong there," he says. "I never did. But what can I do? It's a waiting game. Everyone's waiting to get laid off. The salary and the benefits are too good to just quit."

I nod.

"You can't trust anyone, you know," he says. "No one's really accountable for anything. Everything is done by committee."

I've heard this said before. We have sparkling water with lemon and rosemary chicken and chocolate napoleons for dessert. I've definitely used up my thirty minutes.

Over napoleons, Mr. K explains the pondcrete. The solar ponds. The 903 Pad. The spray irrigation. The leaking plutonium processing line. "The plant is a mess," he says. "When the raid happened, everything just stopped and plutonium was stuck on the production line. Plutonium is stored in various stages all over the plant. It's nothing but a big shell game."

I recall how some of the employees have talked about the big sprinklers used for the spray irrigation. "That water is green!" someone joked. "Maybe it's the guacamole from the cafeteria!" I've heard talk of all the stranded plutonium as well, but Mr. K is starting to sound a little paranoid to me.

"I never go down to the hot areas," I say. "I can't, anyway. I don't have a Q clearance."

He puts down his fork. "You don't plan to make a career of this, do you?"

"No," I say. "I'm just a graduate student who needs a job."

"Good," he says. "With luck we'll both be out of here."

We get back to the office just as the overhead lights start flickering. A voice comes over the PA system. "If the lights go out, do not be nervous," it says. "I repeat. Do not be nervous. Technicians are working."

"It's the commies," Mr. K says, winking.

The PA comes on again. "Thank you for your attention. Have a nice day."

IT'S BEEN a year since I've seen my father. Halloween nears, and Sean and Nathan decide they want to dress up like puppies, with big ears and brown splotches like Heathcliff. One day after school we drive to the craft store so I can buy patches of felt. I'm no seamstress, but I think I can glue brown and black patches of felt onto white sweatpants and T-shirts. With makeshift tails and ears, it might work.

As we pull into the entrance of the shopping center, a yellow cab suddenly appears and veers sharply in front of me. I hit the brakes. For a moment I'm sure we're about to be hit—and then the cab is gone. But not before I catch a glimpse of the driver.

I coast into a parking spot and turn off the engine. My hands are shaking.

"Are you okay, Mom?" Sean asks.

"I'm fine." I rest my head for a moment on the wheel. "I know that person."

"Who? The person who almost hit us?"

"Yes."

"Who was it?"

"That was your grandpa, honey." I regret the words as soon as I say them. "He didn't see us," I say. "He didn't know who we were." I start the engine and we drive home, our errand temporarily forgotten.

DEBRA AND Diane decide that I am a friend.

Debra wears high heels at work—three-inch minimum—but she keeps a pair of tennis shoes under her desk. On her lunch break, if the weather is nice enough, she walks briskly around the plant for exercise, all the way down the hill, past the 300 and 700 buildings with their chain-link fences and razor wire, and up the other side, where the buildings are more open. "Join me," she says.

It feels like an invitation to a secret club.

On sunny autumn days, it's a breathtaking view. On one side lie the mountains; on the other, a landscape dotted with houses that stretch all the way to Denver. "The air is so clean here," Debra says. "It comes down right off the mountains." We catch glimpses of rabbits and groundhogs in the grass. A pair of bald eagles has been sighted near Standley Lake.

We walk past a large, flat graveled area cordoned off with what looks like yellow police tape. A few oil barrels stand upright, and parts of the area are under a tent. It looks like they're preparing for a wedding or a rock concert.

"What's that?" I ask.

"Oh, that," Debra says. "That's the 903 Pad." She walks quickly, arms moving up and down to keep her heart rate up. The thousands of barrels are gone and parts of the area are covered with gravel and asphalt.

"Why is it roped off?"

"There's some plutonium that leaked out there."

I reach out and touch the yellow ribbon. I'm struck by the memory of my sister Karma and me, riding our horses around the perimeter of the plant, kicking the No Trespassing signs with the toes of our cowboy boots.

"What's the difference between one side of the ribbon and the other?"

"Oh, we don't have to worry about that," Debra assured me. "They say this side is safe."

"How does the plutonium know to stay on that side of the line?"

"It knows. Plutonium doesn't travel."

When I return to my desk, Anne—the secretary who greeted me on the first day, with a photo of her daughter on her desk—is ruefully watching the phone lines. "I'm holding down the fort," she chirps. "Everyone's still at lunch." Anne has warmed up to me, too. I've also discovered she's a little more subversive than the others. She asks whether I've been out walking and I say yes. It still feels odd to be wearing a skirt with white socks and tennis shoes, but there's no place to change clothes. A little bit of sunshine at lunch makes it easier to sit in my cubicle all afternoon. It helps keep up my energy, which has been lagging lately.

A week earlier after class, I stopped in the student health center to meet with a doctor. "I don't know what's wrong with me," I said. "I don't feel right. I'm always tired, and it's been going on for a while."

They take some blood tests. I'm waiting for the results.

"Did you see any of those Preble mice?" Anne asks. This is a running joke in the company. Recently the EPA started a petition to protect the tiny Preble's meadow jumping mouse, possibly the rarest small mammal in North America, which apparently likes to live in the Rocky Flats buffer zone.

"I guess they're too small to see," I joke.

Anne's not joking. She leans forward. "Here's the thing," she whispers. "They're more concerned about protecting some damned rodent than they are about protecting people."

I DISCOVER a kindred spirit at Rocky Flats. Patricia is also a graduate student at the University of Denver, working as an administrative assistant. She plans to quit at Christmas. "I'm out of here," she says. "This place is looney tunes. But it's good money." She's smart and funny and, in her dark-rimmed glasses, already looks like an English professor. We meet for lunch on the patio outside the administration building and gossip about our departments.

One day she brings a friend along, a technical writer who's here for only a few months. He's tall and skinny and his dark-rimmed glasses match Patricia's. We break into our brown-bag lunches and start talking about what really goes on at Rocky Flats.

"It's not actually a bomb," Patricia says.

"Right," I say.

"Well, what is it then?" he asks.

"A pit," I say.

"That's a bomb," he says.

"No, it's not," I say. I should know. I've been typing pages and pages about pits.

"A pit is only a critical component of a nuclear bomb," says Patricia with authority. She's been typing pages and pages, too. "It's not the bomb itself."

He laughs. "Are you girls kidding me?" He cracks open a soda.

"No," we reply in unison.

"That's like saying that water is only a critical component of the ocean. Or that the planets are merely critical components of the solar system." He pauses. "There's no bomb without the pit," he says somberly. "The pit is it."

Patricia shoots me a look: he takes things a little too seriously, doesn't he?

We finish our sandwiches, toss our crumpled bags in the trash, and go back to work.

ONE MORNING, one of the managers stops by to say hello. "Here's a heads-up, Kristen," he says. "Be prepared. You might come in some morning and be told to go home because there's no money. It's budget time, and there might not be enough funding through procurement."

Be prepared? I'm living paycheck to paycheck.

On November 4, a memo from the EG&G manager of Rocky Flats warns that if negotiations between the private company that provides security services for the plant and the union do not come to a mutually agreeable conclusion, "a guard force work stoppage could occur." There's no mutual agreement. The guards go on strike. People work on staggered schedules and double shifts. Some guards are in favor of the strike; some are against it. No one knows if they'll get hired back.

All of us are nervous about our jobs in one way or another. I read in the company newsletter that four hundred to seven hundred people will "voluntarily or involuntarily" leave their jobs by the end of 1995. The blow is softened somewhat for some permanent employees by severance packages and educational and training benefits, including a two-and-a-half-hour seminar addressing the phenomenon of LCS, or "layoff casualty syndrome."

In the women's bathroom, someone's taped to the mirror a newspaper ad for the Denver Rescue Mission that says, "Give the Gift of Food This Christmas: Buy a Hot Meal for a Homeless Person." Over the photo of a man with a plate of food, the words "Rocky Flats Employee Picture Here" are written in large black letters.

I look at myself in the mirror. My hair is pulled back and my face looks harsh in the white light. There are deep circles under my eyes. My mother often tells me how tired I look.

The guard strike ends, but no one seems happy about it.

There's some good news, however. The plant is buzzing with a report that Hazel O'Leary, the first woman and first African American to serve as secretary of energy, will visit the facility. I hope to see her; under her directorship, the Clinton administration has released millions of previously classified documents related to the Cold War—a move that's been met with skepticism by my co-workers. Outside the plant, though, she's a folk hero of sorts, famous for saying, "This is not your father's DOE." But on the day she visits, although most of the managerial and engineering staff get to hear her speech, I'm left behind to handle the phones.

"Hey, don't sweat it," my turkey-sandwich lunch friend says in consolation. "She says all the right things, but she's still administration. Do you think she'll pay a visit to the hot zone and see how things really are? No. The managers, the administration, they never go down there. They don't go down in the bowels of the plant. They keep their hands clean."

I'm never sure what's truth and what's hearsay.

November turns into December. At the end of a long day I drive home and pick up Sean and Nathan from the babysitter's. Her name is Jennifer and she's in high school; I pay her to watch the boys for a couple of hours after school. Today she's frazzled. The boys are tired and fussy and a little wild; it's been a long afternoon. She hands them over to me with no small sense of exasperation.

They fuss and wiggle and refuse to get in their car seats. I'm tired, too. I lay down the law. "Sean and Nathan! If you don't settle down, I'm going to take you to the zoo to live with the wild animals!"

They get in their seats. But nothing will settle them down until we arrive home and have plates of spaghetti with meatballs and chocolate milk, and their eyes grow big and sleepy. Sean does his arithmetic problems and I help Nathan practice his spelling while he takes his bubble bath. He soaps his hair up into a spiky Mohawk as I sit on the floor and read him his words.

I tuck them into bed. They're too sleepy for a story, and I'm too tired to read to them.

"Mom?" Sean asks as I turn off the light. "Did it take you this long to grow up when you were my age?"

I pause. He's five years old.

He doesn't wait for an answer. He has another question. "You're not really going to send us to the zoo to live with the wild animals, are you?" He looks like he's given this some serious thought.

"No, sweetie," I say. I feel a catch in my throat, and I kiss his forehead. "We're all staying right here."

I go downstairs and take off my shoes. I make a cup of tea, stretch out on the couch, and turn on the television. I'll give myself a few minutes before I go to bed.

Suddenly I sit bolt upright. Rocky Flats is on television.

ABC Nightline is interviewing people I know. The narrator, Dave Marash, talks about years of contamination at Rocky Flats, and how production was halted after the 1989 FBI raid. Since then, Rocky Flats has been in a state of limbo, wanting to resume building nuclear weapons while trying to deal with environmental regulations the plant had been able to avoid in the past. Rocky Flats has five of the nation's top ten most dangerous buildings in the country, Marash says. Building 771 is number one. Building 776 is number two.

Marash reports that an internal memo shows that as much as 13.2 metric tons (or 14.5 U.S. tons) of plutonium may be stockpiled around the plant, including more than five thousand sealed containers of waste, many containing a buildup of hydrogen gas that can cause a container to rupture and scatter plutonium. Cans that were not supposed to be stored for more than a year have been stored for five. Mark Silverman, the DOE manager at Rocky Flats, appears onscreen. I know his voice well from the PA system at work. "We know, for example, it's in the vents. It's in the ductwork. We know it's in the glove boxes, in the lathes. We know it's in the walls and ceilings. We just can't tell you exactly how much is at any given location in a lot of places." Silverman adds that there was very poor record-keeping at Rocky Flats, and "we do not have as-built drawings. So a building was built, and then added onto, and we literally don't know where every pipe is or every line is."

Have a nice day, I think. I grab my journal and start scribbling their words.

"This may look," Marash says, "like an anonymous stretch of

asphalt." My heart jumps. That's the 903 Pad that I walk by on my lunch hour. "From here," he continues, "contaminated groundwater leaked down the ridge towards the plain and the northern and western suburbs of Denver. Some of the barrels rusted and started leaking . . . and the migration of toxic waste can be traced on a map of drainage patterns in the Rocky Flats area. Walnut Creek drains down and dumps into Great Western Reservoir, and then Woman Creek comes down and feeds the Standley Lake reservoir. Samples from the bottoms of both reservoirs show deposits of plutonium. The plutonium traveled through the water and through the air." Dr. Gale Biggs appears on screen, noting that, according to the findings of Dr. Harvey Nichols in a report to the Department of Energy, "the plutonium levels do not drop off as you go farther away from the plant."

Marash interviews Jim Kelly—the longtime worker at Rocky Flats who was on the roof during the Mother's Day fire—who shakes his head. "It was production, production, production," Kelly says. "Safety was a word. It wasn't really practiced. The job was to get the product out the door, and if you got it done safely, okay, and if you didn't, they'd turn their head."

Marash then talks about the grand jury investigation that began in 1989 after the FBI raid. He interviews Ryan Ross (also known as Bryan Abas), the journalist who broke the story of the runaway grand jury to the press. "The jurors thought that anybody who'd committed a crime should be held accountable for it," Ross says. "They didn't care whether they worked in the federal government, or in the private sector, or how high up in the government they were." He notes that a dozen sections were taken out of the jury report. "Almost all of them had to do with the conclusions of the jury that the illegal conduct they found that Rockwell was engaged in was continuing to be done under the successor contractor [EG&G]."

One of the grand jurors appears onscreen. "I had nightmares, you know. I couldn't sleep at night, thinking about what I had heard for a whole week in that jury room."

Paula Elofson-Gardine, a resident who's lived downwind from the plant since 1964, notes that housing development around Rocky Flats continues to grow as home developers lobby the county planning commission. "The greed of developers," she says, "is matched only by their customers, homeowners who are kept ignorant thanks to the sealed grand jury report."

I pace the dark living room for an hour before putting on my nightgown. So many of the things I feared, or were afraid to even think about, are true. It's real, and it's still going on. I take up my journal again. *I just saw Rocky Flats on* ABC Nightline. *Oh my God. I can't sleep.*

I turn off the light and wait for the morning.

THE NEXT morning I drive into work expecting the world to have changed somehow, and it has not. Deer are grazing close to the road and I glimpse the orange tags on their ears. A light dusting of snow covers the grass, pink in contrast, and the clouds overhead are dark in the morning sky. The mountains are deep blue, almost black. On days like this the beauty of the land, of what is now the buffer zone, is stunning.

People are quiet at work. There's the comforting click-clack of fingers on keyboards, a slight scent of nail polish in the air. A box of doughnuts stands open on one desk and a copy of *People* magazine peeks out from a stack of papers on another. The managers' doors are predictably closed.

But later that afternoon, when many of the managers are out of the building, people gather around Anne's desk. Her desk occupies a semi-neutral zone, where people are freer with their comments. It also has a straight-line view of the front door, so we can scatter quickly if needed. Everyone has seen the *Nightline* report.

"It's true that 771 is a mess," says one of the project engineers. "It's an old building, and a dangerous building. You have to be real careful down there in the hot zone. The hourly workers get some compensation, but what's the point? Those guys take a lot of that stuff home with them, whether they know it or not."

"If someone's afraid to be plantside, or afraid to go down to the 700 area, then they shouldn't be working here," a secretary snorts. "It's more dangerous working at a Federal Express office than here. We have strict safety rules—"

"The media?" someone interrupts. "You believe them? Oh right.

Some bubble-headed bleached blonde, airing everyone's laundry. I don't believe a word of the media."

"I don't know what to believe," Anne says. She pauses. "How is the public supposed to know what to believe?"

"There is a lot of waste at Rocky Flats," adds an older woman with beehive hair. "A lot of time and energy wasted, too. As a Christian, that's hard for me to deal with. It's the taxpayers' money, after all. But someone's got to have this job and these benefits. Someone's got to do it. Why not me?"

Another guy thinks that the whole thing has been orchestrated. "They manipulate the press to get more money out here," he says. "Now there will be fewer layoffs and more money from Washington. It's good for us in the long run."

I don't say anything. I'm afraid to open my mouth. Inside I am shaking with anger and fear.

"Well," Anne says, "I guess the only certain and eternal things in life are taxes, death, and Rocky Flats." Everyone laughs. It's a saying that's often repeated at the plant.

LATER THAT afternoon, just before I leave for the day, Debra catches me in the hallway. "I have something for you!" she whispers. Her eyes are dancing. "It's in my car. I'll meet you in the parking lot."

For once there's no wind. The air is sharp with the slightly metallic scent of snow.

"Here!" Debra announces. From her backseat she extracts a large platter, black faux-marble plastic, wrapped in cellophane. It's stacked with Christmas goodies: jam cookies, peanut butter thumbprints, sugar cookies with brightly colored frosting, braided bread, and six or seven other things, all tied with crimson ribbon. "I made them all myself!" she declares.

I am overcome. It must have taken her weeks to do all that baking. "Thank you, Debra." I feel bad. I hadn't thought of anything for her.

"I want you and your boys to have a nice Christmas," she trills. She jumps in her car and waves out the window.

I pick Sean and Nathan up from Jennifer's and they jump with delight at the tray that takes up the entire front seat. We get home and I set it on the kitchen counter. I let Heathcliff in from the backyard and take the boys upstairs to wash up for dinner.

I come back downstairs and the tray has vanished.

"What's this?" I exclaim. It's disappeared into thin air. I walk into the living room, where Heathcliff is sprawled on the couch, half asleep. He's no longer a puppy; his basset hound barrel of a body takes up half the sofa. One paw drops languidly off the edge.

It takes me a moment to realize that the tray—with not one crumb remaining—is on the floor next to him.

The next day I see Debra at her desk. "How did they like it?" she asks. "Did the boys like the cookies?"

I can't bear to tell her the truth, or even think about what the boys' faces looked like when they came downstairs for their dinner.

"They were—are—delicious," I say. "Everyone in my household just loved them!"

CHRISTMAS ARRIVES, and with it—just days before—a paycheck big enough for me to buy gifts. The Sunnyside Temp Agency is raffling off holiday turkeys. I don't win, but I get a coupon for ten dollars off a turkey at the local grocery store, and that helps. We have a real Christmas tree that takes up half the living room, and Sean and Nathan argue about whether or not the dove on the top of the tree is really a chicken. Jasmine shimmies up the tree trunk, hiding in the branches, and occasionally a paw shoots out to take a swipe at a Christmas ornament.

All that negative stuff about Rocky Flats is just a bad dream.

And then I have a real dream about Mark. I haven't thought about him in a long time. The dream takes place on a dark, windy night, and for some reason I'm working late at Rocky Flats. The wind is intense, and I go out to crack the windows on my car, which I often have to do in real life at the plant to prevent the windows from blowing out. I see someone else in the dark parking lot, walking toward a white van. Is it another worker? The person turns, and I see it's Mark.

I've aged, and he hasn't. He's still in his early twenties. I look into his face, into his eyes. His face is the clearest I've ever seen it in a dream. I reach out and hug him, and it feels exactly as it always felt, with my hipbones just below his leather belt.

He pushes back and looks at me. He wants to tell me something, but before he can speak, the dream ends.

FIRE, AGAIN

1991–1996

When Randy Sullivan began work at Rocky Flats in 1991, he discovered a world unto its own.

For his first eighteen months, he worked in the fire prevention department, learning the layouts of the various buildings. It was confusing at first. The confusion was intentional, he learned; many of the plutonium buildings had been designed so as to slow down anyone—terrorists, for example—who didn't belong there. It was like an old medieval village with twisting streets meant to prevent invaders from getting to the palace. But that made it difficult for employees and firefighters, too. Building 881 was a serpentine maze of curves and twists. Building 371 had three underground levels with multiple staircases that led to different points in the building. All the buildings were filled with large machinery in close quarters, which made it difficult to move around. The fire protection systems were all located at the top.

There were fourteen firefighters and three officers on each shift. Randy's schedule was exhausting: twenty-four hours on, twenty-four hours off, twenty-four hours on, twenty-four hours off, twenty-four hours on, and then four days off. Like the guards, he and the other fire-fighters were expected to work out at the company gym and keep themselves in shape.

Randy's education in hazardous materials included special training in how to extinguish a fire in a glove box. Since water was off limits, he was taught to use a glove-box entry horn, a fire extinguisher with a special fitting, like a plastic bag, that tucked around the part of the box that held the arms of the lead-lined gloves. Randy learned to use a knife to cut out the lead gloves, push the horn inside the glove box, and discharge the CO_2 extinguisher.

Randy was also trained as a fire medic, which meant he could respond to cardiac arrests and major traumas. Occasionally the Rocky Flats firefighters were called to handle car accidents on Highway 93, sometimes fatal, on the west side of the plant, even though it was off-site. Highway 93 was particularly treacherous due to the high winds and extreme weather conditions in the area. In the wintertime, the highway could turn into a skating rink of black ice, and winds of one hundred miles per hour were not uncommon. Employees' cars parked in the parking lot often looked like they'd been sandblasted.

Fires at the plant, though, were his main concern. There had been more than two hundred over the years, and he'd heard stories about the 1957 and 1969 fires and how close the plant came to a significant radioactive release. No one really seemed to know the facts. "Can you imagine what would have happened if we'd had a release in this kind of wind?" he said to a friend on a particularly windy day. "There'd be no one left from here to New Mexico!"

There was a tendency to downplay the fires to the public, particularly in the late 1970s. When a reporter learned of a fire on a loading dock outside a plutonium waste-processing building, Rockwell's director of information services, Felix Owen, told the press it hadn't been reported because it was small and "people are scared of fires at Rocky Flats. I don't need to upset them about a little trash fire on a dock." Sometimes firefighters from other firehouses were brought in to fight fires as well. The Arvada and Fairmount fire districts responded repeatedly to fires when thousands of leaking drums stood out on a windswept field at the Rocky Flats Industrial Park, near the plant. Firefighters fought those fires not knowing what was inside the drums or even knowing if it was safe to use water. Acrid smoke from the fires rose into the sky and floated over nearby communities.

Despite the dangers, Randy enjoyed his job. He liked the camaraderie with the other firefighters and the guards. Each year they had a volleyball tournament at the site, the "hose draggers" versus the "pistol monkeys." Despite the intense competition, the guards and firefighters were like brothers. They looked out for one another on the job. At the firehouse, the firefighters cooked meals together. They had a garden just outside the firehouse, with roses that "nothing can kill," but they didn't bother growing tomatoes—who knew if it was safe to eat them?

The mood at the plant could be prickly. During the Gulf War in 1991, it could take as long as an hour and a half to get through the security gate while the guards conducted car searches. Randy didn't mind—he knew it was necessary. And not everyone supported the mission of Rocky Flats. Once when they drove the fire truck through a residential area in Boulder, people threw eggs.

On the other side of the situation were the employees who weren't happy when it became clear that the plant would not resume production of plutonium triggers. Randy had started work two years after the FBI raid shut down most production at the plant. Hiring, ironically, was at an all-time high. Then the news came down that the military mission had been canceled and Rocky Flats might be torn down. There was talk about trying to clean up the site. After all he'd learned, and all he'd heard, Randy didn't think it was possible. He and his fellow firefighters thought it would be somewhere between thirty and fifty years before any serious cleanup could occur. There was no way they could close the entire place down, clean up all the buildings, and tear 'em down and make it prairie again, he said. His buddies agreed.

But despite all the complaining, most people did the same work, kept the same salary, got the same benefits. It was the same job. Just a different mission.

And things were looking up for Randy. He was promoted. Now he was a captain like his father, just a different kind of captain. He was proud of the work he did. He knew all those buildings like the back of his hand.

ONE DAY Diane asks me to lunch.

She's been planning it for days. "Let's go to Boulder," she says, and I'm surprised. I didn't know she had off-site lunch privileges. But she's my supervisor, so I agree, and we drive to a Mexican restaurant. She orders a small margarita for each of us—"Just one!" she says, laughing—and we talk about our kids. It feels odd to be suddenly so familiar with her. She's treating me like an old friend. I'm not quite sure what to think of it. She's tough, almost harsh, and yet fiercely loyal to her family and even to Rocky Flats. When she speaks of her daughter, her voice is tender.

Our enchiladas arrive, and the margaritas start to take effect. "You know," she says, leaning forward, "I might not be here in a year."

"Really?"

"I need to get a different job." She waves her fingers in the air. "It's my husband. He's taking voluntary separation from the plant." Diane's husband works in one of the more dangerous areas at Rocky Flats. "He wants to leave. He wants us both to leave." She sighs. "But it's different for me on the outside. I'll never be able to get a job like this." It's not just the pay but the benefits. Once you've worked at Rocky Flats, it's not uncommon for other companies to consider you a health risk.

"I don't believe that," I say. "I'm sure you'll find something."

The waiter fills our water glasses and brings extra sour cream. I've ordered blue corn enchiladas with black bean relish and sopapillas, my favorite. Diane's plate, too, is filled to overflowing. It will be hard to go back to work.

"I'll end up being a secretary in a real estate office or something. At what? Five bucks an hour?" She shakes her head and laughs in disbelief. When she leans toward me again, her eyes are friendly but intense.

"Listen," she says. "There's a lot of unethical stuff going on down in 771."

"Really? Like what?"

"Bad stuff. Immoral stuff."

I think, *Do I really want to know this?*

"There's no responsibility," she continues.

"What do you mean?"

"Covering things up. CYA stuff."

"For what? Accidents?"

"They cover things up and make them look better than they are," she says. Her face is flushed. "The whole plant. You know what I mean? I'm not supposed to talk. My husband's not supposed to talk, either."

I look down at my plate.

"It's all about money," she says. "No one wants truth."

I nod. I'm there for the money, too.

"Listen," she says again. "You've got options. I don't. You're in school, you'll get your degree." She pushes back her empty margarita glass. "If I were you, I'd get out of here," she says. "Understand?"

The check arrives, and Diane pays. We rise, and by the time we reach the door, it's as if the conversation has never happened. She turns to me and smiles her professional Rocky Flats smile. "Thanks so much for having lunch with me today, Kristen."

I thank her, too, and we go back to the office.

Later that night, I record everything in my journal. *I hate it at Rocky Flats so badly I can hardly stand it.* I have applied for part-time teaching jobs—I won't graduate until the spring—but nothing has turned up. And I still don't feel well. Normally a night owl, I've been going to bed at eight, right after I tuck the boys in. I take vitamins and protein powder and do my yoga exercises. Nothing seems to help. I see yet another doctor, and they can't figure out what's wrong. My glands hurt. My lymph nodes are swollen. I'm always tired.

Life feels like a treadmill.

TAMARA SMITH, the girl who grew up near our house in Bridledale, is several years younger than I, but our lives have followed similar paths. She rode with the same local riding club, and graduated from the same high school, Pomona High. She'd always dreamed of being a teacher, and she attended Brigham Young University in Utah. In addition to getting her degree, she met her future husband, David Meza.

As a child and a teenager, she was plagued with severe allergies, but her health seemed to improve slightly when she was in college. In 2000 that changes. She's always had headaches, off and on, but these seem different. She's just started teaching, and she thinks the headaches might be from fatigue. She makes light of them with her friends. "Do *you* have the medicine that will take this headache away?" she asks. Nothing ever seems to work. There's also something odd about her hands, which are covered with some kind of eczema, like a burn. When she's teaching a class, sometimes people come up to her afterward and ask, "What happened to you?"

Nothing has happened to her.

Tamara comes from a family that doesn't go to doctors. They still live off the land on Standley Lake, grow their own vegetables, raise their own beef cattle, and like to do things their own way according to their faith.

But after a long period of health problems and intensifying headaches, Tamara goes to see a doctor in 2000. Soon she's going every month. At first she's told it's just asthma. They tell her it's stress, it's fatigue, it's her imagination. It's pneumonia, then bronchitis, then mono. But the tests all come back negative.

There are several frantic trips to the emergency room when Tamara can't breathe or faints repeatedly. She's diagnosed with a thyroid problem. After another trip to the emergency room, her husband insists on a CT scan of her brain.

They find a tumor in Tamara's brain the size of a large lemon.

PETER NORDBERG began his new job as a young attorney in Philadelphia in 1990. He was thirty-four years old, just a few years out of law school. Little did he know that he was going to spend nearly every waking hour of the rest of his life on the Rocky Flats class-action lawsuit.

Following the FBI raid on Rocky Flats, in the summer of 1989, a group of deeply concerned residents from Arvada and the surrounding area gathered to discuss whether it might be feasible to sue Rockwell and Dow Chemical over the migration of plutonium onto their properties. They narrowed their concerns to two issues: a request for medical monitoring for people who lived near the plant, which was the higher priority, and the nuisance or trespass issue regarding the effects on the properties of local homeowners due to plutonium and other contaminants that have escaped from the plant. Property values in surrounding neighborhoods have decreased substantially because of their proximity to Rocky Flats. Of the class representatives, Merilyn Cook is the one to lend her name to the case. *Cook v. Rockwell International Corporation* is filed in January 1990.

The lead trial counsel will be Merrill Davidoff and Peter Nordberg from the Berger & Montague firm in Philadelphia, with Davidoff as the Berger lead attorney, and Louise Rosell of Waite, Schneider, Bayless & Chesley in Cincinnati. Local counsel Silver & DeBoskey of Denver also provide trial assistance. The defense immediately begins filing a wave of motions to dismiss and for summary judgment. Much of the information needed for the case is classified, and the DOE controls most of the millions of pages of documents relevant to operations at Rocky Flats as well as to the litigation. After a year of published decisions, opinions, and orders, the judge denies the defendants' motions in most significant respects, including Rockwell's position that there are a number of statute-of-limitations issues that vary with each defendant. He rules that because there is ongoing contamination and an ongoing threat of future contamination, the statute of limitations does not apply. In 1991, however, the court disallows the medical monitoring issue, so the only issue at stake is property values.

It's the kind of case Peter has been waiting for all his life.

Peter's family is Scandinavian, his father a professor, his mother a second-grade schoolteacher. He was born in Washington, D.C., and grew up in a Milwaukee suburb. He attended Catholic parochial school and then Shorewood High, a public high school, where he served as editor-in-chief of the student newspaper. He liked to imagine himself as the next Bob Woodward or Carl Bernstein, a fantasy bolstered by an AP award for best editorial in a Wisconsin high school newspaper.

Peter earned a bachelor's degree from Harvard, and then a JD from the University of Pennsylvania Law School in 1985. One of his many awards was for the highest grade in a criminal procedure class. His first work as an attorney was at the Hanford site, where he worked with people who believed that their diseases or health conditions might have resulted from Hanford exposures. He and the other attorneys met with local residents to hear their stories. One story in particular stuck in his mind. He interviewed a man with a tumor in his stomach, his belly so distended that it looked like he'd swallowed two basketballs. The man worked on a local farm near the Hanford site, shearing sheep. When a Geiger counter was placed next to the sheep's wool after he'd just sheared it, the Geiger counter "went crazy."

From that job, Peter took the position with Berger & Montague in Philadelphia in 1990. His first day of work was the very day they filed *Cook v. Rockwell International Corporation.*

Peter is put on the case. The first year is consumed by motions to dismiss and, due to the high level of secrecy and the sealed records related to the FBI raid and grand jury investigation that began in 1989, factual information is hard to come by.

Soon Peter is working eighteen hours a day on the case. There are no vacations, and not much time for a social life. His Philadelphia apartment is on the same block as the law firm, and he can see the door of the law firm from his window. It doesn't matter if he's at home or at work. He's always working on Rocky Flats.

The months drag on. To Peter, it feels like a cat-and-mouse game with the defendants. The delays are endless, and it's always one step forward, two steps back. The DOE is not very cooperative. They don't seem concerned about time or money. The Price-Anderson Act, which indemnifies private contractors from potential damages, also protects them from legal fees and expenses. Dow and Rockwell are the defendants in this case, but their legal costs—which will eventually exceed $60 million—are paid by the DOE (in other words, the taxpayer). At one point,

Colorado U.S. district judge John Kane finds the entire DOE in contempt of court for delays in turning over documents to the plaintiffs. And there are problems with the documents that have been released: thousands of pages have information blacked out—or simply blanked out—for what the government says are national security reasons. Documentation from the FBI raid and grand jury testimony is completely sealed.

Peter is particularly mystified by the DOE's approach to MUF. More than a ton of plutonium is apparently missing. The defendants claim that the plutonium is not actually *missing*—that there are errors on paper, accounting miscalculations, plutonium caught in ducts and vents, and so on—but Peter finds this preposterous. The DOE wants citizens to trust their measurements of plutonium, and its risks, down to the tiniest fraction of a gram, yet they can't account for thousands of pounds of this same material? It would be funny if the circumstances weren't so dire.

On the night of January 29, 1997—Peter's forty-first birthday—his life takes an unexpected turn. As usual, he's alone in his apartment, working on Rocky Flats. Mykaila, an employee of America Online, is working as a late-night Internet monitor and sends him a happy-birthday message. Every night at midnight, as part of her job, she sets her computer so she can't receive any return messages and then she sends an instant message to the thousands of customers who have a birthday that day. She enjoys being the first person to wish them a happy birthday, compliments of America Online.

This night, for some reason, she forgets to turn off the instant messaging program. Peter is working on the web. He gets the birthday wish. *What a nice surprise,* he thinks. He texts back. *Thank you,* he says.

You're welcome, she replies.

Thousands of miles separate them, but there's an instant intuitive bond in the flurry of messages that follows. They text again the next night, and the night after that. It's an almost uncanny connection. For months Peter and Mykaila carry on an Internet and telephone relationship, talking for hours each night. They finally meet for the first time in person in April, and marry in May. At the wedding Peter says, "It took half my life to find you," and Mykaila feels the same way.

But there is no honeymoon. Peter had been working on the Rocky Flats case for six years before he met Mykaila, and she's heard all about it. She's just as passionate about the Rocky Flats case as Peter is. She moves to Philadelphia from Texas with her two daughters, and soon she's working side by side with Peter.

After their marriage, Peter works mostly from home, and one entire wing of the house becomes his office. Mykaila and the children—her two daughters and Peter's son from an earlier marriage—spread out a picnic dinner on the office floor while Peter eats at his desk. Soon the children are almost as conversant with levels of plutonium and contamination and criminal trespass as Peter is.

Peter is an engaged but somewhat unconventional parent. The rest of the household adjusts, more or less, to his round-the-clock schedule. Mykaila finds him in the kitchen with the girls, teaching them to do the twist to Chubby Checker, or playing basketball in the driveway in the middle of the night with his son. And then he goes back to work.

Eventually he adopts the girls, and Mykaila and Peter have another son, a special-needs child named Brinkley.

The Rocky Flats class-action lawsuit, and the more than twelve thousand people it represents, is the center of the household. Peter works on birthdays and Christmas Eve, and even on Christmas Day if Mykaila lets him get away with it. But Mykaila's never had a single regret. She understands what drives his work, and she feels he's dealing with attorneys who don't have the same moral code that he does. He cares about Hanford and Rocky Flats, and the people whose properties were devalued and health was threatened. "If one person dies because of Rocky Flats," he tells her, "we can't let that person die in vain. I want to make sure their kids and their grandkids know that once upon a time, someone put up one hell of a fight."

After years of marriage, her heart is still in her throat every single time she sees him.

One evening, in the midst of pretrial preparation, Peter picks up the book *A Civil Action* by Jonathan Harr, the story of a dramatic lawsuit by a group of citizens in a Boston suburb against two corporate giants that had secretly polluted the local water, causing leukemia in children. He reads page after page in horror. In many ways the story mirrors the Rocky Flats saga, and it's a strong parallel to his own situation. He thinks about the plant and its history and all the people, all the stories, everything that's happened, the broad range of characters and the scope of human tragedy. *It's Dickensian,* he thinks.

I THINK about the orange-tagged deer grazing in the buffer zone and the beef cattle on the other side of the fence. The rabbits with their radioactive feet and the tiny Preble's meadow jumping mouse—are his little feet radioactive, too? The birds who chirp in the trees where I sit outside to eat my lunch and the prairie dogs who pop their heads up from their burrows when I walk around the site with Debra. Animals don't heed boundary signs. What do those birds carry with them when they fly off into the sky? How far down do prairie dogs go when they burrow underground? What about the rabbits in my old backyard? Were they hot?

Three weeks after my lunch with Diane, I quit. I'm offered a part-time job at the Colorado School of Mines, an engineering college in Golden, teaching literature to freshmen. It will carry me through until I graduate in May. There's no doubt in my mind. The money isn't as good and I'm not sure how I'll keep the bills paid, but my time at Rocky Flats is over.

"So you're going to be teaching Shakespeare and Dickens to engineering students?" Mr. K laughs when he stops by to say good-bye. "Good luck with that!"

I'm going to miss the people I work with, including Mr. K; George, the turkey-sandwich guy; Patricia and her geeky tech-writer pal; even Debra and Diane. I'll miss the early-morning drives out to the east gate with the sun rising above the dark blue mountains.

On the afternoon of my last day, I stop by the guard shack to turn in my badge. I've heard stories about the mountains of paperwork employees have to sign when they quit. But there's nothing for me to sign.

"Nothing?" I ask.

"Nope," the guard snaps. He's in a rush. I'm not the only person in line. "You're an employee of the Sunnyside Temp Agency, not EG&G or the Department of Energy. Your agreement is with them."

My agreement with them is over.

But my real relationship with Rocky Flats has only just begun. I have boxes of notes, employee newsletters, newspaper articles, my journals, and a burning desire to research the full story of the plant. Now that I've been on the inside of Rocky Flats and I'm beginning to understand it from both sides of the fence—the workers and the activists, the government and the local residents—I want to write.

I call Karma. "I'm going to write a book about Rocky Flats," I say.

There's a long pause. "Big subject," she says. "Are you scared? No one says anything about Rocky Flats."

"I want to write about the two things that have frightened me most in life," I say. "Rocky Flats, and Dad's alcoholism." I can't tell the story of the plant without telling the story of my family. It all seems connected. The ironic thing about all of this, I think, is that I spent years in Europe, traveling around and looking for things to write about. Nothing had ever happened to me in Arvada, Colorado, I thought, that would be interesting to anyone. It's turned out that the most important story to tell is quite literally in my own backyard.

IN JANUARY 1996 I start my new teaching job and spend evenings finishing my Ph.D. dissertation. I'm supposed to graduate in May, but my health worsens.

One weekend I go hiking with a friend. The snow has melted and spring flowers are just beginning to appear in the high country. We head up to Crested Butte and plan to hike up Refrigerator Pass, an area famous for its wildflowers. We've only been on the trail for an hour when I have to stop. The peak still lies ahead; we're in a long, low valley filled with red and blue buds peeking through the winter grass, and the path is easy and flat. "I need to lie down," I say. I feel very faint. "You're running a fever," my friend says, his hand on my forehead. My heart is racing. My whole body feels swollen. We turn back.

I go back to the doctor, a new one, and am diagnosed with chronic fatigue syndrome. "It's mostly in your head," he says. How can this be in my head? The fever is constant. I have no energy. I'm scared I won't graduate.

On April 16, I turn in my dissertation in the morning and see yet another doctor in the afternoon. The left lymph node in my neck feels as big as a football. I'm supposed to begin preparing for my oral defense in May, but it's hard to think straight. I'm referred to two more specialists. The next morning I write in my journal: *I dreamt I was a medical experiment.*

On April 29 I meet with an oncologist. He schedules a biopsy, and tells me that it could be lymphoma or Hodgkin's disease. I want to wait for the surgery until after my dissertation defense, two weeks away, but he says the surgery can't wait. At the end of our meeting he tells me to go home and think about who could raise my sons if something happened to me.

That night, Sean wakes from a bad dream and crawls into my bed. I've been careful not to tell the boys all that is happening, but they seem to sense something is wrong. Sean curls up into a warm ball by my side. It isn't long before Nathan and Heathcliff join us.

We sleep.

PETER NORDBERG and his colleagues work steadily, despite the fact that many of the thousands of class members have died or moved elsewhere. This is a crucially important case, not only to the thousands of people whose lives and properties were affected or might be affected in the future, but to the history and legacy of Rocky Flats. The DOE, Rockwell, and Dow Chemical want the full story of Rocky Flats to be suppressed and quickly forgotten.

There has never been any health monitoring of people living around Rocky Flats. Peter believes that the lives and experiences of these people should not be in vain. If the DOE and Rockwell prevail, it will be easier to make people believe that some plutonium is acceptable, never mind any of the other toxic and radioactive elements released into the environment. Rocky Flats could become a sort of poster child for other contaminated areas around the country that the government wants to turn into wildlife refuges open to the public, potentially putting local communities at risk.

Peter does his research, carries things around in his head, and puts off writing a brief until the last possible second. Then crunch time begins. He and Mykaila are a team. She works with Peter and reads every opinion, every word of every brief. She keeps the coffeepot full twenty-four hours a day, and the glass of cranberry juice filled on his desk. Food is a no-forks affair; Peter, if he eats at all, eats with one hand and keeps writing with the other. The Nordbergs are sociable with their friends and neighbors, but during crunch time no one comes to the house. The kids are more or less on their own. They know that their parents, and especially their dad, are intensely focused, and they know what's at stake. The whole household stays up with Peter in his near-trancelike state.

When the brief is finally finished, it's like a holiday. Peter pushes back from his desk, plays Pinball Wizard for a while, finishes his last glass of cranberry juice, and sleeps.

The case drags on.

ON MAY 4, 1996, I have surgery to remove the left lymph node in my neck. I'm just days away from my dissertation defense and graduation. I haven't even considered the question the doctor asked me: Who will raise your children if you have lymphoma? It's an unfair and impossible question.

I come home with bandages and a stiff white brace around my neck, and sleep for an entire day.

On May 6, the results come back.

No cancer.

I cry with relief. My mother brings the boys back home and we spend the afternoon planting spring flowers in our front yard.

The following day, May 7, I pass my dissertation defense with the brace and bandages still around my neck. The professors who make up my committee politely refrain from asking about my health, yet afterward they're not only congratulatory but relieved when I tell them it isn't cancer. My mother is proud that I'll have a Ph.D.; I'm one of the few people in my family to go to graduate school. Karma and Karin will follow soon after.

A few days later I meet with the oncologist again. He tells me that my body is definitely fighting something, and fighting hard, but that he doesn't know what it is.

I ask what the next step is. I'm greatly relieved not to have cancer, but the symptoms haven't gone away.

"I don't know," he says. "I treat cancer, and it's not cancer. I don't know what it is. I wish you the best." He escorts me to the door.

When the neck brace comes off, I have a scar running down the side of my neck. I tell Sean and Nathan it's a pirate scar. One of my friends at school tells me it's a "downwinder scar." I ask what she means. She tells me there's a stretch near Hanford, Washington, that people call "death mile." Hanford is, like Rocky Flats, a nuclear production complex, and residents who live downwind from the facility claim there have been unusually high levels of cancer deaths, and many have neck scars from thyroid operations they blame on radioactive releases from the plant.

My brother and sisters, especially Karma and Kurt, have similar symptoms, particularly chronic fatigue, fever, and swollen lymph nodes. No one has any answers for us. But we're Norwegian. It's not acceptable to complain. If you wait long enough, my mother says, just about anything will get better or go away. And our health problems are minor compared to those of others we know.

IF TAMARA Smith Meza learned anything from growing up in a strong family, it was how to make decisions in her own way. She's not afraid to buck convention.

After the surgery for her tumor, the doctor recommends radiation and chemotherapy. What, Tamara asks, will be the benefit or outcome? She's told that the radiation will likely cause her to lose her eyesight, as the tumor is behind one eye. Further, the doctor says, people with her type of tumor have an 85–90 percent chance that the tumor will return, even after radiation and chemo.

"Then I don't want it," she says. "If we don't know it's really going to help or not, then I'm going to forgo it."

The doctor informs her that her chance of surviving beyond five years is very slim. It could be less. The treatment might help. It's worth taking a chance.

"Well," she says, "what's the point of doing radiation and chemotherapy if I'm still going to die? What are my other treatment options?"

There are no other treatment options.

With her family's help, Tamara finds a doctor in New York who offers alternative treatment for her type of cancer, including radical diet changes. Slowly Tamara's health improves. It's not until her third visit with her New York doctor that she asks him if he's ever heard of Rocky Flats. "Of course!" he says. She tells him that she moved out to Standley Lake when she was four and has lived near the plant all her life. For a moment he's speechless. He tells her that there is extensive evidence that shows people who have brain cancer have often had some type of exposure to radioactivity. "In my opinion," he says, "I'm sure that your brain cancer is related in some way to growing up by Rocky Flats."

Tamara's not surprised, but discovering the cause of her cancer isn't what matters most to her. What matters are her faith and her family, and making the best of whatever time she's got left. She continues to have regular MRIs. After three years, there's no sign of another tumor.

FOLLOWING THE raid, the DOE began looking for a contractor that could handle cleanup at the plant without going overboard on cost, and in April 1995, Kaiser-Hill Inc. won the contract to begin to coordinate the Rocky Flats cleanup. EG&G continued to manage the plant. About a year later, in April 1996, Mark Silverman stepped down from his position as Rocky Flats manager for the DOE. He was fifty-seven, and his time at Rocky Flats had been notable for a more honest dialogue with the media, a reversal of the DOE's long-standing policy of rewarding contractors for work attempted (rather than work completed), and for the hosting of the first delegation of Russians to visit Rocky Flats. The job, he said, was taking too big a toll on his personal life. He'd spent three years getting to his office by 6:00 a.m., working into the evening, and then working long into the night from his computer at home. Sometimes he couldn't sleep anyway.

He was also discouraged by what he saw as growing apathy in local communities, and worried that citizen indifference might hurt the planned cleanup. "We may have done too good a job of convincing the public that we're doing things safely at the site," he said. "As a result, people aren't so concerned about the site. If the people don't care, the elected officials don't care. And if the elected officials don't care, we can't get the funding to do the job out here."

Four years after he quits, Silverman is diagnosed with an inoperable brain tumor, which he will fight for another four years.

The DOE said in 1995 that the technology did not exist to clean up the plant adequately, and estimated the project would take seventy years and cost $36 billion. In 2000, the same year when Silverman's cancer is diagnosed, Kaiser-Hill wins a second contract to complete the closure and cleanup of the entire 6,245-acre site when they agree to do it in less than six years on a budget of $3.96 billion. EG&G is out. Where will the corners be cut? The public won't find out until after the ink is dry.

DESPITE ALL his training, Randy's never had to fight an actual plutonium fire. He's fought a grass fire or two, along with fires caused by lightning, a fairly common occurrence. One time he'd been standing at the west gate with another firefighter, picking up a pizza for dinner, when they saw lightning strike the ground. They saw the flames and called it in; the fire ended up burning 110 acres in the buffer zone before it was contained.

Some of the grass fires are intentional. In April 2000, Rocky Flats wants to do a "prescribed burn" of five hundred acres. Since the 1950s, the DOE has used mowers to control vegetation at and around its nuclear facilities. But in 1999 there is a policy change. The DOE plans to conduct prescribed burns of large areas of land not only at Rocky Flats but at other DOE nuclear facilities, including Hanford and Los Alamos, to keep down vegetation. Local residents and independent scientists protest that burning contaminated grass and plants will release radioactive smoke, easily inhaled, that will expose people living in the area to plutonium and other contaminants. Dr. Harvey Nichols and others suggest a less-than-perfect solution: allowing goats to graze on the land and keep the vegetation in check. Unfortunately the goats—like the cattle and the deer—would likely ingest contaminants, but at least the contamination would be contained and not floating over the Denver area. The DOE rejects the idea.

After pressure from local citizens and the press, the DOE reduces the burn from five hundred acres to a "test burn" of fifty acres on April 6, 2000. The fire creates a large cloud of smoke. Paula Elofson-Gardine, a local resident and executive director of the Environmental Information Network, alerts the media and has a KMGH Channel 7 news crew at her home. Paula has a Radalert Geiger counter, a real-time handheld radiation monitor that measures alpha and beta particles as well as gamma and X-rays, and she and the news team track the cloud and its effect. In less than forty minutes the cloud travels fourteen miles around the metro Denver area, and is visible from Paula's second-floor window. She and the news crew can smell the metallic odor of the smoke. Before the burn, Paula measures background radiation at 8 to 15 counts per minute. During the burn, the radiation readings quickly reach the highest standard of detection on the Radalert, 19,999 counts per minute, an extraordinarily high level by any standard. The next day the reading goes down to 1,147 counts per minute, and the rate slowly declines over the next few weeks. It stays at about 10 counts per minute above background level for a full year.

"There is no official evidence of what exactly was in that smoke," Paula says to the press. Whatever it was, it wasn't good. But few people in the Denver area seem to be paying much attention. They've been told that Rocky Flats is being cleaned up, and they believe everything is fine.

Randy knows that everything is not fine at Rocky Flats. Over the years he's responded to all sorts of situations on and off the plant site. But not much in the plutonium department. He's been lucky.

But on this May morning in 2003, he's not so lucky.

It's not quite Mother's Day, but spring is in the air. Randy reports for work. He's captain and his buddy Paul Kuhn is shift commander, and they both are hungry. Because the plant is in the process of shutting down, many of the cafeterias are closed and employees now depend on roach coaches that drive around selling honey buns, sandwiches, and soda. The one that stops at the fire station sells pretty good burritos.

"I'll buy breakfast," Paul says, and hands Randy five bucks. Paul is a stocky guy, solid, originally from Czechoslovakia. Both men are in their forties; they've worked together for years.

Randy takes the money, tucks it in his wallet, and steps out into the sunlight. He's wearing his dress blues. It's a fine morning. He reaches for a burrito, and just then his radio goes off.

"Fire response," the voice says. A "pyrophoric incident," as Kaiser-Hill and the DOE like to call it. "Building 371."

Randy's heart jumps. "What do you do when you hear something's happened in 371?" the men often joke. "You pucker."

At this point, Building 371 is the most active plutonium building on-site. That's where they try to stabilize plutonium before shipping it off to the Savannah River Site in South Carolina for storage. All the weapons-grade plutonium on-site ends up at Building 371 sooner or later. And the building is huge, a maze of corridors and underground levels and stairwells. There are many steps, and it takes firefighters a long time to get into the building and then get out again, especially with all their cumbersome equipment.

Breakfast is forgotten. Randy sprints back inside. He and Paul gear up, jump in the fire truck, and head down to the Zone. The radio crackles again. "Fire confirmed." No word on how big it is, but Randy knows the size doesn't really matter. If it's a plutonium fire, big or small, it's bad.

They pull the fire truck up to the front of the building. A shrill alarm fills the air. They begin strapping on their air tanks. Evacuation has begun. Workers pour through the doors and stand nervously outside in the sunlight. Randy races through his mental checklist: dosimeter, badge, regulator, mask, full tank of air. All good. Both men are tense, but they keep their emotions in check.

Suddenly the building manager emerges. "Wait," he shouts. "Wait! There was a fire, but they put it out."

"What?" Randy says. It's hard to hear with all the noise. The manager repeats his words.

"It's out. The fire is out."

Paul and Randy exchange looks. It takes a moment to realize what he's saying, and their relief is palpable.

Randy shifts gears, from fourth gear to second. "So, okay," he says to Paul and the other fire personnel. "It's not that big a deal now. But we still need to go in and investigate. We need to make sure that even though the fire is out, conditions are safe."

Paul nods. Time is still critical. It's agreed that Randy will go interior and Paul will stay exterior. They'll keep in radio contact as best as they can. Many of the buildings at Rocky Flats are so deep underground and so complex that radio communication doesn't always work.

Randy begins descending the steps with two more firefighters following behind. The fire has been reported in the sub-basement, and it's a long way down. They're halfway there when Randy's radio crackles again.

The fire has rekindled.

Crap, he thinks. *Guess I have to change gears again.* And now the stakes are back up. Randy knows that when you go in to fight an active plutonium fire, you're going to get crapped up with radiation. You get hot.

He takes a deep breath from his tank, clambers down the rest of the stairway, and pushes open the door, not knowing what to expect.

Randy's first thought is, *What the hell are all these people doing in this room?* There are people spilling out of the room into the hallway, some dressed only in common bib overalls. Everyone is panicked. It's a confined space, and there is too much equipment and too many bodies. The air is filled with particulate from the fire extinguishers. He estimates visibility to be about six inches. He can barely make out anything in the chaos. A couple of workers are still trying to discharge a dry chemical fire extinguisher in the general direction of the flames.

The criticality alarm is going off and the noise is deafening.

"The fire's in a glove box!" a worker shouts. Randy can't see it.

Building 371 is in deconstruction and decontamination mode, and the glove box that's supposedly on fire has been tented off. Workers have constructed a shroud around it to keep all the contamination inside while they take it apart and try to decontaminate it. *It figures,* Randy thinks. The fire department has discussed the fact that there is a greater chance of something going wrong when workers are tearing things down than when the plant is in actual production. The pressure is on to get the place shut down fast and under budget. People are pushing hard to finish the mission. Sometimes they take shortcuts. Sometimes they make mistakes.

This is a full-size glove box, as big as a refrigerator. And it's a special glove box, called a "guillotine" box, designed with a spring-loaded trap-door that's set to slam down in the event of a fire or any unauthorized entry. Anyone or anything coming in through that trapdoor could be literally cut in half. A person could lose an arm or even be decapitated.

No one informs Randy or Paul that the glove box is an armed guillotine box.

The criticality alarm blares, red light throbbing, indicating that there is plutonium contamination in the air. Randy starts pulling people out of the room. "Get out of here now!" he yells. "Go! Go! Get out!"

When the room is clear he convenes briefly with the other two fire-fighters in the exterior hallway. *This time,* he thinks, *the shit has really hit the fan.* He calls Paul, who is still manning the outside of the building, and fills him in. "It's bad. I need more CO_2 cans. I need more dry chem extinguishers. Fast." He instructs one firefighter to stay in the hallway and the second to follow closely behind him.

He goes back in. The room is a fog. He's shocked that there are still workers in the room. "Get the hell out of here!" he shouts. But they seem to know where the fire is. "Here!" One points. "Here it is!"

"Where?" Randy shouts. His voice sounds like Darth Vader's. All the firefighters sound like Darth Vader when they're wearing their respirators. It's almost impossible to communicate clearly.

"Over here!" the worker shouts again, his voice unimpeded by a respirator. Why isn't he wearing a respirator? The workers are supposed to be familiar with the dangers of plutonium fires and the proper procedures to follow when they occur. If a plutonium button starts to glow, the worker grabs a coffee can and sprinkles a little magnesium oxide powder or sand to smother it. Anything more than that, they're supposed to call the fire department and get out of there.

The workers in this room, though, are scared. Something's gone awry. And they haven't evacuated like they're supposed to.

Randy still can't see the fire. His vision is distorted from all the chemical particulate in the air. He goes around to the back of the glove box. "No, up here!" the worker yells at the front. Randy follows his lead but still can't see it. "Get out!" he yells. "You guys get out of here!"

He goes to the back of the glove box again. And there it is.

The flames reach as high as three or four feet. And they're a funny color, a kind of metallic blue that he knows is not a good sign. Not good for his body. Not good for his lungs.

But he doesn't have time to think about it. At any rate, he's packed up and sealed about as tight as anyone could be.

He examines the glove box. This particular one is a trash glove box, used to collect radioactive refuse from all the other glove boxes. Chemical wipes and all sorts of junk end up here. There's no way to tell how much plutonium might be in there. A cut has been made high on the side, and it looks like a piece of metal has fallen off, creating a spark and igniting the trash.

Randy still can't find the source of the fire. His partner hands him a dry chemical fire extinguisher and he gets it in place and engages the flames.

It has no effect.

Randy glances down at the floor of the tent, and what he sees almost stops him in his tracks. There are eight empty fire extinguishers down there. *That's ridiculous,* he thinks. *How can that be?* Workers are trained to use one fire extinguisher, maybe two at the most. If one doesn't work, they're supposed to back out and immediately call the fire squad.

He appreciates the fact that they tried to contain the fire. But eight canisters? What were they trying to do? *My God,* he thinks. *Everything has gone wrong with this fire.*

The criticality alarm blares so loudly he can't hear his radio. He can barely see his hands in front of his face.

Go easy, he tells himself. *Go easy.*

His partner hands him another can. This time Randy knocks the flames down slightly. He pollutes the air even more with dry chemical—visibility is almost zero—but it's all he's got to work with.

There's no time to waste, but he decides to use a new technology the firefighters have just received—a thermal imaging camera clipped to his belt. Everything in the glove box is shrouded in a dry chemical fog. He leans deep inside the box and starts taking pictures. He pulls out, checks the camera, leans back in and shoots again. Finally he sees the genesis of the fire, deep at the back of the glove box and buried in trash. There are filters inside, encased in wood, and the wood is smoldering and burning. No matter how much dry chemical he uses, he's not going to be able to put that out.

"The dry chem's not making it!" he yells. He grabs a pipe pole and starts digging, trying to grab the material and pull it out of the box. It's unbearably hot. The material is heavy. Randy leans into the glove box again and again, working as quickly as possible. He knows there's plutonium in there. How much, he doesn't know.

He thinks about using water. There's no water line established, of course, because water isn't supposed to be an option. Randy knows they used water on the Mother's Day fire, but that was a last-ditch effort to save the plant—and the city—from an all-out nuclear holocaust. And he's heard that there was a chain of unlikely events that made the situation come out a lot better than it probably should have. He's heard stories that a fire engine hit a power line and knocked out the power, which helped the firefighters and kept the roof intact.

If that roof had melted, he thinks, Colorado Springs would now be the capital of Colorado.

Will he, too, get a lucky break?

He can't get this fire under control. The dry chem isn't working, and there's no water line. His radio is silent. There's no communication coming down from above.

It's unclear what he should do. He's running out of time.

Outside the building, where Paul is trying to monitor Randy's actions, they're setting up a full-scale decontamination station.

Randy checks his tank. His air is low. The two firefighters assisting him are also running low. They're hot, the air is full of vapor from the dry chem, and they need fresh tanks of air.

"Paul!" Randy barks into his radio. "The dry chem isn't working. Can we use water?" He says it again with more emphasis. "Can we use water, Paul?"

There's a long pause, and then the word comes down. "Yes," Paul says. "We're bringing a line in now."

The moments pass interminably. Finally a firefighter appears at the bottom of the stairs. They establish the line and the standpipe, and then snake the line into the glove-box room. Randy keeps pulling pieces of trash, radioactive and otherwise, out of the glove box, while the man behind him holds the water line. "Just knock it gently," Randy says. "Put a little water on the burning charcoal. Real slow." Suddenly he pulls out a plastic liter bottle. "That's not good," he mutters. Plastic liter bottles are used to store plutonium-laced liquids.

The water seems to be working. The two men are rapidly gulping air, going through one tank, then another. The room is intensely hot, the work physically and emotionally draining, and it's taking longer than expected. Randy can't read his partner's face behind the mask, but he knows they're both nearing exhaustion.

"We need another crew!" Randy yells into his radio. "We need someone to spell us. It's getting pretty difficult in here."

There's no response. He's not sure anyone's heard him. Finally the new operations chief shows up in a face mask and regulator, with a fresh crew behind him. The entire firehouse crew, as a matter of fact. This is what firefighters call "dumping the house."

By the time they arrive, though, the fire is essentially out.

Randy orders his exhausted crew out of the building, with instructions to fill their air tanks before they start up the stairs and head for decon. "We have a long way to hump out," he shouts. "Top off your tanks." But they don't listen. They just want out. As the two men head up the stairs, Randy hears the men's personal air alarms go off. They're anxious to get out and think they've got enough air to get up the stairs, or they want to leave more air for the new crew. "Crap," Randy mutters. "It's like herding cats."

Nonetheless, he keeps going until he's sure the fire's out. He looks around the room. It's a mess. Dry chem and water and God knows what. Visibility is still poor. But at least the fire is under control. The new crew can begin to clean things up. As a captain, Randy is always the first in and the last out, and now it's time for him to get out. He takes pride in protecting his crew. He waits until his last guy is up the stairs, and then he begins the climb.

Abruptly his own air alarm goes off. He looks down at his air tank gauge. It's red.

He keeps going. He reaches the top of the stairs and bursts out of the building, where a radiation monitor hands him a fresh respirator. He breathes deeply. A staging center for decontamination has been set up right in front of the building. They're trying to keep everything contained. Randy's crew is already getting scrubbed.

He's told that most of his crew likely suffered not much more than alpha radiation. But his case is different. He was in that room for a long time, with all that dry chemical in the air. Plutonium and other contaminants bond to airborne material. And plutonium particles are microscopically small. It doesn't take much to cause a problem. Even though he was wrapped up as tightly as possible, he's got internal contamination.

He takes off his gear, peels off his clothes, and steps into the decontamination room wearing nothing but his boxers and a respirator. The radiation monitor passes a wand closely over his entire body.

"You're pretty hot, sir." The man steps back a little.

"How hot?"

"Your skin contamination is high. High, high, high."

This is not good, Randy thinks. *Definitely not good*. It's a warm afternoon but he's shivering.

The radiation monitor leaves and Randy hears voices in the other room. A doctor emerges. "I'm sorry," he says. "We can't do you here, Captain. Your counts are too high to decon you here." Only basic decontamination could be done at the site of the fire. He'll have to be taken up to the decon room at the medical building, a special trailer for decontamination consisting of one big room with three separate chambers behind glass doors. The room is filtered, with a shower and a cot in case a contaminated employee has to spend the night.

Randy shivers and waits. Some of the men from his own crew, freshly scrubbed and decontaminated, come into the room. They truss him up in tiebacks, pull gloves on his hands, and tape him up so that no part of his body is exposed. He is isolated and quarantined within his own skin. The mood is somber and no one speaks. An ambulance pulls up. They put him on a gurney, roll him into the back, and drive him, red lights flashing, to the medical building.

Once again Randy is stripped down, but this time all he gets to keep is the respirator. A special medical team shows up, a team with long experience working with contaminated people in Building 771. A doctor stands by with a DTPA syringe, which Randy and his co-workers know as "the big honking needle." DTPA, or diethylene triamine pentacetic acid, is a chelating agent that isolates internal contamination from radioactive materials—plutonium, americium, or curium—and binds with the radioactive material or poison and then passes it from the body in urine. DTPA can't reverse health effects caused by the radioactive material, but it helps decrease the amount of time it takes the radioactive poison to leave the body. He gets scrubbed from head to toe with a bleach solution until he feels raw all over. The radiation monitor checks him again. He's still hot. Screaming hot, especially around the belly, from having leaned repeatedly into the glove-box door.

They scrub, check, scrub, check, scrub, check. This goes on for nearly three hours. Randy hears the doctors talking about his radiation counts in the other room. Then they come in again and scrub him some more.

There's talk about whether or not his entire body should be shaved.

Hair can sometimes be a trap for radioactive particles. "At least I'm a Cherokee," Randy jokes. "Not much body hair."

He escapes being shaved. Finally they stop scrubbing. But it's not over yet. Now they begin to try to measure internal contamination more specifically and determine how much soluble plutonium is passing through Randy's body. Insoluble deposits in lymph nodes, the liver, or the bone—plutonium that stays in the body—are very difficult to measure or detect. Randy gets a nasal swipe and lung count, then has to provide fecal and urine samples.

When Randy finally gets back to the fire station, it's past dinnertime. He calls his wife. It's been a long day, and it will be a few more hours before they'll let him go home. He worries, as always, whether or not he's bringing any radioactivity home with him.

The tests for internal contamination will go on for weeks and months, and he'll need to bring in fecal samples regularly. The other guys joke with him about it. "You get to do the Cool Whip container thing!"

"Yeah," he says. It's like being a member of a special Rocky Flats club. He thinks back to how the day began, with the warm morning sun and a breakfast burrito. "This turned into a whole full-blown thing, didn't it? A damn big deal."

Later, when he's interviewed by DOE officials, Randy learns that the glove box he'd been leaning into was a guillotine box, poised to mechanically spring and sever anything that intruded—a hand, an arm, or a head. Why hadn't it operated properly? No one seems to know. Why wasn't he informed? Apparently the workers were panicked trying to control the fire.

"That disturbs me," he remarks to Paul. He tries not to think about it.

On September 19, 2005, Tamara Smith Meza returns from church with her husband and is talking on the phone with a friend. She's had to cut back on her teaching, of course, but at her last checkup her doctor gave her a clean bill of health. Three years have passed since the brain tumor. Both she and her husband are relieved.

Suddenly a sharp, searing pain shoots through her head. Tamara gasps. "I don't know what's going on," she says to her friend, "but my head hurts so bad I think I better go. I gotta go." She hangs up the phone. "David!" she yells. "David!" He rushes into the room to find her kneeling on the living room floor, touching her head and then looking at her hands as if she were looking for blood on her fingers. She begins to cry, and then vomits. "I can't feel my left side!" she cries, and then passes out.

David picks her up and lays her on the couch. He calls 911. He hangs up the phone and sees that Tamara has stopped breathing. He can't see any vital signs. He clears her airway and starts doing CPR, and continues until the ambulance arrives.

The tumor has returned. Tamara has surgery again. This time the side effects are more severe: she has trouble with walking, balance, and memory. But she recovers.

Four months later, a third brain tumor appears. The first tumor grew in the memory region of her brain, the second in the speech and language section, and the new tumor is in the cavity left by the removal of the first tumor.

Tamara Smith Meza's surgeries and health problems make it impossible for her to continue teaching. She and her husband return to Colorado and buy a house just a few minutes away from the house she grew up in, where her parents still live.

Sometimes people ask her if she thinks Rocky Flats might have affected her health or caused her condition. She doesn't think too much about it. She's focused on her treatments, on her diet, on her relationship with her family and with God. But Tamara's husband, David, does think about whether Rocky Flats might have had something to do with it. There's no way to prove it, of course. It's all invisible, unseen, unverifiable.

8

WHAT LIES BENEATH

1996–2011

S ilence is an easy habit. But it doesn't come naturally. Silence has to be cultivated, enforced by implication and innuendo, looks and glances, hints of dark consequence. Silence is greedy. It insists upon its own necessity. It transcends generations.

Silence is almost always well-intentioned. What parent hasn't scolded their child? *We don't talk about things like that. Just look the other way. Keep your thoughts to yourself. This is just for our family to know. You can forget this ever happened. Let's not upset anyone. If you can't say anything nice, don't say anything at all.*

The cost of silence and the secrets it contains is high, but you don't learn the price until later. Secrets depend upon the smooth façade of silence, on the calm flat water that hides the darker depths.

In the weeks and months that follow my surgery, I try to forget about Rocky Flats. My symptoms come and go, and I continue to see various physicians. When I mention to one doctor that I grew up near Rocky Flats and briefly worked at the plant, and I wonder if there might be a connection of some kind, he laughs. "You can't worry about things like that," he says.

And it isn't cancer, after all. I should feel relieved.

What does it matter, anyway, where my symptoms come from? It could be anything. Allergies, viruses, flu, exhaustion, bad weather, a bad day at the office. Maybe it's all in my imagination. The uncertainty is frustrating.

But my siblings share many of my symptoms. And it's still there, the lingering feeling that this chapter wasn't supposed to be part of my story, or my family's story, or anyone's story. Governments aren't supposed to poison their own people.

We weren't supposed to know about Rocky Flats during the production years, and now we are supposed to forget it ever existed.

ONE DAY, not long after I quit my job at the plant, I'm having lunch with a friend when she mentions a woman named Ann White. "You'd like her," she says. "And she knows a lot about Rocky Flats."

Ann White turns out to be a very energetic woman with gray hair and a quick smile. And she does know a lot about the plant. She's long been involved with citizen groups that are concerned about ongoing contamination issues at Rocky Flats, and she tells me how she was arrested for protesting.

"They say they're going to clean it up," she says. "Cheap."

When I worked at Rocky Flats, the DOE projected that it would take until 2065—nearly one hundred years from when my family moved to Bridledale—and more than $36 billion to clean up the nuclear waste. There was a question of whether they even had the technology to clean it up, and some non-DOE scientists talked about making the Rocky Flats site a "national sacrifice zone," completely closed off to the public.

She told me how in 1996, the year after I quit Rocky Flats, Mark Silverman, the DOE manager at Rocky Flats—whose voice I had known well from the crackling loudspeakers over my head at work—had acknowledged to her and a group of citizens that Rocky Flats was facing a crisis. Tons of plutonium were unsafely stored in proximity to more than two million residents. He talked about an Accelerated Site Action Plan (ASAP) that would include a 130-acre cement cap over the high-security, highly contaminated processing areas. By 2010, he said, most of the major buildings would be demolished and plutonium would be consolidated and sealed behind thick layers of concrete.

"They're just going to cover it up?" I ask.

"Apparently so," Ann replies. "Parts of it, anyway."

"By 2010?"

Ann nods. "Can you believe it?"

I think about what might be under that cement cap. The Infinity Rooms, for one thing.

Infinity Rooms—called that because the radioactivity is so high it can't be measured by standard equipment—are so profoundly contaminated after four decades of pit manufacturing that no one can access them without "extraordinary safety measures," including full-body suits and oxygen tanks. Workers have to wear body suits and respirators just to peek in the door. With a total area of 4,500 square feet—about the size of a professional basketball court—they are scattered throughout three separate buildings. Back in 1968, one of the rooms was welded shut, a worker's tools left inside, like an abandoned workers' cave at the bottom of a mineshaft. Except this cave was lethal—forever. Infinity Rooms are legendary in plant gossip.

Then there were the "Pac-Man" rooms, named after the video game because they "gobbled up" and stored contaminated equipment from around the site. A worker had to go through four separate airlock chambers to enter a Pac-Man room. There were the barrels of radio-active waste—roughly a thousand of them—awaiting shipment to other federal facilities. There was Building 371, which had never functioned properly and operated for only two years before it was shut down due to safety issues. In Building 371 there were seventeen different Infinity Rooms on three floors—2,200 square feet of plutonium-contaminated space.

"They're hoping we'll just forget about it," Ann says. "If we can't see it, it can't be real, right?"

☢

STANDLEY LAKE was a magnet for me and my siblings when the afternoons were long and sleepy, and we could sit in the tall grass and watch the water ripple and dance as if it were tapped by invisible fingertips. We felt small in the face of the lake and the long fields leading up to the blue foothills and the gray hulking mountains beyond. The mud along the water's edge was thick and pungent and clung to our skin and clothes like mottled glue.

I return to the lake even though the tall grass no longer makes me feel small. My sisters and brother do, too. Maybe it reminds us of a happier, peaceful time. The lake no longer looks the same; the surrounding pastures are now filled with housing developments, and an RV campsite and boat dock stand where once I galloped my horse.

Everything has changed.

One afternoon my brother, Kurt, who still lives near Arvada, is walking along the lake's edge with his wife, Cindy. Their two golden retrievers run and leap ahead of them, happy to be off their leashes. Kurt reaches down and grabs two sticks and throws them, one by one, out into the water.

"Uh-oh," Cindy says. "Here comes trouble."

A patrol boat pulls up with two men inside. They have a bullhorn. "Get your dogs out of the water," a voice calls.

"What?" Cindy laughs. She turns to Kurt. They've always walked along the lake and let the dogs play in the water.

"Now," the man demands. "Get your dogs out of the water now." He means it.

Kurt calls and the dogs come swimming back, noses high, front paws paddling. They climb up on the bank and shake themselves thoroughly, drops flying.

"Keep your dogs out of the lake. No one gets in the lake," the man barks. The boat turns to go back.

"Wait!" Kurt yells. "Why?"

The boat turns slightly. The man drops the bullhorn. "It's drinking water," he yells. "No one can get in the water."

"What?" Kurt looks around. Several speedboats are making big circles on the lake. "People are waterskiing. And swimming." He laughs. They're worried about a few dog hairs getting in the drinking supply?

"You can ski in the middle of the lake, or you and your dogs can swim in the middle of the lake, but you can't wade in or kick up any sediment," the man says.

"Why?" Kurt asks.

"Because it's drinking water," he replies.

Kurt turns to Cindy. "That doesn't make any sense."

Lake officials—as well as the DOE, the Colorado Department of Public Health and the Environment (CDPHE), and even the EPA—are counting on people not to question their logic.

The fact is, there's plutonium in that sediment.

Kurt doesn't trust anything about Rocky Flats. He's started delivering pizzas at night for a little extra income for his family. One of his coworkers is a previous Rocky Flats worker. "I have nothing left of my life," he tells Kurt. His arms are covered shoulder to wrist with tattoos, an attempt to hide the scars he says he carries from being scrubbed down with steel wool after contamination "incidents" at Rocky Flats. His health is poor and he's living on pizza delivery paychecks and tips.

Kurt's surprised. Most of the Rocky Flats workers he knows about are the parents of his friends. Many of them have died. But this guy is much younger.

MANY STUDIES have been done on Rocky Flats over the years. In 1996 a Boston University epidemiologist, Richard Clapp, found a disproportionate rate of lung and bone cancers in areas around Rocky Flats. There was good reason, he concluded, to continue to survey the incidence of cancer and other diseases in exposed communities and to monitor public health.

But no health testing or medical monitoring has ever been done for people living near Rocky Flats, although it has been done elsewhere in the United States. In 1989 a class-action lawsuit by residents living near the Fernald uranium processing facility near Cincinnati, Ohio, settled for $78 million and established the Fernald Medical Monitoring Program, providing comprehensive monitoring of the health of roughly 9,500 people. Another class-action lawsuit, brought by former workers at Fernald, ended in August 1994 with a $15 million settlement plus lifelong medical monitoring for workers.

Following Clapp's study, in 1998 the Colorado Cancer Registry of the CDPHE released a report asserting that there was no evidence of adverse health effects directly attributable to the plant in residential areas near Rocky Flats compared to other parts of the greater Denver area. Some citizens were relieved. Others questioned the methodology, especially given the fact that contamination from the fires at Rocky Flats, and the 1957 fire in particular, blanketed the broader Denver metropolitan area. A radiation health specialist, Bernd Franke, noted that the report was meant to "calm people down for public relations purposes."

The year 1999 marked the end of a decade-long study commissioned by Governor Roy Romer and administered by the CDPHE. Because the court documents from the FBI raid were—and still are—sealed, and no actual testing or monitoring of the health of individuals had been performed, the report depended primarily on available records, research and analysis of past releases of radioactive materials and chemicals, and "dose reconstruction," a method that involves calculating how much plutonium might have been released from the plant and how people might have been affected, including the potential for an increased risk of cancer. Even relying on these methods, the Rocky Flats Historical Public Exposures Study discovered shocking levels of contamination.

The study identified plutonium and carbon tetrachloride as the most significant contaminants from the plant. Further, between 1,100 and 5,400 tons of carbon tetrachloride, a solvent used for cleaning and degreasing, were released. Other contaminants included beryllium, dioxin, uranium, and tritium. People who lived near the plant and led "active, outdoor lifestyles" had the highest level of exposure to airborne plutonium. (The study itself focuses on "ranch workers," but this would have applied to children playing outside, too—and children are even more vulnerable than adults.) The largest amounts of plutonium released from Rocky Flats into surrounding neighborhoods and communities came from the 1957 fire and, in the late 1960s, from a waste-oil storage area—the leaking barrels. Soil sampling showed that the highest off-site plutonium concentrations in soil were mostly east of the plant. My old stomping ground. People were exposed to contaminants by drinking water and ingesting vegetables and meat and through skin contact, but, the report explained, those exposures were found to be significantly smaller than exposures from breathing plutonium.

With respect to water, the study found that plutonium and tritium were the key radioactive materials in water samples from Walnut Creek and Woman Creek, both of which flowed off-site from Rocky Flats. The two creeks eventually flowed into local reservoirs that supplied, or continue to supply, drinking water to several local cities. As a result of this contamination, in 1997 the city of Broomfield built a new drinking water supply, and Great Western Reservoir was no longer used for city drinking water.

Standley Lake, however, continues to provide drinking water for local communities, and the lake is used extensively for recreation. Woman Creek—the creek where I kissed Adam so long ago—flows directly from the Rocky Flats site into Standley Lake. In 1996, partly due to public protest, Woman Creek Reservoir was built to try to stem the flow of Rocky Flats contaminants into Standley Lake. No attempt was made, however, to clean up the already existing contaminants.

The report stated that most of the plutonium from Rocky Flats was in a form that wouldn't readily dissolve in water, and thus tended to sink to the bottom of the streambeds of Walnut and Woman Creeks and into the sediment of Standley Lake. The DOE stated that drinking water contained only "trace elements" of contamination and that "average contamination levels . . . have never exceeded drinking water standards or relevant health guidelines, even for plutonium." The report conceded that residents using Standley Lake for recreational purposes "might incidentally ingest sediment and water that contains, or previously contained, site-related contaminants," but blandly asserted that such contact "is not expected to be detrimental to one's health" and concluded that there is "no apparent public health hazard." The report did note, however, that further study was needed regarding risks to the public from environmental plutonium exposure, giving as an example the plutonium deposited in the silt of Standley Lake.

There's good reason not to kick up that sediment. You wouldn't know it, though. There are no signs mentioning the presence of plutonium in Standley Lake.

The main risk of inhaled plutonium, the study noted, was cancer of the lung, liver, bone, and bone marrow. Nonetheless, the study stated that the risk of getting cancer was minimal, and that all remaining contaminants in the air, water, and soil beyond plant boundaries did not exceed safe levels established by the government. Of the $8.7 million of federal funds used for the study, more than $1 million was spent on public relations to reassure people that they had nothing to worry about. This included guest newspaper columns and letters to the editor written by a public relations firm and signed by health department officials, as well as "polishing the speaking styles" of health department consultants and setting up talks and luncheons with local community groups.

The study did not change anything. Despite ongoing requests from Colorado citizens for further health testing and monitoring, the DOE and CDPHE say that due to population changes, low levels of exposure, and the fact that no disease can be attributed solely to plutonium, it is not feasible to perform an epidemiological study of residents around Rocky Flats.

EACH YEAR my siblings and I return with our families to our mother's house in Arvada for the holidays. It's just days before Christmas in 2004, and we're preparing for our usual Christmas Eve family gathering. Sean and Nathan, who have always been tall for their age, nearly reach my shoulders. My mother has ordered lefse—no lutefisk, thank goodness—and a chocolate Yule log with all the trimmings, and we discuss where to buy the turkey now that Jackson's Turkey Farm is long gone. Ever since

I worked at Rocky Flats, I can't help but think of pondcrete when I see the white, gelatinous lutefisk that my mother loves. I don't mention this to her; that's one joke she probably wouldn't appreciate.

My mother's townhouse is beautifully decorated for the holidays. She loves parties and holidays and extravaganzas. There's a big wreath on the door, and evergreen boughs are strewn across the fireplace mantel. Holiday potpourri simmers in a pot, and Christmas carols blare on the stereo. She's had her hair done up in blond curls and she wears a bright red sweatshirt with sequined appliqués of Santas and reindeer, mostly for the grandchildren, I think. On Christmas Eve she'll don a more tasteful pair of slacks and a red cashmere sweater with a matching necklace and earrings.

Some Christmas Eves, my father slips in quietly with a box of chocolates in hand and joins us silently for dinner. He rarely speaks and never stays long. His exits are as stealthy as his entrances. We're not even sure why he comes.

It's been years since that's happened. My father now lives in a small untidy apartment, a mausoleum of unwashed dishes and crumpled clothing, old newspapers, books with worn covers, and an ancient television with no knobs.

I stand in the kitchen in front of the oven, pulling cookies out on hot baking sheets and rolling them in powdered sugar. I don't know that this will be our last real Christmas together as a family. The years ahead will bring a host of health problems for my mother, including Alzheimer's. Her health will decline so quickly we will hardly know how to deal with it.

But on this day she is cheerful and chatty and looks radiant. The phone rings. She answers it in the other room. I hear her voice go on for a moment or two and then she comes into the kitchen, receiver in hand, with a meaningful look in her eye. "It's your father."

"Oh, thanks, Mom," I say, more than a little sardonically. I still carry the image of him from my childhood: neglectful and neglecting, neglected himself, no doubt, this man who never showed me much affection or attention.

What is a daughter's responsibility to a father? Is a child obligated to love, care for, and respect a parent in the face of indifference? Indifference that is perhaps more devastating than conflict or anger? He has been absent, tormented, darkly destructive, angry, sometimes threatening. I am invisible to him, yet my father looms so large in my memory and imagination that I can't seem to knock him loose from my head or my heart.

"Take it," my mother says, handing me the phone. "You should talk to him."

I take the phone and say hello flatly.

"Kris!" His voice sounds overly loud and a little forced. "How are you?"

"Okay."

We have nothing to say to each other, I think. *We don't even know each other.*

He wants to meet for coffee.

"I don't know—"

"Go," my mother says. She seems to have radar and picks up the faintest details of any phone conversation in the house. She waves both her hands in the air, as if shooing me away like a fly. "Go. Go!"

"All right."

We agree to meet at a Starbucks just south of Arvada. "Just for a few minutes," my dad says. "I'll be driving my cab. I'm on shift."

I arrive first. I'm late, but I know he'll be later. That's a time-honored tradition in our family. We're always late. Five minutes pass, then ten. I order a coffee for myself and sit down. Another ten minutes pass. Maybe, I think, he's not going to show up.

There's a time-honored tradition of that, too.

Eventually a sun-worn yellow cab pulls into the parking lot, moving slowly, carefully. My father climbs out. He holds his hand up against the sun and pushes against the door of the coffee shop. "Kris?"

He's aged. So have I, I suppose, though I can't recall the last time we saw each other. His hair is uncut, gray and very straight and a little too long, reaching past his collar. He wears a corduroy shirt the color of rust, well-made but worn, perhaps bought at a secondhand store. The shirt is rumpled and covered with lint. As always, he seems completely unaware of his personal appearance. His skin is pale and as thin as paper, sagging a bit under the jaw. He looks at me with a slightly unfocused look. I don't know whether to hug him or shake his hand or do nothing. I smile.

"I have to get new glasses," he says. "I'm sorry I can't see you very well. My puppy chewed up my glasses. And my shoes and my furniture, too." The corners of his mouth tease up into a smile.

"What kind of puppy?"

"A sheltie. His name is Dusty."

"Are you going to keep him?"

"Of course," he says. "I'm thinking of getting a second one to keep him company."

"Can I get you a coffee?" I ask.

"No. I'll get it." He reaches into his pocket and pulls out two crumpled one-dollar bills. "I'll take the usual," he says to the guy behind the counter.

"Fine, sir." They seem to know him here. It's odd to think of my father hanging around Starbucks. I put my money back in my pocket.

"Cream?" I ask.

"No," he says darkly, as if I should know better. Danes drink their coffee black. Norwegians, too. My mother never understood why I put cream in my coffee.

"Well," he says, settling into a chair. "It's been a rough day."

I wonder if it's stressful, this idea of coming to meet with me. "A bad day?"

"Yes indeed," he says. "As of six o'clock this morning, I have given up smoking."

"Given up smoking?" Both my parents have smoked since they were thirteen or fourteen. That's what kids did on the farm, they say.

"Doctor's orders," he sighs. He pats his chest. "I'm having a little heart trouble."

"Is it serious?"

"It might be. Who knows?" He sips his coffee. "I see doctors all the time but I don't like them very much."

My mother doesn't like them much, either.

"Christmas is coming," he says. Despite everything he's always had a good appetite, and I know he appreciates the family cooking. "I asked the doctor if I could have a glass of wine with Christmas dinner, and he said no. Not even that. But it doesn't matter. I'm not much of a drinker these days."

We sit in silence.

"The boys are doing well," I say. "Sean and Nathan."

"That's good. They're big now?"

"Yes."

"Good."

I'm not sure what else to say. I pause. "I'm writing a book about Rocky Flats, Dad. Did you know that?" I've discussed it with my mother and siblings. They're curious and supportive, although my mother wonders if there's much of anything to write about. She doesn't believe there really was any contamination. She wanted to raise her children in a perfect environment, and except for a few problems with my dad, it was, to her mind, perfect.

He seems to know that I'm writing a book.

"Well," he says, "I never thought about it much back then, when you kids were small, but I guess I'm glad about the water."

"What do you mean?"

"You probably don't remember this, but I tried to dig a well out there. Went down to two hundred feet. Nothing. Absolutely nothing. The neighbor across the way went down thirty feet and found water right away. But we just couldn't get it. We ended up going on city water." He looks at me fuzzily. "Maybe a good thing, eh?" He laughs.

I sip my coffee, nearly gone. He's scarcely touched his. Now he looks at his watch. "Gosh, Kris, I've got to run."

But he doesn't get up.

"I didn't mean to be late today. I'm sorry to make you wait."

"That's okay, Dad." It's the first time he's ever apologized to me for anything.

"I was going to call you on your cell to tell you I'd be fifteen minutes late. And then I thought I'd just call your mother. But the time got away from me."

"It's fine."

"No, let me tell you what happened." He leans forward. My dad is a great storyteller, and he loves to talk to people. He likes driving a cab for that reason alone. His stories are usually exchanged with strangers, but this time he has a story for me.

"Right after I talked to you at your mother's, a woman called and said she needed a ride to the airport. An airport trip is a good gig, you know. Forty bucks."

I nod.

"She was young. Mid-twenties." He pauses as if recalling for a moment what it felt like to be in his mid-twenties. "So I show up. She's a mess. She'd had a fight with her boyfriend the night before, and she'd piled up all her stuff on the curb. Everything she owned. She had suitcases and clothes, loose clothes, and all these teddy bears. I've never seen so many teddy bears. I helped her pile them into the trunk and the backseat."

The guy behind the counter is listening to us with curiosity.

"Jesus. I've never seen so many teddy bears." He pauses to wipe his forehead. "So we get almost all the way to the airport. And I know you're waiting for me." It's a long trip from Arvada to the airport, forty minutes if there's no traffic. "And she says, all of a sudden, take me back." He looks at me.

"Take me back?"

"Yes." He looks out the window at his cab.

The interior of the coffee shop is dark and cool. We both look out to the parking lot, where the cars, including the cab, glare almost painfully in the sun.

"So I take her back," he says. "I had to unload all those teddy bears." He shakes his head from side to side, like a slightly bemused judge before a jury. Then he grows serious again. "So that's why I was thinking about calling you. I knew you were waiting. But I had to take her back. And I made it. I made it in time."

I sit in silence for a moment, and then we both stand.

"It was good to see you," he says.

"I love you, Dad." The words slip out. I can't recall the last time I said that.

"I love you too, Kris," he says.

He turns and walks out to his cab.

THE CLASS-ACTION lawsuit *Cook v. Rockwell International Corporation*, delayed in the courts for more than sixteen years from when it was first filed in January 1990, is reassigned to three different district court judges before it finally goes to trial in 2005. The defendants, represented by the one-thousand-plus-member law firm of Kirkland & Ellis in Chicago, argue that scientists had determined that even though plutonium had settled on the plaintiffs' land, it was of "no consequence or concern." Lead counsel for the plaintiffs, attorneys Davidoff, Nordberg, and Sorensen, along with local counsel, argue that Dow and Rockwell were reckless in handling radioactive materials known to be dangerous, allowing plutonium to escape the boundaries of the plant, and that much of the defense was an attempt to continue to suppress and "spin the facts" of what happened at Rocky Flats and its effect on local communities.

Peter Nordberg has put sixteen years of his life into this case. *That's longer than it takes to get a doctorate,* he muses. Four times longer, in fact, than the time it took to fight World War II. He thinks about Mykaila, who has talked with him about this case every single day for the past ten years, since the day they first met. He's been mostly living in a hotel room since the trial began four months ago. Mykaila is back in Philadelphia taking care of the children. They've both been waiting for this moment.

In a Denver courtroom on February 14, 2006—Valentine's Day— lead trial counsel Merrill Davidoff is tied up with another case, and it's Peter who sits at the table in front of the jurors with the jury form in front of him. His armpits are damp under his suit, and he hopes no one will notice. He tries to keep his face expressionless. The courtroom is full, and many in the room are local residents.

Peter looks over at the defendant's lead attorney, David Bernick, a man that *Forbes* magazine describes as a "five-foot-seven dynamo." The *New York Times* calls him a "quiet corporate rescue artist." He's defended tobacco companies, breast-implant makers, major auto companies, and asbestos cases. He's quick to jump to his feet and not afraid to shout in the courtroom. Bernick doesn't look the slightest bit nervous.

Peter shifts in his chair. He tries to appear calm, though his heart is pounding. He's the one who made the final oral argument, and he thinks it was the high point of his career. Years of research and reasoning and passion went into it. One of the court exhibits showed a chart of MUF over the years. Even today, nearly three thousand pounds of plutonium are unaccounted for. The defendants, Rockwell and Dow, said that these were accounting errors and bookkeeping problems. You're making a big deal out of nothing, they said. Much of the plutonium had been shuffled around the plant; Rocky Flats wasn't as sophisticated back in the 1950s and 1960s; and even up to the present, they said, plutonium was a difficult thing to track.

The response of the plaintiff's attorneys, including Peter, was, "Are you kidding?" For one thing, some of that plutonium is in the backyards of the local residents they're representing.

The numbers seem almost ludicrous. Three thousand pounds of plutonium is enough to make hundreds of Nagasaki bombs. When a millionth of a gram of plutonium, breathed into the lungs or ingested into the body, can cause cancer, the potential effects of one and a half tons of lost plutonium is difficult to fathom.

Local rancher Bini Abbott was called in to show a film clip of the extraordinarily high winds at her ranch—wind that likely carries particles of plutonium. Dr. Shawn Smallwood, the scientist who conducted soil movement studies at Rocky Flats, showed a video of energetic groundhogs digging and cavorting in the soil at Rocky Flats, bringing soil material to the surface.

Rockwell and Dow argued that scientists had given Rocky Flats a clean bill of health and local citizens had suffered no ill effects from the plant's operations. Nordberg and his colleagues presented testimony and evidence demonstrating that the DOE controlled or influenced much of the scientific and epidemiological research on the health effects of plutonium.

Expert witnesses testified on both sides. Jon Lipsky described the facts the FBI uncovered during the raid. Mike Norton testified for the defense, but, surprisingly, even he concurred that serious environmental crimes occurred at Rocky Flats. Rockwell and Dow maintained that no harm occurred and that their studies proved this.

Peter isn't one to blame the workers at Rocky Flats. Most were conscientious. And besides, it was the Cold War. People thought it was important to have a nuclear deterrent. No one had ever made plutonium for nuclear weapons in bulk before, and Washington had been bearing down on the plant, saying we needed plutonium pits, we needed more, and we needed them immediately. Maybe, Peter thinks, we had to build the bombs. Maybe we had to build them in the numbers we did. And maybe some information had to be classified.

The problem was the blanket secrecy, starting in Washington and trickling down to the contractor at the site. Any scrutiny of the plant's activity was viewed with suspicion, and it became convenient to invoke the need for secrecy in order to evade accountability. Soon secrecy was institutionalized. When plutonium escaped the site and endangered local populations, the government stonewalled and provided nothing but bland reassurances. To Peter's mind, this case wasn't just about plutonium in people's backyards, property values, or health issues. It was about an abuse of the power of classified information, an abuse of secrecy, with devastating consequences for the public.

The jury files back into the courtroom, four men and six women. The judge takes his seat and begins to read the verdict form, one question, one answer, one question, one answer, one after the other.

A civil trial does not require a unanimous verdict, so the jury can find the defendants liable even with two dissenting votes. The judge continues to go through every item on the thirty-page verdict form. The jurors have voted unanimously on nearly every point. Both defendants are found liable for trespass and nuisance. By the time the judge comes to the part about punitive damages, it begins to hit Peter that the jury has taken their side. There is not a single dissenting vote when it comes to the amount of damages. Have they, in fact, won?

The courtroom is perfectly still. The judge's final words hang in the air for a silent moment before the room erupts.

The jury finds that the DOE, the Dow Chemical Company, and the former Rockwell International Corp. have been negligent and caused damage to properties surrounding Rocky Flats. They award punitive damages of $110.8 million against Dow Chemical and $89.4 million against Rockwell, with an additional $177 million in actual damages levied against each company, for a total judgment against the companies of $554.2 million.

The other plaintiffs' attorneys rise and turn to hug their clients. Nordberg is stunned. His spirits soar. It's a David-and-Goliath moment.

David Bernick jumps to his feet, filled with indignation. He begins shouting at the jurors, questioning their motives and values. For a moment the noise stops and the courtroom is shocked. This is not normal procedure. The jurors don't know how to respond. Peter recovers his composure and stands to object. The judge sides with him, quiets the courtroom, and tells the jurors that they are excused.

Some of the original parties to the litigation have died. Others have moved away. Those who remain, those who have just heard the verdict, can hardly believe their ears. They were all present for every day of the trial. "It was a no-brainer," says Bini Abbott. "I was worried about the health situation. I had some crippled animals born, and I wanted to find out if Rocky Flats was responsible." Bini herself has had cancer. She'd known Lloyd Mixon, who owned the infamous crippled pig named Scooter and had since moved away.

The press is waiting for them outside on the courtroom steps, cameras flashing. Merilyn Cook, the local ranch owner whose name is on the case, smiles broadly into the bright Colorado sunshine. "It's a tremendous verdict," she says. "It just feels wonderful." Merrill Davidoff, the Berger lawyer who led the case for the plaintiffs, tells the press, "I think the jury was outraged at the contempt which Dow and Rockwell showed for the community."

Dow's general counsel, Charles J. Kalil, releases a formal statement saying, "The jury's verdict is disappointing as numerous independent scientific studies have concluded that there is no past, present or future public-health threat." David Bernick is abrupt in his comments to the press. "We don't think that this was at all a fitting end to a very long and important legal process," he says. "And we're going to appeal."

No one seems to be listening to him.

One person who's not present to hear the verdict is plant manager Mark Silverman, who was awarded the Presidential Distinguished Rank Award by President Clinton before he died of an inoperable brain tumor in May 2005.

Peter walks out of the courtroom and calls Mykaila. She's not home, so he leaves a message—a message that she tells him later "sounds much more ebullient than your usual Scandinavian self!" She's thrilled that the case has ended well, and happy that Peter will finally be coming home.

There is talk among the plaintiffs and attorneys of going out for a glass of champagne to celebrate twenty years of work. But people are tired, physically and emotionally. Many of them trickle off, one by one, to their homes and hotel rooms. Peter stays to have a drink or two with those who remain, and then flies home, finally, to Philadelphia. The euphoric feeling never goes away.

I can't physically be in that courtroom, but I'm glued to the television and news reports. My search for a university teaching position has taken me to Tennessee, where I'm teaching at the University of Memphis. Sean and Nathan are in high school. Rocky Flats feels far away and yet just as near as ever.

My family is ineligible for compensation. My parents sold our house—for a fraction of its value—just months short of the parameter date established by the lawsuit. We will never see a cent. But that doesn't matter. My siblings and I never believed this day would actually come.

I call my sister Karma, who's now living in New Mexico and working as an environmental soil scientist. There's a moment of silence on the phone as we both fight back tears.

✪

AFTER ESTIMATING in 1995 that it would cost upwards of $36 billion and take seventy years to clean up Rocky Flats, the DOE is faced with a dilemma. The technology for a thorough cleanup is limited, and the amount of money and time it will take to clean up Rocky Flats to anything close to "background level," or anything resembling the natural condition of the site before the facility was built, is much more than they're willing to pay.

Simply admitting that mistakes have been made and closing the site permanently to public access isn't an option, either, partly for political reasons. It's prime development land, surrounded by even more prime development land, just minutes away from Denver and Boulder, with a great view of the mountains. New homes, communities, malls, and roads are springing up all around Rocky Flats. No one wants to think about plutonium anymore. And the DOE isn't interested in designating any of the former nuclear weapons sites as "national sacrifice zones," as some scientists have recommended.

So the DOE, the EPA, and the CDPHE change their minds. Or, rather, they change their standards. In 1995 the DOE signs a contract with Kaiser-Hill to begin cleanup at Rocky Flats. EG&G will be phased out. They make a deal with Congress to impose limits on the cleanup. This agreement, renewed in 2000, sets a deadline of December 2006 and dictates that cleanup and closure activities cannot exceed $7 billion. Instead of basing site cleanup levels on what would be considered safe for people to live or work on, they choose, as a model, a person who would most likely spend only a few hours a week on a wildlife refuge site: a wildlife refuge worker. This mature adult would have very limited exposure to the air, water, and soil. The equation doesn't take into consideration refuge visitors, or residents or businesspeople, or small children—children are especially vulnerable—who might spend longer periods in the area. Surely, they say, we can clean the site up well enough that a part-time seasonal wildlife worker's chance of growing ill or developing cancer might be a little higher than average, but not unacceptable, all things considered.

But how clean is clean? Exactly how much residual plutonium will remain in the soil? In 1996, as part of the Rocky Flats Clean-Up Agreement—a legally binding agreement for site remediation—the DOE, EPA, and CDPHE put forth interim radionuclide soil action levels (RSALs) for plutonium and other radionuclides. The RSALs will determine how much radioactive material will have to be removed and how much can remain on the site. If contamination levels are below the RSAL, no action will be taken.

Even the most cynical observers are surprised when the RSALs are announced. The legally binding standard the government intends to adopt for cleaning up plutonium in the soil on the Rocky Flats site is 651 picocuries per gram of soil, significantly higher than at any other site in the United States.

There will be no cleanup of off-site soil.

How does 651 picocuries per gram of soil at Rocky Flats compare to other plutonium-contaminated sites? The bomb test sites at Enewetak Atoll and Johnston Atoll in the Pacific Ocean are lower, at 40 and 14 picocuries per gram of soil, respectively. The Livermore National Lab in California is at 10 picocuries per gram of soil. Fort Dix in New Jersey is at 8 picocuries per gram of soil. A portion of the Hanford, Washington, site, one of the most contaminated sites in the country, is at 34. Only the Nevada Test Site, where the nuclear bombs were tested, is higher, at 200 picocuries per gram of soil—and then only for part of the site that is not open to the public.

Citizen and watchdog groups are deeply alarmed. They call for an independent assessment of the RSALs and how they were determined. In response, in 1998 the DOE agrees to fund an independent assessment of the cleanup standards. After fifteen months of study, Risk Assessment Corporation, a well-known team of scientists, releases a peer-reviewed study that recommends a level of no more than 35 picocuries per gram of soil, a reduction of 95 percent from what the DOE, EPA, and CDPHE have originally agreed to as the legally binding standard. The Soil Action Level Oversight Panel sends this recommendation to the DOE.

The DOE never formally responds.

Instead, the DOE, EPA, and CDPHE come up with a compromise of their own. The top three feet of soil at Rocky Flats will be cleaned up to 50 picocuries per gram. Soil from three to six feet below the surface will be cleaned up to a level of 1,000 to 7,000 picocuries per gram. There will be no limit on the amount of plutonium that will remain in the soil six feet or more below the surface, despite the fact that a great deal of material and contaminants remains below the surface, in parts of the lower floors and basements of buildings and the extensive piping that connected them.

Critics note that just cleaning up the top contamination level does not take into consideration soil movement due to weather, erosion, or—perhaps most worrisome—burrowing animals. A 1996 study of burrowing animals present at Rocky Flats shows that they constantly redistribute soil and its contents. Animals dig to depths of ten to sixteen feet and disturb as much as 11 to 12 percent of surface soil in any given year. Wind and water actions contribute further to soil movement.

In 2000, Congress proposes transforming the Rocky Flats Environmental Technology Site into a wildlife refuge, and the Rocky Flats National Wildlife Refuge Act passes in 2001. Kaiser-Hill promises they can do the cleanup fast, and for the agreed-upon price. The graduated standard for residual plutonium in soil is officially adopted in 2003. Of the $7 billion slated for cleanup, most will go to site security, relocation of weapons-grade material, removal of bomb-production waste, and demolition of buildings. Only 7 percent of the total—roughly $473 million—will go for actual soil and water cleanup.

Rocky Flats will become a plutonium graveyard, with a few sprinkles on top.

Government officials claim that background levels for plutonium—due to fallout from the atmospheric testing of nuclear weapons from 1945 to 1963—make the whole argument moot anyway. In 1963 many countries signed the Limited Test Ban Treaty, promising to stop testing nuclear weapons in the atmosphere, underwater, or in outer space. Underground nuclear testing was still permitted. The ban met with mixed results: France continued atmospheric testing until 1974 and China until 1980. Underground testing continued in the Soviet Union until 1990, in the United Kingdom until 1991, in the United States until 1992, and in both China and France until 1996. These countries promised to discontinue all nuclear weapons testing after the Comprehensive Test Ban Treaty in 1996 (a treaty that has never been ratified by the U.S. Congress). Some countries that did not sign the treaty have continued testing. India and Pakistan last tested nuclear weapons in 1998, North Korea in 2009.

The average background level for plutonium along the Front Range of the Rockies resulting from global fallout is 0.04 picocuries per gram of soil. Although measurements vary, most levels at the Rocky Flats National Wildlife Refuge, and some of the off-site areas around Rocky Flats, are dozens of times higher.

The wildlife refuge is designed to protect wildlife and habitat as well as provide opportunities for public hiking, biking, and possibly hunting. In December 2004, the U.S. Fish and Wildlife Service announces the results of a study to determine if the deer at Rocky Flats will be safe for human consumption. Tissue is collected from twenty-six deer and a control deer. Only the deer from Rocky Flats have detectable levels of plutonium in bone samples. The government determines that the risk level from eating deer with "tissue activity level" falls within an acceptable range.

"Close it, fence it, pave it over," implores an environmental engineer at a forum in Boulder a few months later. But the government has no intention of closing off the land. In their view, turning former nuclear sites into nature reserves is "thrifty environmentalism. It would cost a fortune to clean up the site so people could live there . . . but making it safe for 'wildlife-dependent' public use is more affordable."

ROCKY FLATS tends to run in families. During the Cold War years, it was common for entire families to work at the plant. Workers spread the word about good jobs and great pay, and government officials found it easier to do security clearances on relatives of employees who had already been cleared.

Many workers, and sometimes members of the same family, became ill. Rocky Flats was supposed to keep track of on-the-job exposures, but their record-keeping was inconsistent and sometimes inaccurate. Records were often lost or misplaced. Dosimeter badges, which employees wore to record daily exposure levels, were sometimes mixed up or zeroed out. Workers did not have full access to their own exposure records.

Sixteen members of the Dobrovolny family worked at Rocky Flats. By 2007, seven were sick or dead from health issues related to Rocky Flats. Mark Dobrovolny worked at Rocky Flats for years. His father worked at the plant and died of lung cancer. Mark, who painted walls and floors at Rocky Flats with a coating designed to prevent spilled radioactive material from getting into the air, has health issues, but is reluctant to request his exposure records. "Actually, I don't want to know," he said. "And I don't know that I would believe the information that was there." Like other workers, he was never sure if he was really safe. "Everyone who worked out there thought they would get cancer," he said. His former wife, Michelle, who also worked at Rocky Flats, has chronic pneumonia, bronchitis, and other health problems. "The hardest thing is watching my family members die around me," she said.

Over the years, a number of studies have been done on worker illnesses at Rocky Flats. A DOE-financed study in 1987 found that workers exposed to plutonium experienced a higher proportion of deaths tied to brain tumors and cancer in blood-producing organs. That same year, the epidemiologist Dr. Gregg Wilkinson published an article finding that some exposed Rocky Flats workers with internal plutonium deposits as low as 5 percent of the permissible "lifetime body burden" developed a variety of cancers in excess of what was normal for unexposed workers. The *New York Times* reported that Wilkinson, whose report had received extensive peer review, had been told to alter the results of his study before publishing it. When he first showed his findings to a physician at Los Alamos, the doctor exclaimed, "Oh, this is terrible. . . . If these findings [are] real, they would shut down the nuclear industry." Wilkinson said a deputy director "stated that I should not write to please peer reviewers, but rather I should write to please the DOE because they provided funding and support." Wilkinson published the results anyway. He was ordered to submit all future work to an assistant director of the Los Alamos lab for approval before publishing any further data, and in protest he left his job.

Wilkinson then went to the State University of New York at Buffalo and conducted a study on the health of women employed at twelve DOE sites around the country, including Rocky Flats. Over the years the DOE had employed more than eighty thousand women, but had never conducted a study of the health effects of nuclear work on women. Wilkinson found that exposure to external ionizing radiation appeared to be associated with an increase in leukemia and brain tumors, and an increased risk for all cancers combined.

In 1990, testing by doctors at the National Jewish Hospital in Denver revealed that many workers were suffering from chronic beryllium disease. Unlike cancer or other diseases that were difficult to tie directly to exposure to plutonium or other radioactive elements, beryllium disease was clearly a result of working conditions at Rocky Flats. Exposure to beryllium dust can lead to a buildup of scar tissue in the lungs that reduces the capacity to process oxygen. Several hundred workers were diagnosed with chronic beryllium disease, which has no cure and is often fatal.

A 2003 study found that Rocky Flats workers who inhaled radioactive particles of plutonium were particularly vulnerable to lung cancer. Rocky Flats workers in general may also be more susceptible to brain cancer, and they appear slightly more susceptible to other cancers such as leukemia. Conversely, in 2009 the CDPHE study concluded that worker cancer deaths were not that much higher than in the general population.

There always seemed to be at least one contradictory study that made it hard to draw firm conclusions.

In 2000, Energy Secretary Bill Richardson acknowledged that workplace exposures had harmed the health of workers in the U.S. nuclear weapons industry. That same year Congress passed the Energy Employees Occupational Illness Compensation Act (EEOICA), a program intended to compensate ailing workers and shift the burden of proof of exposure to the government rather than the worker. The government relies on dose reconstruction, which uses estimates and computer modeling to try to determine past exposure to radiation. The process is expensive, arduous, and prone to error, partly because of industry-wide poor record-keeping.

During the production years at Rocky Flats from 1952 to 1989, more than sixteen thousand people worked at Rocky Flats, and thousands more during the postproduction years. Approximately six thousand of them have applied for compensation on the grounds that their illnesses were caused by radiation or toxic chemicals. In addition to lost records, many have had to deal with repeated delays and complicated exposure formulas. Although hundreds of workers have been helped by the program, by 2011 two of every three who have sought compensation have had their claims denied. Four of the seven members of the Dobrovolny family were denied medical care and compensation.

Charlie Wolf is one of the few managers at Rocky Flats who goes into the hot areas with his workers to coordinate a job or help out in a pinch. Hired in 2003, he is deeply involved in cleanup operations, but he doesn't live to see the final result of his efforts. Just a year later he develops an aggressive brain tumor called a glioma. The doctors originally give him six months to live—he fights for six years, and dies in 2010. His cancer can be tied directly to Rocky Flats, yet he is repeatedly denied compensation. It isn't until after his death that his wife and daughters receive $150,000 in compensation, which barely makes a dent in his medical expenses.

In February 2010, Representative Mark Udall introduces the Charlie Wolf Nuclear Workers Compensation Act, designed to reduce some of the red tape of EEOICA and expand the category of individuals and list of cancers that qualify for compensation. The legislation gains little support.

THE CLEANUP itself generates waste. Every glove, shovel, screwdriver, and wrench used to disassemble Rocky Flats equipment becomes toxic, too. Tents are constructed to try to limit airborne contamination when buildings are demolished. For every single pound of detoxified equipment, another 1.6 pounds of waste is generated.

In August 2004, former grand jury foreman Wes McKinley, whistle-blower Jacque Brever, and FBI agent Jon Lipsky hold a press conference on the steps of the Denver capitol to protest plans to open the Rocky Flats National Wildlife Refuge to the public. On the way to the event, Lipsky receives a call from his employer. When it's his turn to speak to reporters, he says, "I received a call from the FBI ordering me not to talk about the Rocky Flats case, so I can't tell you what I came here to tell you. . . . As a father and a fellow human being, I urge you not to allow recreation at Rocky Flats. I'm sorry I can't tell you more."

On October 13, 2005, the last shipment of transuranic (plutonium-laden) radioactive waste from Rocky Flats heads off to the Waste Isolation Pilot Plant (WIPP) in New Mexico, which opened on March 26, 1999, after twenty-five years of planning and complications, including protests and lawsuits. Based on the compromised cleanup standards, the Rocky Flats cleanup is declared successful. "Our success at Rocky Flats is a great inspiration to those other sites [around the country]," says Clay Sell, deputy secretary of the DOE. "Six years ago, seven years ago, the problems at Rocky seemed insurmountable." The EPA certifies the site on June 13, 2007. In July the DOE transfers nearly 4,000 of the 6,200 acres of the Rocky Flats site to the U.S. Fish and Wildlife Service for eventual use as a public recreation area. Approximately 1,300 acres of deeply contaminated land, including the former industrial zone and the areas capped with cement, are to remain off-limits. The DOE will retain this land for long-term surveillance and monitoring.

When the cleanup is declared complete, 2,600 pounds of plutonium are still missing. Studies demonstrate that vegetation and plants at Rocky Flats absorb plutonium from the soil—called "uptake"—and birds and animals carry it far afield. The control of spreading weeds is a constant problem. Another study shows that small burrowing animals such as groundhogs and even the tiny Preble's meadow jumping mouse regularly bring plutonium to the surface, despite the DOE's computer modeling that claims plutonium will remain stable below the surface.

Despite public opposition and concern, plans move forward for the Rocky Flats National Wildlife Refuge. Visitors, including school groups, will enjoy a visitor station staffed on a "seasonal basis," at least sixteen miles of open trails, parking at trailheads, scenic overlooks, and the opportunity to view wildlife and go hiking and biking. Home development near Rocky Flats continues at a rapid pace, with the closest house within two miles. Nearly two and a half million people live in the Denver metropolitan area within a fifty-mile radius of the former nuclear weapons factory.

In the spring of 2005, Tamara Smith Meza goes in for her regularly scheduled MRI scan. For months the results have been normal, but this time the doctors are concerned. It looks like the tumor has returned and has spread to Tamara's abdomen. The doctor schedules a full-body PET scan, but in preparation she has to undergo an ultrasound test. She usually keeps her emotions in check, but this time she is frightened.

On the day of Tamara's ultrasound, her regular doctor isn't in, but he leaves instructions for the technician to go ahead with the ultrasound in preparation for the PET scan. Tamara and David sit down and answer the usual questions about her health, her list of medications—questions she has answered a million times. One question in particular is always more difficult than the others.

"Are you pregnant?" the technician asks.

"No."

"Are you sure?"

Tamara is very sure. She glances at David. Like her sister, Tamara has always had fertility problems. She and David are now thinking about adopting a child.

Tamara lies down on the table and the technician smears jelly on her abdomen. She begins to move the wand over her skin. David sits nearby, looking intently at the screen. Suddenly the technician laughs. "Well, there it is!"

David looks up in surprise. This seems a little unprofessional. Tamara is stunned by the technician's enthusiastic tone. "What?" she asks. She cranes her head toward the screen. "Is that the tumor?"

"Tumor?" the technician says. "What tumor? That's a baby!"

"A baby?" Tamara looks at the technician as if she were out of her mind. David is speechless.

"Yes." The technician leans forward and moves the pointer on the screen.

"But we can't have children!" Tamara says. "We know. We've tried. I'm sick."

"Well," the technician says, "I'm not your regular doctor, and I don't know what your health issues are. But tumors don't have heartbeats. And this has a heartbeat."

Six months later, Tamara and David are the parents of a healthy baby boy. They name him Isaac.

JIM STONE, the troubleshooter-turned-whistle-blower whose testimony helped shut down Rocky Flats, has waited a long time for his reward. Following the FBI raid in 1989, he filed a whistle-blower fraud case against Rockwell and was awarded $4.2 million in damages. But he never saw a dime of it. Rockwell appealed the case and eventually it went to the U.S. Supreme Court. After his eighteen-year fight, in 2007 the U.S. Supreme Court denies Stone any damages and also makes it more difficult for whistle-blowers to share in proceeds from fraud lawsuits against government contractors. By this time Stone is eighty-one years old and living in a nursing home. Two weeks after the decision, he dies from complications related to Alzheimer's.

"He died with nothing more than the clothes on his back and the love of his family and friends," says his son Bob Stone. "I know if he had it to do all over again, even knowing how it turned out, he would have done it just the same."

IN THE spring of 2010, four years after the decision in *Cook v. Rockwell International Corporation*, the appeal finally goes before a three-judge panel at the 10th Circuit Court of Appeals in Denver. The award, now at $926 million including damages and accrued interest, to more than twelve thousand Colorado homeowners is at stake. The case has been in litigation for nearly twenty years.

Peter Nordberg and his colleagues present their case—again—before the judges. Peter returns home to Mykaila, once again exhausted, but confident that they've done a good job.

My siblings and I tensely wait to hear the judges' decision. No matter what happens, there will be no compensation for us. Even so, not just our troubled years but our best years as a family were in Bridledale, and we loved it. We loved the land and we loved the house. We want some sense of justice.

I am hopeful that the decision will stand. The evidence seems overwhelming. My brother, Kurt—who's always teased me about wearing rose-colored glasses—is cynical. "Are you kidding, Kris?" he asks.

He's right. On September 2, 2010, the decision is overturned. The three-judge panel rules that under the law, the presence of plutonium on properties south and east of the plant at best shows only a risk and not actual damage to residents' health or properties. " 'DNA damage and cell death' do not constitute a bodily injury in the absence of the manifestation of an actual disease or injury," the panel writes, later adding, "Plaintiffs must necessarily establish that plutonium particles released from Rocky Flats caused a detectable level of actual damage." Decreased property values cannot, the panel further states, be counted as damage.

The court also rules that the jury reached its decision based on faulty instructions that incorrectly stated the law.

I am stunned.

Peter Nordberg never hears the verdict.

In April 2010, just after returning from the appeals court in Colorado, Peter buys tickets for Mykaila and their two daughters to see Garth Brooks in Las Vegas. He isn't a fan himself, and as always he has plenty of work to do, but he wants them to have a good time. When Mykaila and the girls return from their trip, Peter is sick. It isn't unusual for him to come back from Colorado tired after a stressful court proceeding; sometimes the altitude bothers him, though mostly he just needs to unwind. This time, though, it seems worse. Mykaila is worried. "I'm fine," he tells her. He doesn't want to go to the doctor; he hates doctors. In all their years of marriage, he's only been to the doctor three times. Mykaila thinks, *Okay, well, he's just tired. He's run a marathon of a case. He just needs rest.*

But on the third day he's ill, Mykaila insists he go to the emergency room. He seems to worsen even as they drive, and by the time they get there, he can't get out of the car and asks for a wheelchair. In the emergency room, Peter's heart rate is above 250 and he can barely breathe.

The doctors stabilize him and get him upstairs to a hospital room. He's told he'll be in the hospital for a day or two of observation. When Peter starts pestering the staff with lots of questions, Mykaila knows he's back to his old self. "I feel a hundred percent better," he says. From his bed, he makes a video for their son Brinkley. "I'm going to have dinner now. I will see you in a couple of days," he says into the camera. "I love you. Good-bye." Mykaila sits with him as he finishes dinner and starts to fall asleep.

Suddenly all his vital signs go critical. His kidneys stop working. His lungs fill with fluid. "I'm drowning!" he cries. He loses consciousness. His heart quits. Doctors perform CPR and get his heart rate back. Over the next twelve hours, they lose him and bring him back six times. On the seventh time, they can't resuscitate him.

Mykaila, who's never left his side, loses him as quickly as she found him.

It turns out that Peter had a heart problem, a birth defect. His heart is four times the size it should be.

Peter was fifty-four years old, and he and Mykaila had been married thirteen years. He left behind four children: Brinkley, age eleven; another son, age twenty-three; and twin eighteen-year-old stepdaughters. He died on his older son's birthday, two weeks before the girls graduated from high school.

Peter taught his children to be strong, and he taught them to be savvy. He told them that no one, not even the government, is above questioning. Ask respectfully, he said to them, but always question. You can't sit and say nothing.

Mykaila thinks about his writing, his research, the briefs that read almost like novels. One reason the trial was initially so successful was that Peter was able to take technical and complicated information and turn it into a story that helped the jurors understand Rocky Flats and all its implications. No one ever knew his health was at risk. For years, she says, "he'd been putting in twenty-hour days with a body that needed to be working three hours a day."

Mykaila is grateful for one thing: Peter will never know that the case he devoted his life to has been overturned, within the space of only a few minutes, by three judges on a sunny Colorado afternoon.

THE DAMAGING effects of high doses of radiation are well documented. The effect of a high dose usually shows up shortly after a person is exposed, and the severity of the symptoms—including dysfunction or death of large numbers of cells, which can lead to cancer—increases with the dose.

By contrast, the health effects of long-term, low-level radioactive contamination are not immediately obvious, and have been under debate ever since John Gofman's 1969 recommendation that the AEC dramatically lower its threshold level. In 2000, however, scientists at Los Alamos wrote of low-level exposures, "Ionizing radiation of any kind can lead to alterations of a living cell's genetic makeup, and sometimes those alterations trigger the uncontrolled growth and multiplication of that cell's progeny, more commonly known as cancer. . . . Moreover, there is a substantial delay between the time of exposure and the appearance of the effect. If the effect is cancer, the delay ranges from several years for leukemia to decades for solid tumors." A recent, three-nation study by the International Agency for Research on Cancer (IARC) shows a relationship between external radiation and leukemia. And new studies by the DOE have found a relationship between external radiation dose and lung cancer at the Fernald site, and leukemia at the Savannah River site.

Back in 1981, Dr. Carl Johnson reported a significantly higher rate of cancer in neighborhoods around Rocky Flats, data that was later confirmed by other studies. His research also showed a higher rate of thyroid cancer, particularly in females.

Schoolmates Stacy and Curtis Bunce grew up down the road from us, in a historic house on Simms Avenue. Like us, they swam in the irrigation ditch. Like Tamara's family, they had a garden in the backyard and their own well for water. The family lived in the house for thirty years.

Stacy's father died of cancer. Her mother has thyroid cancer. Curtis has cancer of the thyroid and neck lymph nodes as well as squamous cell carcinoma. Stacy has polycystic ovarian syndrome (an immune disorder that can cause infertility) and early chronic thyroiditis. "Nearly every family we know in the neighborhood has had some form of cancer or thyroid problem," Curtis says.

"We thought of Rocky Flats just as we thought about the grocery store down the road. It was just there," Stacy adds. "We didn't give it a second thought."

The Dunns' story is similar. In 1984, John and Barbara Dunn moved to Colorado from South Carolina for John's new job as a hotel manager in Boulder. They bought a home in a new housing development four miles downwind of Rocky Flats, not far from Bridledale. Each day when John drove to work, he passed the protesters and workers lining the road to the Rocky Flats entrance. He and Barbara knew little about Rocky Flats, and when Barbara asked people about it, she was told not to worry. When their daughter was born, they named her Kristin after a Norwegian friend. Kristin played in the sandbox and swam in the shallow irrigation ditch that ran across the back of their property. The family had a vegetable garden.

Eventually John was transferred and the family moved away. Years passed. In 2010, Kristin is a student at Michigan State when she comes home from college with what she thinks is a cold or virus. But there is a marble-sized swelling in her neck, and her symptoms worsen. When a biopsy comes back negative, the doctor suggests the mass be removed anyway. The surgeon is stitching her up when the lab calls the operating room with a diagnosis of advanced cancer. When Kristen awakens in the recovery room, her thyroid has been removed.

When Kristin is first diagnosed, John jokes with a friend and co-worker, "I wonder if Rocky Flats has anything to do with it!" To his surprise he learns that another employee of the Boulder hotel, who lived in the very same neighborhood during that same time period, has a fifteen-year-old daughter with thyroid cancer. This girl has also just had her thyroid removed.

Is it Rocky Flats? Tamara Meza's doctor believes her brain tumors are likely caused by Rocky Flats. In November 2011, her cancer returned and she underwent surgery for a fifth brain tumor. And doctors have told the Bunce family that their cancers are not genetic or coincidental; they believe the family was exposed to radiation. Curtis Bunce's doctor recommends that he look into assistance for people who have lived near facilities like Rocky Flats, something similar to the Fernald Medical Monitoring Program in Ohio.

But there is no public health monitoring or medical assistance for people who live or lived near Rocky Flats.

In April 2010, a group of citizens sponsored by the Rocky Mountain Peace and Justice Center independently raise funds to hire a lab to sample and test soil in two locations off the Rocky Flats site, one indoors and one outdoors. The indoor sample is taken from a crawl space beneath a house built in 1960 about one mile southeast of the site, not far from my childhood home. The study finds plutonium in breathable form in the crawl-space sample, and possibly in the outdoor sample.

The group sends letters to pertinent government officials with three recommendations: establish a program for sampling dust in surface soil at and around Rocky Flats for its plutonium content; maintain the wildlife refuge as open space that remains closed to the public; and establish a program to monitor the health of people living in the affected areas. The DOE responds indirectly to the first recommendation and not at all to the others; the Fish and Wildlife Service responds to none of them.

In September 2011 the citizen group hires technical specialists to do another, more extensive sampling along a road next to the site, near where significant home construction is under way, as well as other nearby locations. Testing is done farther out from the site for comparison purposes. The Fish and Wildlife Service declines permission for the group to enter the refuge. As of this writing, results are pending and further analysis continues.

Were we—are we—living under the protection of the bomb, or under its shadow?

WE LIVE with the deadly legacy of nuclear weapons. The graphite bricks used by Enrico Fermi in the first "atomic pile" during the Manhattan Project are now buried in a Cook County forest preserve. The Hanford site in Washington holds the acid used to extract plutonium for the first atomic bomb test in New Mexico. The manufacture of each container of enriched uranium, each reactor fuel element, and each gram of plutonium created a wide range of radioactive waste, all of which must be analyzed, categorized, handled, and stored differently. Much of it is so toxic that it has to be specially treated before it can be stored or disposed of, and it must be isolated for hundreds of centuries.

The technical challenges of storing or dealing with radioactive waste are daunting, partly because radioactive waste can remain toxic for such vast lengths of time—from 10,000 to millions of years. The most problematic elements are neptunium-237, with a half-life of 2 million years, and plutonium-239, with a half-life of 24,000 years. It will be 240,000-plus years before the plutonium we made in the 1940s will approach the end of its radioactive life. Radioactive waste must be stored and managed in a stable place—safe from earthquakes, water, or other weather elements—and be under the constant watchful eye of stable human institutions or governments. Government agencies struggle with how to communicate to future generations how lethal these storage sites are. Signs in English will likely be inadequate, and perhaps are even today. Linguists are working to develop symbols or pictures that will warn of contamination and can last hundreds and thousands of years.

The DOE and the Colorado Department of Public Health and the Environment continue to state that Rocky Flats and the surrounding areas are completely safe. In a 2009 promotional video, "From Weapons to Wildlife," Carl Spreng of the Colorado Department of Health says he looks forward to Rocky Flats opening to the public. "That's an exciting prospect. These lands have been essentially protected for many years because they were part of the buffer zone of Rocky Flats, and it's an example of some pristine prairie areas that will be a valuable asset to the citizens in the area and the state of Colorado. The land is," he emphasizes, "a jewel."

The land is not pristine. Some of the most hazardous materials known to mankind remain on the site. You can't see, smell, or taste any of it. In September 2004, in response to the Draft Environmental Impact Statement for the Rocky Flats Wildlife Refuge, 81 percent of the commenting public rejected public access to the refuge and felt that it should be permanently closed.

Even some workers express concern over the cleanup. Shirley Garcia worked at Rocky Flats for fifteen years, starting out in 1982 in Building 371. She was one of the few women who loaded salt and plutonium into huge furnaces, working in heavy lead-lined gloves and pulling the crucibles out of the furnaces. "It all comes down to cost," she says. "There's just not enough money to get it done. I'm afraid if we don't have a good enough cleanup now, we're not going to get another chance. This is our one and only chance. In the past, with the government, if you messed up, they would throw money at you, and you'd go do it again. This is not going to be the case this time. We're going to have one shot and one shot only."

The DOE believes the cleanup is adequate and the site is safe. "Everybody's allowed to have misconceptions," notes the current DOE site manager at Rocky Flats. "I think that once [people] get out here and experience it, then it will be their decision as to whether or not they want to maintain those old preconceptions or not. I think what they'll find out is that there's nothing out here but great stuff. . . . I don't see there being a risk here."

The DOE readily admits that groundwater at several areas below the Rocky Flats site is contaminated. However, current testing of wells along the Rocky Flats property line "suggest[s] that the groundwater contamination plumes remain on site." John Rampe, a former Energy Department environmental scientist, believes the refuge is safe. "We find occasional plutonium or other contaminants that don't meet the state standards," he says. "When we find it, we remove it. We have removed dozens of miles of soil, scraped off the top layers and sent them to waste facilities." A spokesman from the U.S. Fish and Wildlife Service says, "There is absolutely no reason to warn people about this place. The refuge is safe; it would only scare people."

ON AN August morning in 2010, I return to Colorado to visit Karma and my father, who is now in an assisted-living home. The flight into Denver International Airport is rough; a summer thunderstorm that threatens to turn into a tornado tosses the small plane like a cat tosses a mouse, and all the passengers look a little green. It's a relief to touch ground.

Karma and I decide to take Dad to his favorite pizza place for lunch. He grows tired of cafeteria food. It's a beautiful afternoon with the brisk, clear air of Colorado that's like nowhere else. We sit out on the patio, overlooking a mountain stream. My father's health has declined in recent years, but his mind is sharp. He reads all the books I send him and he wants to talk about them. He pays attention to politics. He asks us both about our work and our lives and seems genuinely interested. We order pizza and a couple of beers. My dad has a Coke.

Ordinarily, we never talk about the past. This time, though, I have a question. I've carried it in my mind for more than thirty years. It's the one question my mother said I must never ask.

"Dad," I ask between bites, "do you remember that car accident we had so long ago?"

He pauses and considers for a moment. "Yes." No wariness. A straight answer.

"Did you know I broke my neck in that accident?"

"Oh." He looks surprised, then his face drops. "Oh, no." He puts down his slice of pizza and looks down for a long moment. I'm not sure how to read the silence. He furrows his brow.

Then he looks up. "No, I didn't know." His eyes behind his glasses are moist but direct. "I'm sorry, Kris."

I find it hard to speak.

We break the silence by each reaching for another slice. "I broke my back in that accident," he adds as a kind of afterthought. "Still gives me trouble."

Karma suggests dessert, and we move on to happier subjects. When we get up to leave, Dad quickly wraps the last three pieces of pizza in a napkin and buttons them inside his shirt.

"Why don't we get a box for that?" I ask.

"No, no," he says, shaking his head and grinning. "It's better this way. I feel like I'm getting away with something."

THE NEXT day, I rent a car and drive the sixteen miles from Denver out to what used to be Rocky Flats. There are new roads where no roads were before, houses and strip malls and office buildings where once there were fields. I drive down Indiana Street, to the spot where I used to wait at the east gate in a long early-morning line of cars. There's nothing here now. The land is empty and still except for a number of unmarked monitoring devices and several No Trespassing signs. Legislation that would have required additional signage informing visitors of what happened here, and why it might still be dangerous, has twice been defeated. The wide paved road leading to the guard checkpoint is gone, the asphalt dug up and covered with soil and low grasses that sway slightly in the wind.

I drive past Standley Lake, where a few boats bob on the water, and into Bridledale. Many of the same neighbors are still there, although attitudes have changed. The golden, guileless optimism of the 1970s is gone. Bridledale, it turns out, was not the heaven my mother imagined. The economy declined. People divorced. Their kids married and divorced.

No one defends Rocky Flats anymore. Every family has a cancer story or knows of one. Still, if people are critical, it's in a whisper.

The evergreen trees my mother planted along our long driveway are monstrous now, too large for the narrow strip of earth that anchors them. The rail fence is worn and sways like the back of an old horse. Our house looks weathered, a little shabby, the concrete of the front porch cracked. I knock on the door and the man who answers the door kindly listens to my request. He lets me wander around the house, all gold and avocado—

little has changed—and even the purple walls of my bedroom are the same, except with a different bed, a different bedspread. In the den a well-stocked wood fire roars, although the temperature outside is barely cool. The house feels warm and safe, a sanctuary. From the kitchen window, from which I could once see the Rocky Flats water tower, there is nothing but open space punctuated by housing developments and the deep blue mountains in the distance.

I shake the owner's warm, damp hand and climb back into my car. It doesn't take long to reach the cemetery in old Arvada, the rows and rows of crumbling markers on the small rise just above the square brick block that was our first house. The neighborhood has fallen into decline. The Arvada Beauty Academy, where my mother spent long afternoons getting her beehive hairdos, is gone, but the Arvada Pizza Parlor remains, the paint on the façade faint, the lettering faded. The customers don't need a sign.

I pass through the cemetery gate with the stone marker carved with the year 1863, and drive down the pebbled road to where my grandparents are buried. Someday my father will be buried here, too. When my siblings and I helped him move out of his apartment, he gave me a worn cigar box filled with photos dating back to the forties. "You're the family historian," he said. "You keep all the stories." My mother—whose death three years ago from a stroke is still so fresh it grips my heart—requested that her ashes be buried in her beloved Minnesota. Colorado was never really her home.

I sit down on the ground between the two markers and look toward the mountains. It all feels familiar. The rough grass scratches my ankles and here, above all the houses, the strip malls, the checkered patterns of housing developments that stretch as far as I can see, a meadowlark sings. Will I be buried here as well? Will a meadowlark still sing?

I inventory my body. It's a sturdy one, all things considered. Maybe some of what my mother used to say about our Scandinavian ancestry— farmers, most of them—is true. Working the land made us strong. I'm not a farmer, or even a very good gardener. But I love the land; I love this land.

In the geography of land and the geography of the body, some things are seen and some are unseen. My stubby toes remember dim-witted Chappie, who artlessly stomped on my feet with his iron shoes, and sly Tonka, who did the same but was more intentional. My left knee has a long scar, a reminder of my rock-climbing days. Four inches below my belly button is the tiny pink smile of the C-section that brought Sean into the world. My neck bears two scars. One is a small puncture from the handlebar of my first tricycle. The second scar, more visible, is a vertical line along my neck that takes the place of my left lymph node. It keeps me mindful of the scare I had with lymphoma and the doctor who told me I had better figure out who was going to raise Sean and Nathan because it probably wasn't going to be me.

The body is an organ of memory, holding traces of all our experiences. The land, too, carries the burden of all its changes. To truly see and understand a landscape is to see its depth as well as its smooth surfaces, its beauty and its scars.

I have spots in my lungs. I've never smoked. A speck of plutonium in the lung looks like a tiny starburst, a punctuation point of energy reaching out in infinitesimal pointy spires to the surrounding tissue. My spots, so far, are harmless. But other problems persist. I changed my diet and lifestyle to try to manage the symptoms I've struggled with my entire life, with no clear diagnosis. Karma has had several bouts of cancer related to her reproductive organs. Kurt has rheumatoid arthritis. Kurt, Karma, and I all have ongoing problems that seem to be related to immune system deficiencies: chronic fatigue, muscle ache, swollen lymph nodes, and a high white blood cell count.

"It appears that my body has no control of its immune system," Kurt wrote recently, after a fresh round of blood tests, "but the reason behind it is undetermined. They've tested for just about every known disease or cancer, but thankfully all are negative. Crazy stuff, huh? Rocky Flats strikes again?"

We'll never know for sure. Karin's health seems to be good, but she holds fast to the Scandinavian family doctrine of never complaining and never going to the doctor.

We're the lucky ones. Nearly every family we grew up with has been affected by cancer in some way. Some of those illnesses and deaths can be linked, directly or indirectly, to Rocky Flats.

Everything else is, as the government likes to say, nothing but conjecture. Speculation. Exaggeration. Media hype. They say there is no direct link between Rocky Flats and health effects in the surrounding communities. All contaminants are at levels that have been declared safe by the government.

It's hard to imagine that everyone in the government believes this.

I think of the day I sat next to Mark's body, as still as stone, at the mortuary in Boulder. How quickly time passes, and how quickly things change. Yet the emotion remains the same. I wish I had been able to say good-bye. I wonder if my father ever thinks about the time he stood at my bedroom door and I wasn't able to open it. Would my life, his life, be any different if I had risen from that bed? I remember being a child and the sweet freedom of galloping through the fields around our house and out to the lake. That time seems very long ago.

When Kristen Haag died in Bridledale at age eleven, her father, the man who built our house, considered taking Rocky Flats and the DOE to court. Money can't take away grief, the neighbors said. You can't make the government into a scapegoat. It's easy to feel paranoid about things you can't control. Sometimes people just die of cancer. Sometimes even a child dies. These things are in God's hands.

We don't talk about plutonium. It's bad for business. It reminds us of what we don't want to acknowledge about ourselves. We built nuclear bombs, and we poisoned ourselves in the process. Where does the fault lie? Atomic secrecy, the Cold War culture, bureaucratic indifference, corporate greed, a complacent citizenry, a failed democracy? What is a culture but a group of individuals acting on the basis of shared values?

In less than a generation we have almost forgotten what happened at Rocky Flats, and why it must never happen again. In a few years it will be completely forgotten, as if it never occurred at all. Will those who walk on the trails and pitch their tents to watch the stars know what the land can't forget? Years and decades will pass; governments and government agencies will change. People will build homes and businesses and roads and parks on land tainted by an invisible and invincible demon. And no one will know.

In early 2011, following the reversal of the jury verdict, the plaintiffs for the Rocky Flats class-action lawsuit requested that an entire panel of circuit-court judges hear the case, a type of request that is usually granted if there is a case with broad legal impact or if there are contradictory decisions from different courts. The court denied this request. As of this writing, the Supreme Court is deciding whether or not it will review the case.

The Charlie Wolf Act has never passed.

All documentation from the 1989 FBI raid on Rocky Flats and the subsequent grand jury investigation is, after twenty-two years, still sealed. Jurors are still not allowed to speak. The crimes committed at Rocky Flats, and the full story of the environmental contamination that resulted from those crimes, are still sealed in the grand jury vault.

Who can imagine our culture, our human lives, 24,000 years in the future? The Cold War will be just one of many wars that my grandchildren will study in school. For their grandchildren, Rocky Flats will be a tiny footnote in history. Four, five, six generations are nothing compared to 24,000 years.

The bones of my grandparents have turned to dust, and all that remains are their names carved in granite. Someday I, too, will have a stone marker. I think of Mark, and the tall tree that now stands where his ashes are buried.

There will be no markers at Rocky Flats. What lies beneath will remain. A few light drops on my cheek make me turn my face to the sky. The air is suddenly acrid with the scent of raindrops and soil, a clean sharp smell, and it's time for me to go. I watch a growling mass of clouds, as white as marshmallows just moments ago but now dark and mottled, move toward me from the line of mountains. The drops quicken, and I can see sheets of rain moving across the flat plain of rooftops. The granite stones grow wet and the ground turns to mud.

I have always loved the many moods of the sky at Rocky Flats. Turquoise and teal in summer, fiery red at sunset, iron gray when snow is on the way. The land rolls in waves of tall prairie grass bowed to the wind, or sprawling mantles of white frosted with a thin sheath of ice in winter.

But the serenity of the landscape belies the battles that still wage over who controls the land, how dangerous the levels of contamination are, and what's to be done about it. Roughly one-third of the site is permanently fenced off due to high levels of contamination. The opening of the rest of the area as a wildlife refuge and public recreation area—about five thousand acres—is temporarily on hold. The U.S. Fish and Wildlife Service (FWS) says it lacks the $200,000 a year it would take to build walking paths and hire staff. Many citizens remain unconvinced that the refuge is safe for public recreation.

The controversy over land surrounding Rocky Flats continues as well. Government agencies claim that the Rocky Flats National Wildlife Refuge is safe and nearby areas are fine for homes, businesses, and recreation. Yet at the same time a 1970 study by Atomic Energy Commission scientists showed that land now held by the FWS on the eastern portion of the Rocky Flats site—as well as an adjacent off-site area of roughly thirty square miles—had been heavily contaminated with plutonium released from the plant. The FWS is currently considering several proposals for the use of some of this contaminated land. The privately financed Jefferson Parkway Authority proposes a major toll highway on land along the eastern edge of the Rocky Flats National Wildlife Refuge that would link up with an existing beltway around the city of Denver. Completion of this project would spur even more residential and commercial development near Rocky Flats. The city of Golden, which opposes the highway, wants to build a bikeway along the same strip of land. Hundreds of local citizens signed a petition demanding that the FWS first determine the quantity, depth, and extent of plutonium contamination in the land proposed for the highway or bikeway, noting that construction itself churns up plutonium-laden dust that could pose a risk to construction workers and residents. In the fall of 2011 the FWS announced that it would not implement such a study.

The problems faced at Rocky Flats are shared by former nuclear sites around the United States and around the world. Hanford, which housed nine nuclear reactors, is one of the most heavily polluted places on earth. Parts of the buffer zone are now the Hanford Reach National Monument, established in 2000 by President Bill Clinton. Only certain areas are open to the public. Fernald, the former uranium processing facility in Ohio, released millions of pounds of uranium dust into the air and contaminated surrounding areas with radioactivity. The site is permanently closed. Nobody can ever safely live here, federal scientists concede, and Fernald will have to be closely monitored essentially forever.

On March 11, 2011, following a level 9.0 earthquake and consequent tsunami, the meltdown of three nuclear reactors at the Fukushima Daiichi nuclear power plant in Japan led to the release of large amounts of radioactive material into the air and into the ocean. It was the world's worst nuclear disaster since Chernobyl. More than 100,000 Japanese residents in surrounding communities were forced to flee. Throughout

Japan, radioactive substances were found not only in beef, milk, spinach, and tea leaves, but also in rice, an essential part of the Japanese diet. In the United States, special monitors deployed for a short time by the EPA following the accident picked up reportedly low levels of radiation from Japan all along the California, Oregon, and Washington coastline. More than a dozen cities in the United States tested positive for fallout from Fukushima in their water supplies. Scientists found radiation from Japan in milk from Arizona to Arkansas to Vermont.

Japanese officials downplayed the accident. Initially assessed as Level Four on the International Nuclear Event Scale (INES), the accident was raised to Level Five and eventually to Level Seven, the highest on the INES. Skeptics in Japan and abroad accused the government of "a consistent pattern of official lying, foot-dragging and concealment." At an antinuclear protest in Tokyo on September 19, 2011, attended by sixty thousand people, Fukushima resident Ruiko Muto compared the people of Fukushima to *hibakusha*, the name for survivors of the bombings of Nagasaki and Hiroshima. "Day after day, many inescapable decisions were forced upon us. To flee, or not to flee? To eat, or not to eat? To make our children wear masks, or not to make them? To speak out, or to remain silent?" she asked. Many of the *hibakusha* have long been opposed to nuclear weapons, but because the Japanese government maintained that nuclear weapons and nuclear power were two separate and unrelated issues, few Japanese have opposed nuclear power plants. By the fall of 2011, that had changed.

Next to Fukushima and Chernobyl, the explosion in 1957 of the underground nuclear waste tank at the Mayak plant, near Kyshtym, Russia, is considered the third-worst nuclear disaster in history, and it reveals the same troubling pattern of government silence and misinformation. The nuclear accident at Chernobyl, classified as a Level Seven on the INES, caused the evacuation of 135,000 people and the release of radioactive material four hundred times higher than what had been released by the Hiroshima bomb. It was only after radiation levels set off alarms in Sweden that Soviet officials allowed publicly that a disaster had occurred.

At Fukushima, a twelve-mile exclusion zone for the most highly contaminated land remains in place around the plant, and officials are considering further expansion of this zone. The estimated cost to clean up the "vast areas" contaminated by the Fukushima accident is at least $13 billion. The accident at Chernobyl contaminated approximately 100,000 square kilometers (roughly 62,000 square miles) with fallout, and levels of contamination were detected all over Europe. At Mayak, hundreds of square miles around the plant are uninhabitable. Today Mayak reprocesses waste from foreign nuclear reactors for profit.

In the United States we currently have approximately 25,000 plutonium pits in our stockpile: roughly 10,000 in nuclear warheads, 5,000 in "strategic reserve," and more than 10,000 "surplus" pits at the Pantex plant near Amarillo, Texas. When production was halted at Rocky Flats after the 1989 FBI raid, the DOE lost the ability to produce plutonium pits. During its production years, Rocky Flats produced more than 1,800 pits per year. In 1998, nine years after the raid, the production of plutonium pits began again at the Los Alamos National Laboratory, with only a few pits produced per year. The DOE says that aging plutonium pits may be unreliable and new pit production is necessary to maintain our stockpile—although many specialists believe plutonium pits are stable for at least half a century, and recent studies suggest an even longer shelf life. Nonetheless, the National Nuclear Security Administration (NNSA) currently seeks to build a modern pit facility capable of producing 450 or more pits per year. Total construction cost of this facility is estimated at more than $2 billion. So far, NNSA has failed to gain full congressional support.

Many inescapable decisions have been forced upon us—decisions about nuclear weapons and nuclear energy that will have far-reaching consequences with sometimes dangerous and unintended results. To speak out or to remain silent is the first and most crucial decision we can make.

PLUTONIAN ODE

by ALLEN GINSBERG

I

What new element before us unborn in nature? Is there
    a new thing under the Sun?
At last inquisitive Whitman a modern epic, detonative,
    Scientific theme
First penned unmindful by Doctor Seaborg with poison-
    ous hand, named for Death's planet through the
    sea beyond Uranus
whose chthonic ore fathers this magma-teared Lord of
    Hades, Sire of avenging Furies, billionaire Hell-
    King worshipped once
with black sheep throats cut, priest's face averted from
    underground mysteries in a single temple at Eleusis,
Spring-green Persephone nuptialed to his inevitable
    Shade, Demeter mother of asphodel weeping dew,
her daughter stored in salty caverns under white snow,
    black hail, grey winter rain or Polar ice, immemor-
    able seasons before
Fish flew in Heaven, before a Ram died by the starry
    bush, before the Bull stamped sky and earth
or Twins inscribed their memories in clay or Crab'd
    flood
washed memory from the skull, or Lion sniffed the
    lilac breeze in Eden—
Before the Great Year began turning its twelve signs,

ere constellations wheeled for twenty-four thousand
    sunny years
slowly round their axis in Sagittarius, one hundred
    sixty-seven thousand times returning to this night

Radioactive Nemesis were you there at the beginning
    black Dumb tongueless unsmelling blast of Disil-
    lusion?
I manifest your Baptismal Word after four billion years
I guess your birthday in Earthling Night, I salute your
    dreadful presence lasting majestic as the Gods,
Sabaot, Jehova, Astapheus, Adonaeus, Elohim, Iao,
    Ialdabaoth, Aeon from Aeon born ignorant in an
    Abyss of Light,
Sophia's reflections glittering thoughtful galaxies, whirl-
    pools of starspume silver-thin as hairs of Einstein!
Father Whitman I celebrate a matter that renders Self
    oblivion!
Grand Subject that annihilates inky hands & pages'
    prayers, old orators' inspired Immortalities,
I begin your chant, openmouthed exhaling into spacious
    sky over silent mills at Hanford, Savannah River,
    Rocky Flats, Pantex, Burlington, Albuquerque
I yell thru Washington, South Carolina, Colorado,
    Texas, Iowa, New Mexico,
where nuclear reactors create a new Thing under the
    Sun, where Rockwell war-plants fabricate this death
    stuff trigger in nitrogen baths,
Hanger-Silas Mason assembles the terrified weapon
    secret by ten thousands, & where Manzano Moun-
    tain boasts to store
its dreadful decay through two hundred forty millennia
    while our Galaxy spirals around its nebulous core.

I enter your secret places with my mind, I speak with
    your presence, I roar your Lion Roar with mortal
    mouth.
One microgram inspired to one lung, ten pounds of
    heavy metal dust adrift slow motion over grey
    Alps
the breadth of the planet, how long before your radiance
    speeds blight and death to sentient beings?
Enter my body or not I carol my spirit inside you,
    Unapproachable Weight,
O heavy heavy Element awakened I vocalize your con-
    sciousness to six worlds
I chant your absolute Vanity. Yeah monster of Anger
    birthed in fear O most
Ignorant matter ever created unnatural to Earth! Delusion
    of metal empires!
Destroyer of lying Scientists! Devourer of covetous
    Generals, Incinerator of Armies & Melter of Wars!
Judgement of judgements, Divine Wind over vengeful
    nations, Molester of Presidents, Death-Scandal of
    Capital politics! Ah civilizations stupidly indus-
    trious!
Canker-Hex on multitudes learned or illiterate! Manu-
    factured Spectre of human reason! O solidified
    imago of practitioners in Black Arts
I dare your Reality, I challenge your very being! I
    publish your cause and effect!
I turn the Wheel of Mind on your three hundred tons!
    Your name enters mankind's ear! I embody your
    ultimate powers!
My oratory advances on your vaunted Mystery! This
    breath dispels your braggart fears! I sing your
    form at last

behind your concrete & iron walls inside your fortress
of rubber & translucent silicon shields in filtered
cabinets and baths of lathe oil,

My voice resounds through robot glove boxes & ingot
cans and echoes in electric vaults inert of atmosphere,

I enter with spirit out loud into your fuel rod drums
underground on soundless thrones and beds of
lead

O density! This weightless anthem trumpets transcendent
through hidden chambers and breaks through
iron doors into the Infernal Room!

Over your dreadful vibration this measured harmony
floats audible, these jubilant tones are honey and
milk and wine-sweet water

Poured on the stone block floor, these syllables are
barley groats I scatter on the Reactor's core,

I call your name with hollow vowels, I psalm your Fate
close by, my breath near deathless ever at your
side

to Spell your destiny, I set this verse prophetic on your
mausoleum walls to seal you up Eternally with
Diamond Truth! O doomed Plutonium.

II

The Bard surveys Plutonian history from midnight
lit with Mercury Vapor streetlamps till in dawn's
early light

he contemplates a tranquil politic spaced out between
Nations' thought-forms proliferating bureaucratic

& horrific arm'd, Satanic industries projected sudden
with Five Hundred Billion Dollar Strength

around the world same time this text is set in Boulder,
    Colorado before front range of Rocky Mountains
twelve miles north of Rocky Flats Nuclear Facility in
    United States of North America, Western Hemi-
    sphere
of planet Earth six months and fourteen days around
    our Solar System in a Spiral Galaxy
the local year after Dominion of the last God nineteen
    hundred seventy eight
Completed as yellow hazed dawn clouds brighten East,
    Denver city white below
Blue sky transparent rising empty deep & spacious to a
    morning star high over the balcony
above some autos sat with wheels to curb downhill
    from Flatiron's jagged pine ridge,
sunlit mountain meadows sloped to rust-red sandstone
    cliffs above brick townhouse roofs
as sparrows waked whistling through Marine Street's
    summer green leafed trees.

III

This ode to you O Poets and Orators to come, you
    father Whitman as I join your side, you Congress
    and American people,
you present meditators, spiritual friends & teachers,
    you O Master of the Diamond Arts,
Take this wheel of syllables in hand, these vowels and
    consonants to breath's end
take this inhalation of black poison to your heart, breathe
    out this blessing from your breast on our creation
forests cities oceans deserts rocky flats and mountains
    in the Ten Directions pacify with this exhalation,

enrich this Plutonian Ode to explode its empty thunder
    through earthen thought-worlds
Magnetize this howl with heartless compassion, destroy
    this mountain of Plutonium with ordinary mind
    and body speech,
thus empower this Mind-guard spirit gone out, gone
    out, gone beyond, gone beyond me, Wake space,
    so Ah!

—July 14, 1978

ACKNOWLEDGMENTS

I HAVE many people to thank in the process of researching this book, including neighbors, old classmates, scientists, Rocky Flats workers, activists, attorneys, journalists, physicians, and developers. In addition to the many interviews I conducted myself, I depended heavily on the remarkable archive of Rocky Flats interviews at the Maria Rogers Oral History Program at the Carnegie Library in Boulder, Colorado, where I owe special thanks to Dorothy Ciarlo, Hannah Nordhaus, and Susan Becker. Thanks to J. Wendel Cox and Jennifer Dewey at the Denver Public Library, Western History and Genealogy department, and David M. Hays at the University of Colorado at Boulder Libraries, Archives department.

I am grateful to all of the people who granted me interviews and were very generous with their time. Particular thanks go to workers Randy Sullivan, Stan Skinger, Charlie Wolf, Doug Parker, Laura and Jeff Schultz, Dr. Robert Rothe, and Debby Clark. Charlie Wolf is missed by many. Neighbors and residents include Tamara Smith Meza and her family (as well as physician Nicholas Gonzales); Ann White; the Kirstin Dunn family; Stacy Gardalen (née Bunce), Curtis Bunce, and Patricia Bunce; Bini Abbott; and the Duane Hart family. Dr. LeRoy Moore with the Rocky Mountain Peace and Justice Center was an invaluable resource and I am deeply grateful for his assistance. Thanks to Representative Wes McKinley; Dr. Harvey Nichols, professor emeritus of biology at the University of Colorado at Boulder; and Len Ackland, author of *Making a Real Killing: Rocky Flats and the Nuclear West*. Attorney Peter Nordberg and his wife, Mykaila, shared remarkable stories, and I am grateful to Karen Markert for her assistance with court documentation. Peter Nordberg is deeply missed. Thanks to those who have been involved in the Rocky Flats story in so many ways and shared their stories: Patrick Malone, Shirley Garcia, Hildegard Hix, Mary Harlow, Jack Cohen-Joppa, Pam Solo, Judy Danielson, Paula Elofson-Gardine, Anne Guilfoile, Ellen Klaver, Bob McFarland, Chet Tchozewski, Kenneth Nova, Elene Rosenfeld, Jyoti Wind, and Bob Kinsey. Thanks to Rex Haag and particularly Charles C. McKay, who shared stories of his grandparents and great-grandparents. Thanks to investigative journalists Ryan Ross, Eileen Welsome, and Patricia Calhoun. I am grateful to my dear friend Christie Smith, who sent useful newspaper clippings for years, and Theron Britt for helpful commentary. Alex Stein offered insightful comments and unflagging faith that I would finish this project. Marge and Joe Meek were ever supportive of this work. Warm gratitude to Roberta and Rick Robertson. Photographer Arin Billings shared her remarkable photographs of Rocky Flats workers, and I continue to be inspired by the photography of Robert Del Tredici as well as by Robert Adams and his photos of people living near Rocky Flats.

Thanks to physicist Robert Philbin, who helped me understand the complexities of radioactivity, and Ross Proctor, lieutenant with the Memphis Police Department, for helping me understand issues of domestic preparedness and nuclear weapons.

I am grateful to the research assistants who helped with transcriptions and footnotes: Wendy Sumner Winter, JD Wilson, Andrew Sall, Matt Martin, John Schulze, Derek Gray, Sean Meek, and especially, in the final stretch, Colleen Pawling and Tom Useted. Gwendolyn Ashbaugh Mooney and Greg Larson read the manuscript closely. Thanks to my colleagues in the MFA program at the University of Memphis for their encouragement and support, particularly Richard Bausch, Rebecca Skloot, Sonja Livingston, and Aram Goudsouzian. I'm grateful to Stephen Usery at Book Talk and Corey Mesler at Burke's Books. They are the heart of the literary community in Memphis.

I am indebted to Grant and Peggy Pound at Colorado Art Ranch and the remarkable residents of Trinidad and Libre, Colorado, for two writing residencies that gave me the time and quiet space to complete the final stages of this project. Thanks to the San Jose Literary Arts Council and the University of Memphis for grants, and Denver International

Airport and Colorado Art Ranch for exhibiting photos and text from this book over the summer of 2010.

I owe deep gratitude to the people who made this book happen. John Glusman was extraordinarily enthusiastic, and his comments in the early stages were invaluable. I am extremely fortunate to have the talent and energy of Rachel Klayman, my editor at Crown, as well as the other wonderful people at Crown, including Mark Birkey, Chris Brand, Julie Cepler, Stephanie Chan, Michael Gentile, Leila Lee, Annsley Rosner, Jay Sones, and Barbara Sturman. Publisher Molly Stern has been hugely supportive of this book. A very special thank-you to my agent, Ellen Levine.

Molly Giles believed in this book—and me—from the very first sentence. Heartfelt gratitude—and lefse and lutefisk—to Greg Larson, for love, support, and late-night editing. It turns out that a Norwegian and a Swede make a pretty good team. Thanks, most of all, to my remarkable family: my two sons, Sean and Nathan; my mother, now gone; my father; and my beloved siblings, Karin, Karma, and Kurt.

ROCKY FLATS TIMELINE*

1942   The Manhattan Project begins.

1945   The U.S. Army conducts its first nuclear weapons test on July 16, 1945 in New Mexico. The weapon is referred to as "the Gadget." On August 6, the "Little Boy" atomic bomb is dropped on Hiroshima, Japan. Three days later, the "Fat Man" nuclear bomb, an implosion-design plutonium device, is dropped on Nagasaki.

1946   President Harry S. Truman signs the Atomic Energy Act, creating the Atomic Energy Commission (AEC).

1951   The *Denver Post* reports, "There Is Good News Today: U.S. to Build $45 Million A-Plant near Denver." The site for the plant lies sixteen miles from central Denver and nine miles from Boulder, and site plans rely erroneously on wind pattern reports from Stapleton Airport, not the high mesa of Rocky Flats. Dow Chemical is chosen as the operating contractor.

1957   A major fire occurs in plutonium processing Building 771. Despite the spread of radioactive and toxic contamination to the Denver metropolitan area, residents are not told about the fire until 1970.

1962   The Cuban missile crisis between the United States and the Soviet Union brings the world to the brink of nuclear war.

1969   A major fire in plutonium processing Buildings 776 and 777 becomes the costliest industrial accident in the United States at that time. Cleanup takes two years. The public is largely unaware of the fire.

1970   After a team of independent scientists discovers plutonium at off-site areas around Rocky Flats, the AEC admits to the contamination but announces that it is not a result of the 1969 fire, but rather of the 1957 fire—of which the public was never informed—and of thousands of drums that have been leaking radioactive and toxic materials since the 1960s.

1972   The AEC expands the buffer zone around Rocky Flats and Congress spends $6 million to purchase an additional 4,600 acres, bringing the Rocky Flats site to a total of approximately 6,400 acres.

*Adapted from a timeline prepared by the Rocky Flats Stewardship Council.

1973 The Colorado Department of Health discovers tritium in drinking water downstream of Rocky Flats, but does not alert local officials for five months. The AEC initially denies the presence of tritium.

1974 Governor Richard Lamm and Representative Timothy Wirth establish the Lamm-Wirth Task Force on Rocky Flats to help determine the future of Rocky Flats, given its proximity to the Denver metropolitan area. The task force concludes that nuclear-weapons work should be ended at Rocky Flats and moved to another location.

1975 Rockwell International replaces Dow Chemical as managing contractor of Rocky Flats.

1978 Large-scale protests begin at Rocky Flats. Protesters set up camp on railroad tracks leading into the plant site and remain on the tracks from April until January 1979.

1979 A core meltdown of Unit 2 at the Three-Mile Island nuclear power plant near Harrisburg, Pennsylvania, occurs on March 28. On April 28, thousands of protestors rally at Rocky Flats, including Daniel Ellsberg, Allen Ginsberg, Bonnie Raitt, and Jackson Browne. A counterdemonstration is held by pro–Rocky Flats workers and the United Steelworkers of America.

1983 More than fifteen thousand protesters link hands and nearly encircle the seventeen-mile perimeter of the plant on October 15.

1984 The first Rocky Flats worker is diagnosed with chronic beryllium disease, an incurable illness irrefutably linked with work conditions at the plant.

1986 The Department of Energy (DOE), the Colorado Department of Health, and the Environmental Protection Agency (EPA) sign an agreement to allow, for the first time, partial regulation of radioactive and hazardous waste at Rocky Flats. That same year, the Chernobyl disaster on April 26 at the Chernobyl Nuclear Power Plant in Ukraine releases a large amount of radioactive contamination that spreads over much of Europe. At the time, it is considered the worst nuclear accident in history.

1987 The Rocky Flats Environmental Monitoring Council, a community oversight organization, is formed.

1989 The FBI raids Rocky Flats to collect evidence of alleged environmental lawbreaking at the plant. Production of plutonium triggers ends. A federal grand jury is impaneled to review the evidence and embarks on a nearly three-year-long investigation, hearing hundreds of witnesses and examining thousands of documents.

1990 EG&G takes over from Rockwell as the operator of Rocky Flats. A class-action lawsuit, *Cook v. Rockwell International Corporation*, is filed on behalf of thousands of residents living downwind of the plant. The suit alleges that Dow and Rockwell allowed plutonium from Rocky Flats to contaminate residents' land.

1991 The Soviet Union is dissolved. The Cold War ends. This same year, the Rocky Flats Beryllium Health Surveillance Program is initiated.

1992 The U.S. Attorney and Department of Justice bypass the grand jury and negotiate an out-of-court settlement with Rockwell in which the company pleads guilty to ten violations of the Clean Water Act and federal hazardous waste laws, including illegal storage of hazardous wastes. Rockwell pays a fine of $18.5 million. Outraged grand jurors refuse to be dismissed and write their own report detailing ongoing contamination and calling for the indictment and trial of several Rockwell and DOE officials. Though the report is sealed by the judge and jurors are forbidden to speak about the case, someone leaks a redacted version of the report to *Westword*, a Denver weekly. Meanwhile, President George H. W. Bush announces the end of the W88 warhead program, effectively ending production at Rocky Flats.

1993 The Rocky Flats Citizens Advisory Board replaces the Rocky Flats Environmental Monitoring Council, its mission to provide ongoing community and local government oversight of the cleanup at Rocky Flats.

1994 The Rocky Flats Nuclear Weapons Plant is renamed the Rocky Flats Environmental Technology Site.

1995 In the ongoing class-action lawsuit *Cook v. Rockwell International*, a U.S. district judge holds the DOE in contempt of court for failure to release millions of pages of documentation regarding missing plutonium, health issues, and other information about the plant. The DOE estimates that it would take seventy years and $36 billion to clean up Rocky Flats, and says that the technology to do an adequate cleanup may not exist. Also, the DOE, the EPA, and the Colorado Department of Public Health and the Environment sign the Rocky Flats Cleanup Agreement, which specifies cleanup levels for soils contaminated with radioactive materials at Rocky Flats. Local residents and scientists protest that the levels are too high.

1999 Shipments of transuranic (i.e., plutonium-laden) nuclear waste from Rocky Flats to the Waste Isolation Pilot Plant, east of Carlsbad, New Mexico, begin.

2000   The Energy Employees Occupational Illness Compensation Act is established to compensate nuclear industry workers whose health may have been harmed by workplace exposure to radioactive and chemical toxins. Due to missing and inaccurate records, many workers find it difficult to prove exposure.

2001   The Rocky Flats National Wildlife Refuge Act is signed into law. Kaiser-Hill LLC agrees to clean up the Rocky Flats site for an estimated cost of $7.3 billion and sets a target completion date of 2010.

2003   DOE, EPA, and CDPHE revise the Rocky Flats Cleanup Agreement, setting new cleanup levels for radioactive materials in the soil at the site.

2004   The U.S. Fish and Wildlife Service announces that public recreation will be allowed at the proposed Rocky Flats National Wildlife Refuge.

2005   Kaiser-Hill announces that it has completed the cleanup of Rocky Flats more than fourteen months ahead of schedule. *Cook v. Rockwell International* goes to trial. It is the largest environmental class-action lawsuit in Colorado history. Property owners seek $500 million in damages.

2006   The jury in *Cook v. Rockwell International* awards the plaintiffs almost $554 million. The Rocky Flats Stewardship Council is formed to provide ongoing local government and community oversight of the postclosure management of the Rocky Flats site.

2007   Nearly four thousand acres of the former Rocky Flats Nuclear Weapons Site are transferred to the Department of the Interior for management by the U.S. Fish and Wildlife Service as the Rocky Flats National Wildlife Refuge. As of the publication of this book, the refuge has not been opened to the public.

2008   The judge in *Cook v. Rockwell International* issues a final award of $926 million, including compensatory damages, interest, and exemplary damages.

2010   A three-judge appeals court at the Tenth Circuit Court of Appeals in Denver overturns the decision in *Cook v. Rockwell International* and throws out the award.

2011   Following an earthquake and tsunami, on March 11 three nuclear reactors undergo a full meltdown at the Fukushima I Nuclear Power Plant in Japan, leading to extensive radioactive contamination on the level of the Chernobyl disaster. Radioactivity from Fukushima is measurable on the West Coast of the United States.

NOTES

WRITING THIS book was the work of twelve years. Each chapter is closely based on primary and secondary sources, including books, newspaper articles, journals, technical reports, government reports, and court documentation. In addition, I conducted extensive personal interviews and drew on the more than 150 interviews at the Maria Rogers Oral History Program on Rocky Flats at the Carnegie Branch Library for Local History in Boulder, Colorado. Many of these resources are available on the Internet.

All dialogue is as close to verbatim as possible based on interviews, newspaper articles, audio and video documentation, and other sources. I am thankful for the work of investigative journalists with the *Rocky Mountain News,* the *Denver Post, Westword,* the *New York Times,* the *Los Angeles Times,* the *Washington Post,* and the *Bulletin of the Atomic Scientists,* and particularly grateful to have been able to consult Len Ackland's book, *Making a Real Killing: Rocky Flats and the Nuclear West* (University of New Mexico Press, 1999), which was vital to nearly every chapter but particularly the section on the 1969 Mother's Day fire. Interviews with Bill Dennison, Stan Skinger, and Willie Warling were also essential to that section.

My account of the FBI raid and subsequent grand jury investigation is based on articles, court documentation, and many interviews with Jim Stone, Jon Lipsky, Wes McKinley, Jacque Brever, Peter Nordberg, and others. The book *The Ambushed Grand Jury: How the Justice Department Covered Up Nuclear Crime: And How We Caught Them Red-Handed* by Wes McKinley and Caron Balkany (Apex Press, 2004) was useful.

Sections describing protests at Rocky Flats are based on newspaper articles and firsthand accounts as well as interviews conducted by me or the Maria Rogers Oral History Program with Daniel Ellsberg, Debby Clark, Pam Solo, and others, and a very illuminating article by Edward Abbey.

Technical information is based on extensive DOE documentation (and that of its subcontractors) as cited below, as well as the work of Ed Martell, Carl Johnson, Gregg Wilkinson, Shawn Smallwood, Marco Kaltofen, and others, and several articles by LeRoy Moore. Numerous interviews with Tamara Smith Meza, Peter and Mykaila Nordberg, Ann White, Laura and Jeff Schultz, Charlie Wolf, Charles McKay, Pat McCormick, and Randy Sullivan were particularly helpful, and I am grateful for their willingness to share their lives for this book.

Chapter 1. Mother's Day

PAGE

4 **From 1952 to 1989, Rocky Flats manufactures:** H. Josef Hebert, "Quality of Replacement Plutonium Triggers for Aging Nuclear Warheads Questioned," Associated Press, January 20, 2008.

The creation of each gram of plutonium: Linda Rothstein, "Nothing Clean About 'Cleanup,'" *Bulletin of the Atomic Scientists* (May/June 1995): 34–41.

This is a secret operation: Tamara Jones, "U.S. Vows to Lift 30-Year Veil of Secrecy at Weapons Plants," *Los Angeles Times*, June 17, 1989.

5 **Announcement of the plant:** Robert Perkin, "Denver Gets Atom Plant," *Rocky Mountain News*, March 24, 1951.

The Rocky Flats Nuclear Weapons Plant will become: These sites are the Nevada Test Site, Hanford Reservation, Lawrence Livermore National Labs, Los Alamos National Lab, Sandia National Lab, Waste Isolation Pilot Plant, Pantex Plant, Pinellas Plant, Savannah River Site, Oak Ridge Reservation, Mound Plant, Food Materials Production Center (Fernald), and the Kansas City Plant.

Not even the governor: "Boulder Leaders Cheer Atom Plant," *Boulder Daily Camera*, March 24, 1951.

Colorado's top elected officials are not informed: *Citizen's Guide to Rocky Flats: Colorado's Nuclear Bomb Factory* (Boulder, CO: Rocky Mountain Peace Center, 1992).

Contractors, the local power plant, and local businesses: Robert L. Perkin, "PSC Expected to Get $45 Million Atomic Plant Contract," *Rocky Mountain News*, October 28, 1951.

6 **The newspaper reports that workers on the project:** Tamara Jones and Dan Morain, "Federal Probers Sound Alarms: Rocky Flats Boon Turns into Ecological Nightmare," *Los Angeles Times*, June 20, 1989.

The plant site in Jefferson County: "Colorado Will Get New Atomic Plant," *New York Times*, March 24, 1951.

Officials from the AEC emphasize: Joseph Givando, "Denver A-Plant Plans Shrouded in Strict Secrecy," *Denver Post*, March 24, 1951.

When questioned further by reporters: Len Ackland, *Making a Real Killing: Rocky Flats and the Nuclear West* (Albuquerque: University of New Mexico Press, 1999), 52.

7 **He warns against the location:** Laura Frank and Ann Imse, "Rocky Flats Whistle-blower Dies at 82," *Rocky Mountain News*, April 12, 2007.

"The housing situation is rough here": Sam Lusky, "Atom Plant Workers to Increase State's Housing Problems," *Rocky Mountain News*, March 25, 1951.

8 **Solid and liquid waste is packed into:** Mark Bearwald, "Sprawling Rocky Flats Keeps Its AEC Secret," *Denver Post*, January 16, 1958.

What spews from the smokestacks: "Rocky Flats Stack Spews Higher Level of Uranium," *Denver Post*, April 30, 1953.

18 **More than 7,640 pounds of plutonium:** Ackland, *Making a Real Killing*, 152.

20 **Rock climbing, biking:** Stan Skinger, interviews by author, March 24, 2007, February 28, 2008, and May 19, 2008, and by Dorothy Ciarlo, March 9, 2005 (Maria Rogers Oral History Program, OH 1300V A-B).

21 **felt "somewhat divorced from the actual nuclear weapon":** Robert Rothe, interviews by Hannah Nordhaus, November 10, 2003, and November 13, 2003 (Maria Rogers Oral History Program, OH 1179), and interview by author, February 25, 2012.

A few days later: Bill Dennison, interview by Dorothy Ciarlo, March 1, 2001 (Maria Rogers Oral History Program, OH 1066 A-B).

23 **It's the core of the plant:** Willie Warling, interview by Dorothy Ciarlo, May 27, 1998 (Maria Rogers Oral History Program, OH 0927).

Internal alpha emitters like plutonium: Helen A. Grogan et al., "Assessing Risk of Exposure to Plutonium," Risk Assessment Corporation (February 2000), 6.27–6.39.

29 **The filters had not been replaced:** "Rocky Flats Revisited: Carl Johnson Responds," *Ambio* 2, no. 6 (1982): 376–77.

30 **A spokesman from the AEC:** "Atomic Plant Fire Causes $50,000 Loss," *Denver Post*, September 12, 1957.

30 **When pressed for more information:** "There's No Atomic Blast Danger at Rocky Flats," *Rocky Mountain News*, June 1, 1954.

 Based on soil and water testing: Howard Holme, interview by Hannah Nordhaus, September 1, 2005 (Maria Rogers Oral History Program, OH 1369).

39 **Later an AEC fire investigator will report:** Ackland, *Making a Real Killing*, 153. See also Atomic Energy Commission, "Report on Investigation—1969 Fire," vol. 1.

40 **A plant spokesman states:** "Fire Is Reported at Rocky Flats," *Rocky Mountain News*, May 12, 1969.

42 **Production is halted temporarily:** "Year's Delay Possibility, Probers Told," *Denver Post*, June 24, 1969.

 Due to pressure from concerned scientists: "AEC Admits Plutonium Release at Rocky Flats," *Denver Post*, February 19, 1970.

47 **The contracting laboratory for Rockwell:** *Dark Circle*, 1983, reissued 2007, directed by Judy Irving, Chris Beaver, and Ruth Landy. See also Harvey Wasserman and Norman Solomon, *Killing Our Own: The Disaster of America's Experiment with Atomic Radiation* (New York: Delacorte, 1982).

 Ultimately, Rex comes to believe: Rex Haag, interview by author, Arvada, Colorado, August 10, 2002.

Chapter 2. Drums and Bunnies

48 **One Rocky Flats worker and then another:** Citizen Summary, Rocky Flats Historical Public Exposures Studies, 903 Area, www.cdphe.state.co.us/rf/903area.htm.

49 **The rabbit, dissected for analysis:** Dow Chemical memo, "Contaminated Rabbit," January 3, 1962. Refers to a rabbit dissected on July 24, 1961.

 Occasionally demonstrators begin: Anne Guilfoile, interview by author, Arvada, Colorado, October 29, 2006. The first documented protest at Rocky Flats was in 1968, according to LeRoy Moore.

51 **Around the same time, a Los Angeles construction company:** Tamara Jones and Dan Morain, "Federal Probers Sound Alarms: Rocky Flats Boon Turns into Ecological Nightmare," *Los Angeles Times*, June 20, 1989.

53 **"Liar, liar, plant's on fire!":** Ann Breese White, interview by author, Denver, Colorado, July 31, 2004. See also unpublished essay by Ann Breese White, "Paying the Piper: The Nuclear Legacy of Rocky Flats," January 16, 1996, held in archives at Western History Room, Denver Public Library.

54 **On the west side of town, not far from Rocky Flats:** Pat McCormick,

interview by author, Denver, Colorado, November 30, 2006. See also Pat McCormick, interview by Dorothy Ciarlo, January 13, 2007 (Maria Rogers Oral History Program, OH 1454V).

58 **The show home with a built-in bomb shelter:** "Ode to the Family Fallout Shelter," September 14, 2010, http://dscriber.com/3290-ode-to-the -family-fallout-shelter.

60 **The AEC reports, "There is no evidence":** Anthony Ripley, "Colorado Atom Plant Is Called Radiation Hazard," *New York Times*, February 11, 1970.
And, he adds, there will be no off-site testing: LeRoy Moore, "Democracy and Public Health at Rocky Flats: The Examples of Edward A. Martell and Carl J. Johnson," in *Tortured Science: Health Studies, Ethics and Nuclear Weapons*, edited by Dianne Quigley, Amy Lowman, and Steve Wing (Amityville, NY: Baywood Publishing, 2011).

63 **One girl, Tina:** Name has been changed.

65 **Comparing these samples:** Moore, "Democracy and Public Health at Rocky Flats," 70.
The chicks have beaks so curled: Judy Danielson, "Talk at Rocky Flats Cold War Museum Event," October 28, 2006.
Bini buys several horses each year: Bini Abbott, interview by author, Arvada, Colorado, February 7, 2005.

66 **Plutonium deposits in the top:** Moore, "Democracy and Public Health at Rocky Flats," 70.
In some places, the level is 1,500 times higher than normal: Fox Butterfield, "Dispute on Wastes Poses Threat to Operations at Weapons Plant," *New York Times*, October 21, 1988.
Deposits are heaviest: Ripley, "Colorado Atom Plant Is Called Radiation Hazard."

67 **For plutonium to be truly dangerous:** Ripley, "Colorado Atom Plant Is Called Radiation Hazard."
The CCEI report states: Ripley, "Colorado Atom Plant Is Called Radiation Hazard."

76 **The amount of plutonium released:** Alan Cunningham, "Putzier: Press Plagues Rocky Flats," *Rocky Mountain News*, December 16, 1971.
Glenn Seaborg, the physicist who: Jeremy Bernstein, *Plutonium: A History of the World's Most Dangerous Element* (Ithaca, NY: Cornell University Press, 2007), 105.

77 **Dow Chemical and the AEC have known:** Keith Schneider, "Weapons Plant Pressed for Accounting of Toll on Environment and Health," *New York Times*, February 15, 1990.

77 **Five particularly powerful windstorms:** These windstorms occurred on December 5, 1968; January 6–7 and 30, 1969; and March 19 and April 7, 1969. Citizen Summary, Rocky Flats Historical Public Exposures Studies, 903 Area.

Major General Giller declares: "Rocky Flats Still Smolders," *Science News* 97, no. 8 (February 21, 1970): 194.

In 1969, Gofman and his colleague: Jeremy Pearce, "John W. Gofman, 88, Scientist and Advocate for Nuclear Safety, Dies," *New York Times*, August 26, 2007. See also John W. Gofman, *Radiation-Induced Cancer from Low-Dose Exposure: An Independent Analysis* (San Francisco: Committee for Nuclear Responsibility, Book Division, 1990).

85 **At the 1970 congressional hearing:** Fred Gillies, "Rocky Flats: It's Always There," *Denver Post*, March 21, 1972.

86 **In a statement to the press:** Gillies, "Rocky Flats: It's Always There."

90 **But no action is taken:** Harvey Nichols, interview by author, Boulder, Colorado, October 27, 2006. See also interview by Hannah Nordhaus, September 14, 2005 (Maria Rogers Oral History Program, OH 1372V A-B).

One afternoon after school: Name has been changed.

96 **When he gets the results:** Al Hazle, interview by Hannah Nordhaus, July 22, 2003 (Rocky Flats Cold War Museum and Maria Rogers Oral History Program, OH 1154V A-B).

99 **Karma's had a crush, too, on Scott:** Name has been changed.

A few years down the road, in 1981: Carl J. Johnson, "Cancer Incidence in an Area Contaminated with Radionuclides Near a Nuclear Installation," *Ambio* 10, no. 4 (1981): 176–82.

100 **Dow Chemical and the AEC don't bother:** Len Ackland, *Making a Real Killing: Rocky Flats and the Nuclear West* (Albuquerque: University of New Mexico Press, 1999), 171.

The Colorado Health Department tests the water: Tad Bartimus and Scott McCartney, *Trinity's Children: Living Along America's Nuclear Highway* (Albuquerque: University of New Mexico Press, 1993), 190.

Several residents, including the new mother: Hazle, interview by Nordhaus.

AEC officials are slow to acknowledge: The Lamm-Wirth Task Force Final Report (1975), 46.

residents are told: James Sterba, "Radiation Traced to Atom Plant in Colorado," *New York Times*, September 27, 1973.

101 **Scientists estimate that 50 to 100 curies:** Lamm-Wirth Report, 47. See also "Investigative Report of the 1973 Tritium Release at the Rocky Flats Plant

in Golden, Colorado," Radiation/Noise Branch, Hazardous Materials Control Division, U.S. Environmental Protection Agency, July 1975.

101 **Rocky Flats maintains there is no threat:** Fred Gillies, "AEC Opens Rocky Flats," *Denver Post*, October 21, 1973.

The Environmental Protection Agency sidesteps: Patricia Buffer, "Rocky Flats History," Department of Energy Rocky Flats Field Office (July 2003), 8. There are several versions of this document, with various dates.

"We won't drink the water": Bill Richards, "Plutonium Taints Their Reservoir: Should U.S. Pay for Denver Suburb's New Water Supply?" *Washington Post*, March 21, 1977.

Rocky Flats officials explain: Colorado Department of Public Health and the Environment, "Technical Topics Papers: Surface Water," 3. www.cdphe .state.co.us/rf/contamin.htm.

A storm of publicity eventually forces: Fred Gillies, "Atom Waste Buried in Tons at Flats Plant," *Denver Post*, August 7, 1973.

102 **The presence of strontium strengthens:** Joel Warner, "Servant of the People," *Boulder Weekly*, January 6–13, 2005.

As an aside to Broomfield's worried city manager: Hazle, interview by Nordhaus.

Al Hazle notes in the article: Fred Gillies, "Curium at N-Plant a Surprise," *Denver Post*, September 21, 1973.

103 **The study will measure:** Grace Lichtenstein, "Housing Near Colorado Nuclear Plant Stirs Fears of Possible Health Hazards," *New York Times*, January 11, 1976.

Several of the readings exceed: Carl Johnson, "Survey of Land Proposed for Residential Development East of Rocky Flats." Report to the Jefferson County Commissioners and the Colorado State Health Department, September 12, 1975.

The readings are much higher: Moore, "Democracy and Public Health at Rocky Flats" (referencing Edward A. Martell, interview by Robert Del Tredici, July 22, 1982).

Johnson feels this creates a potential hazard: Carl J. Johnson, Ronald R. Tidball, Ronald C. Severson, "Plutonium Hazard in Respirable Dust on the Surface of Soil," *Science*, New Series 193, no. 4252 (August 6, 1976): 488–90.

It's decided that no more subdivisions: Timothy Lange, "They Fired Dr. Johnson," *Westword*, May 28, 1981.

104 **"If it were," he says, "I'd be the first":** Lichtenstein, "Housing Near Colorado Nuclear Plant Stirs Fears."

104 **In 1951, when Charlie is nine:** Charles McKay, interview by author, Arvada, Colorado, October 27, 2006.

105 **"It's really gone too far":** Howard Holme, interview by Hannah Nordhaus, September 1, 2005 (Rocky Flats Cold War Museum and Maria Rogers Oral History Program, OH 1369).

The same year Church files his lawsuit: Buffer, "Rocky Flats History."

Chapter 3. Nuns and Pirates

113 **In early December 1974:** "Cattle Near Rocky Flats Show High Plutonium Level," *Rocky Mountain News*, December 5, 1974. See also Steve Wynkoop, "Cattle's Lungs Hold Plutonium," *Denver Post*, December 5, 1974.

An Enviromental Protection Agency (EPA) study has found: Tad Bartimus and Scott McCartney, *Trinity's Children: Living Along America's Nuclear Highway* (Albuquerque: University of New Mexico Press, 1993), 190.

Plutonium, uranium, americium, tritium: D. D. Smith and S. C. Black, "Actinide Concentrations in Tissues from Cattle Grazing Near the Rocky Flats Plant," Environmental Protection Agency, Las Vegas, Nevada, February 1975.

"People don't want to buy": Howard Holme, interview by Hannah Nordhaus, September 1, 2005.

The DOE and officials at Rocky Flats: Jack Olsen Jr., "EPA Reverses Finding on Rocky Flats Cattle," *Rocky Mountain News*, January 27, 1975.

114 **The report concludes, however:** Holme, interview by Nordhaus.

115 **Less cancer was found:** Holme, interview by Nordhaus.

"It is as if the government": Len Ackland, *Making a Real Killing: Rocky Flats and the Nuclear West* (Albuquerque: University of New Mexico Press, 1999), 178.

116 **What they don't know:** "Guards Prepared to 'Shoot to Kill' Are Increased at 14 Nuclear Sites," *New York Times*, October 23, 1976.

123 **Rocky Flats officials contend that the plant:** "Why the Need for Rocky Flats?" *Boulder Daily Camera*, April 20, 1975.

In a press interview, task force member: Todd Phipers, "Report Stresses Flats Potential Hazard," *Denver Post*, 1975. Denver Public Library, Western History and Genealogy Clipping File, U.S. Government, Department of Energy, Colorado, Rocky Flats Plant, 1970–1979. See also "Special Task Force Report Findings Revealed," *Rocky Mountain News*, February 15, 1975.

It also criticizes the Price-Anderson Act: The Lamm-Wirth Task Force Final Report (1975), 23.

124 **Contradicting its own recommendation:** Lamm-Wirth Report, 11.
Some critics claim: Ackland, *Making a Real Killing*, 179.

127 **Sister Pam Solo is the sole woman:** Pam Solo, interview by Dorothy Ciarlo, October 24, 2004 (Maria Rogers Oral History Program, OH 1272), and interview by LeRoy Moore, September 23, 1996 (Maria Rogers Oral History Program, OH 1528).

128 **She worries that things will continue:** Solo, interview by Moore.

130 **In 1975 and 1976 he and his colleagues:** Susan Heller Anderson, "Dr. Carl Johnson Is Dead at 58," *New York Times*, December 30, 1988.
Using data collected by: "Research on Adverse Health Effects Related to Rocky Flats," Rocky Flats Historical Public Exposures Studies, Colorado Department of Public Health and the Environment, www.cdphe.state .co.us/rf/adversheal.htm.
The study involves 154,170 people: Carl J. Johnson, "Cancer Incidence in an Area Contaminated with Radionuclides Near a Nuclear Installation," *Ambio* 10, no. 4 (1981): 176–82.
He finds higher-than-average: Carl Johnson, interview by Robert Del Tredici, July 20, 1982. See also Carl Johnson, "Leukemia Death Rates of Residents of Areas Contaminated with Plutonium," Proceedings of the 105th Annual Meeting of the American Public Health Association, Washington, DC, November 1, 1977; Carl Johnson, "Evaluation of the Hazard to Residents of Areas Contaminated with Plutonium," Proceedings of the Fourth International Congress of the International Radiation Protection Association, Paris, April 24–30, 1977, 1:243–46; and Carl Johnson to Jefferson County Board of Health, "Report on Death Rates from Lung Cancer in the Eight Census Tracts near Rocky Flats and in Golden, and in Nineteen Census Tracts at the South End of Jefferson County," November 20, 1977. See also Stephen Talbot, "The H-Bombs Next Door," *The Nation*, February 7, 1981.
Johnson estimates 491: Rex Weyler, *Greenpeace* (Richmond, BC, Can.: Rodale Books, 2004), 555. See also "Rocky Flats Revisited: Carl Johnson Responds," *Ambio* 2, no. 6 (1982).
He also believes that plutonium: "Radioactive 'Releases' Reported at Colorado Plant," *New York Times*, September 26, 1977.
A panel of international peers: Johnson, "Cancer Incidence in an Area Contaminated with Radionuclides."
A report by ERDA, the Energy Research and Development Administration: In 1975 the Atomic Energy Commission (AEC) was split into two parts, the Energy Research and Development Administration (ERDA) and

the Nuclear Regulatory Commission (NRC). In 1977 ERDA became the Department of Energy.

130 **ERDA argues that samples:** "Radioactive 'Releases' Reported at Colorado Plant."

131 **Some scientists question Johnson's study:** zd. J. van Loon, "Reflections on Cancer and Death Rates in Rocky Flats: A Reply to the *Ambio* Report," *Ambio* 2, no. 6 (1982).

Before Johnson's article reaches print: "Dr. Johnson's Credibility," *Denver Post*, May 11, 1979.

Years later, in 1990: Mark Obmascik and Thomas Graf, "Rockwell Won Bonuses Despite Errors," *Denver Post*, January 7, 1990.

The report also states that trucks: "Radioactive 'Releases' Reported at Colorado Plant."

132 **Records going back six years:** John Ashton, "Official Lying Charged at Flats Protest Trial," *Rocky Mountain News*, November 19, 1978.

I believe it's irrelevant: "Rocky Flats Flights Going On for 24 Years," *Rocky Mountain News*, May 8, 1977.

133 **City water always tastes a little better:** Tamara Smith Meza, interviews by author, October 29, 2006, March 20, 2007, October 8, 2008, and e-mails.

137 **Pulling fake terrorist exercises:** Debby Clark, interview by author, December 7, 2006.

138 **Two weeks after my graduation:** Fred Gillies, "Thorium in Jeffco Horses Revealed as Rocky Flats Item," *Denver Post*, May 23, 1976.

145 **Soil and dust testing:** Glenn Troelstrup, "Review Asked to Find Source of Cesium at Rocky Flats," *Denver Post*, April 10, 1977.

Radioactive cesium and strontium are produced: R. Cowen, "Rocky Flats Radiation Remains Unexplained," *Science News* 135, no. 25 (June 24, 1989): 391.

146 **A Jefferson County commissioner:** Fred Gillies, "Jeffco Commissioner Challenges Flats Cesium Data," *Denver Post*, April 9, 1977.

However, he says, "Even if it [a vote to censure Johnson]": George Lane, "Johnson Censure Hinted over Cesium 'News Leak,'" *Denver Post*, April 10, 1977.

151 **Making note of the ongoing blizzard conditions:** Keith Pope and Joseph Daniel, *Year of Disobedience* (Boulder, CO: Daniel Productions, 1979), 23.

152 **"We'll be back":** "Ellsberg and 19 Others Arrested at Protest Site," *New York Times*, May 9, 1978.

155 **Some drivers honk:** Pope and Daniel, *Year of Disobedience*, 61.

155 The train stops: Edward Abbey, "One Man's Nuclear War," *Harper's*, March 1979.

156 Ann and her husband have a home: Ann White, interview by author, July 31, 2004.

157 The tour guide, a Rockwell employee: L. M. Jendrzejczyk, "The Plutonium Syndrome," *New York Times*, March 30, 1979.
     As he is led off: Pope and Daniel, *Year of Disobedience*, 60.

158 But Judge Goldberger begins by ruling: Molly Ivins, "Colorado Trial Reflects Antinuclear Drive," *New York Times*, November 22, 1978.

159 There is no such thing . . . as a "permissible" dose: Robert Del Tredici, *At Work in the Fields of the Bomb* (New York: Harper & Row, 1987), 134.
     The radioactive sand under the barrels: Bob Reuteman, "Vindication at Last for All Who Feared Rocky Flats," *Rocky Mountain News*, February 18, 2006.

160 Dr. John Gofman, from the University of California, Berkeley: Abbey, "One Man's Nuclear War."
     "Objection sustained": Abbey, "One Man's Nuclear War."

161 She describes the night she was arrested: Pope and Daniel, *Year of Disobedience*, 17.
     The jury is to disregard: Abbey, "One Man's Nuclear War."

162 In November 1978: Karen Newman, "Flats Area Bodies Yield Plutonium," *Rocky Mountain News*, November 11, 1978.
     Hundreds of frozen sex organs: "Sexual Organs Held in Frozen Limbo," *Los Angeles Times*, February 19, 1995.
     When data is finally published: Interviews with John Cobb by Hannah Nordhaus, Albuquerque, New Mexico, December 24, 2003, and February 12, 2004 (Maria Rogers Oral History Program, OH 1180V). See also John Cobb et al., "Plutonium Burdens in People Living Around the Rocky Flats Plant," Environmental Protection Agency summary, March 1983.
     At a court hearing: Keith Schneider, "Data for Nuclear Arms Workers Cast Light on Three Decades of Plutonium Peril," *New York Times*, November 18, 1985. See also Janet Day, "Flats Radiation Killed Worker, Judge Rules," *Rocky Mountain News*, September 29, 1990.

163 Potential homeowners had been asked to sign: Sandra Hubbs, "Impact of Flats FHA Freeze Uncertain," *Denver Post*, November 8, 1978.
     "This notice is to inform you": *Citizen's Guide to Rocky Flats*, 47.
     A spokesperson for the Rocky Flats Monitoring Council: Mark Stevens, "Denver Plutonium Plant Upsetting the Neighbors," *Christian Science Monitor*, October 27, 1978.

Chapter 4. Operation Desert Glow

168 In his published study: Carl J. Johnson, "Cancer Incidence in an Area Contaminated with Radionuclides Near a Nuclear Installation," *Ambio* 10, no. 4 (1981): 176–82.

In the five years after Rocky Flats was built: Margie McAllister, "The Reluctant Radical," *Sunday Camera Magazine*, October 6, 1985.

Birth defects are higher: Fred Gillies, "Higher Birth Defects Tied to Flats Plant," *Denver Post*, November 1, 1978.

In Area 1, which extends: McAllister, "The Reluctant Radical."

Lung and bronchial cancer for males: Johnson, "Cancer Incidence in an Area Contaminated with Radionuclides."

When Johnson turns his attention: McAllister, "The Reluctant Radical."

In January 1980, for the first time: Peggy Strain, "EPA Concedes That Rocky Flats Contamination May Cause Deaths," *Denver Post*, January 13, 1980.

169 He says that the county board of health: Timothy Lange, "They Fired Dr. Johnson," *Westword*, May 28, 1981. LeRoy Moore and several other citizens attended the meeting at the Jefferson County Courthouse at which the board decided to fire Johnson. There was no opportunity for public comment.

170 Unwillingly, after more than seven years: LeRoy Moore, "Democracy and Public Health at Rocky Flats: The Examples of Edward A. Martell and Carl J. Johnson," in *Tortured Science: Health Studies, Ethics and Nuclear Weapons*, edited by Dianne Quigley, Amy Lowman, and Steve Wing (Amityville, NY: Baywood Publishing, 2011), 106.

And a member of the county board: Lange, "They Fired Dr. Johnson." See also LeRoy Moore, letter to Dan Rather suggesting this topic for *Sixty Minutes*, June 3, 1981; Rocky Flats Action Group, "Johnson Fired as Jeffco Health Head," *Action: The Voice of Nuclear Criticism and Education in Colorado* 6, no. 3 (June/July 1981); and Paul Krehbiel, "Johnson Seeks Reinstatement," *Citizens Healthwatch* (January–March 1982).

When the case goes to trial: Carl Johnson and John R. Holland, "Politicization of Public Health," presented at the U.S. Conference of Local Health Officers and the American Public Health Association, Washington, DC, November 18, 1985. Referenced in Moore, "Democracy and Public Health at Rocky Flats."

172 In October 1982, Stone sends: "A Sad Final Note for Flats Whistleblower," *Denver Post*, March 31, 2007.

"Oh, you're going to get it now!": Jim Stone, interviews by Dorothy Ciarlo,

June 10, 1999 (Maria Rogers Oral History Program, OH 0978), June 24, 1999 (OH 1979), and January 6, 2000 (OH 0980).

175 **When asked if enough uranium-235 and -238:** Richard Rhodes, *The Making of the Atomic Bomb* (New York: Simon & Schuster, 1986), 294, 500.

177 **After the Second World War:** Eileen Welsome, *The Plutonium Files: America's Secret Medical Experiments in the Cold War* (New York: Dial Press, 1999), 123.

These studies determined: "Closing the Circle on the Splitting of the Atom," Department of Energy (1995), 38.

Particles of plutonium weighing 10 micrograms: LeRoy Moore, "A dozen reasons why the Rocky Flats National Wildlife Refuge should remain closed to the public," June 2010, http://leroymoore.wordpress.com/2010/06/15/a-dozen-reasons-why-the-rocky-flats-national-wildlife-refuge-should-remain-closed-to-the-public/.

Workers at Los Alamos were already operating: DOE Openness: Advisory Committee on Human Radiation Experiments. The Office of Health, Safety and Security, www.hss.energy.gov/HealthSafety/ohre/roadmap/achre/chap5_1.html.

178 **When they report the problem to management:** Barry Siegel, "Showdown at Rocky Flats: When Federal Agents Take on a Government Nuclear Bomb Plant, Lines of Law and Politics Blur, and Moral Responsibility Is Tested," *Los Angeles Times*, August 8, 1993.

179 **At Rocky Flats, it's common for managers to blindfold:** Siegel, "Showdown at Rocky Flats."

Facilities like Rocky Flats have to break the law: Siegel, "Showdown at Rocky Flats."

But a confidential internal DOE memo: Patricia Calhoun, "Truth Decay," *Westword*, October 13, 2005.

180 **Jim Stone continues to write:** Virginia Culver, "Whistle-blower Helped Shut Flats," *Denver Post*, April 13, 2007.

181 **In the spring of 1979:** "2 Monks Lead Anti-War Unit in N-Protest," *Denver Post*, April 24, 1979.

182 **"We are determined to put an end":** Pam Solo, interview by Dorothy Ciarlo, October 20, 2004 (Maria Rogers Oral History Program, OH 1272V A-B); interview by LeRoy Moore, September 23, 1996 (OH 1528). See also Pam Solo, *From Protest to Policy: Beyond the Freeze to Common Security* (Pensacola, FL: Ballinger, 1988).

Kites flutter in the wind: Jack Cox, "Nuclear Protest Abounds with Color and Characters," *Denver Post*, April 29, 1979.

182 **Now Ellsberg calls Rocky Flats:** Daniel Ellsberg, interview by Dorothy Ciarlo, April 13, 2003 (Maria Rogers Oral History Program, OH 1137V A-B), and interview by LeRoy Moore, April 24, 1998 (Maria Rogers Oral History Program, OH 1530).

183 **The rally is peaceful:** Cox, "Nuclear Protest Abounds with Color and Characters."

On Sunday, 286 men and women: Joseph Seldner, "N-Protesters Fined $1,000," *Denver Post*, June 1, 1979.

184 **The plant bills itself:** Patricia Buffer, "Rocky Flats History," Department of Energy, September 1973.

A year later he transferred: Keith Schneider, "Data for Nuclear Arms Workers Cast Light on 3 Decades of Plutonium Peril," *New York Times*, November 18, 1989.

185 **As to what's in those cans:** Jim Kirksey, "Engineer: No Triggers at Rocky Flats," *Denver Post*, May 18, 1979.

"We aren't dying of cancer": "Pro-Flats Coalition Seeks City's Support," *Denver Post*, May 25, 1979.

186 **"We want the world to know":** Pamela Avery, "Flats Scene of Pro-Nuke Rally," *Rocky Mountain News*, August 27, 1979.

On the cold morning of September 26, 1979: Tom Clark, "Is Denver Safe from Rocky Flats?" *Denver Magazine*, March 1980.

187 **One sunny weekend:** Laurie's name has been changed.

Organizers estimate they need: Joe Garner, Karen Bailey, and Sharon Novotne, "10,000 Protest at Weapons Plant," *Rocky Mountain News*, October 16, 1983.

188 **State Trooper Dave Harper:** Karen Bailey, "Human Chain Was Joined in Spirit," *Rocky Mountain News*, October 16, 1983.

189 **Jack Weaver, a plutonium production manager:** Ackland, *Making a Real Killing*, 194.

The circle does, in fact, fall short: Bill Walker and Virginia Culver, "15,000 Protest Flats' Nuclear Work," *Denver Post*, October 16, 1983.

191 **Pat Mahoney will eventually serve:** Virginia Culver, "Faith Was Flats Protester's Arsenal," *Denver Post*, August 11, 2008.

Pat McCormick begins to think: Pat McCormick, interview by author, November 30, 2006. See also interview by Dorothy Ciarlo, January 13, 2007 (Maria Rogers Oral History Program, OH 1454V).

193 **Rocky Flats guard Debby Clark:** Debby Clark, interview by author, December 20, 2006.

195 **Eighty-five percent of the radioactive waste:** John Leach, "Rocky Flats' Wastes Pose Hazard in Idaho," *Boulder Daily Camera*, August 5, 1979. In 1978 alone, Rocky Flats shipped 9.6 million pounds of low-level waste and 1.4 million pounds of transuranic waste off-site, most of it to Idaho.

196 **Located twenty-six miles east of Carlsbad:** Keith Schneider, "U.S. Seeks to Store Nuclear Waste at Army Bases to Save Plutonium Plant," *New York Times*, November 10, 1989. See also Tamara Jones, "Nuclear Refuse Piles Up; Dump Site Is Delayed," *Los Angeles Times*, November 28, 1988.

"If you can't store it": Jones, "Nuclear Refuse Piles Up."

A spokesman tells the press: Fox Butterfield, "Idaho Firm on Barring Atomic Waste," *New York Times*, October 23, 1988.

Romer chastises officials: Ackland, *Making a Real Killing*, 213.

"The legal grounds are not near as important": Butterfield, "Idaho Firm on Barring Atomic Waste."

"No sale," says Washington governor: "7 States Decline Requests to Take Nuclear Waste," *Los Angeles Times*, October 12, 1989.

197 **Governor Romer proposes a short-term solution:** Schneider, "U.S. Seeks to Store Nuclear Waste at Army Bases to Save Plutonium Plant."

The red boxcar filled with radioactive material: George J. Church, "Playing Atomic NIMBY," *Time*, December 26, 1988.

Designed and built by Rocky Flats workers to burn: Buffer, "Rocky Flats History."

198 **Following a meeting of several hundred local residents:** *Denver Post*, May 19, 1987, Denver Public Library, Western History and Genealogy Clipping File, U.S. Government, Department of Energy, Colorado, Rocky Flats Plant, 1980–1989.

199 **He concedes that the health department's stance:** McAllister, "The Reluctant Radical."

The judge agrees: Fred Gillies, "Judge: Rocky Flats' Link to Cancer Rate Unproven," *Denver Post*, July 4, 1985.

Ultimately the suit is settled: Calhoun, "Truth Decay."

200 **"The purpose for the lawsuit":** Calhoun, "Truth Decay."

Even without the pulpit of his job: McAllister, "The Reluctant Radical."

In fact, he reminds the press: Keith Schneider, "Weapons Plant Pressed for Accounting of Toll on Environment and Health," *New York Times*, February 15, 1990.

Further, Johnson notes, the court system: McAllister, "The Reluctant Radical."

200 **The group Physicians for Social Responsibility calls Rocky Flats:** Gary
    Schmitz, "Doctors: A 'Creeping Chernobyl' Created," *Denver Post*, Octo-
    ber 27, 1988.
    Pat Schroeder insists the study: Jerry Brown, "Rocky Flats Study Should
    Be Given to Outside Experts, Schroeder Says," *Rocky Mountain News*,
    August 18, 1979.
201 **Moving plutonium operations would result:** Buffer, "Rocky Flats History."
    Officials can't agree on what type: Beth Gaeddert, "Obstacles Stall Flats
    Emergency Booklet," *Rocky Mountain News*, November 1, 1979.
202 **However, Johnson states, "Very little":** Clark, "Is Denver Safe from Rocky
    Flats?"
    The plan never moves beyond the drafting: O'Keefe, "What's the Drill: A
    Real Emergency Catches Rocky Flats Napping," *Westword*, September 20,
    1989.
    Niels Schonbeck, the biochemistry professor: O'Keefe, "What's the Drill";
    see also *Citizen's Guide to Rocky Flats: Colorado's Nuclear Bomb Factory* (Boul-
    der, CO: Rocky Mountain Peace Center, 1992), 47.
203 **Nonetheless, a few months later:** Buffer, "Rocky Flats History."
    Full production resumes: *ABC Nightline*, December 20, 1994.
    In March 1982, Bruce Shepard, a Colorado Springs developer: LeRoy
    Moore, "Democracy and Public Health at Rocky Flats: The Examples of
    Edward A. Martell and Carl J. Johnson," in *Tortured Science: Health Stud-
    ies, Ethics and Nuclear Weapons,* edited by Dianne Quigley, Amy Lowman,
    and Steve Wing (Amityville, NY: Baywood Publishing, 2011), p. 106. Also
    see Robert Kowalski, "HUD Official Helped Kill Flats Homebuyer Alert,"
    Denver Post, August 20, 1989. Those two housing developments were
    the Countryside development and the Lake Arbor development, both in
    Westminster.
205 **I meet a boy named Andrew:** Name has been changed.
206 **The DOE isn't unaware of problems:** Janet Day, "DOE Rips Rocky Flats
    Management," *Rocky Mountain News*, May 21, 1988.
    And the government will continue: Gary Schmitz, "Government May
    Abandon Rocky Flats Nuclear Plant," *Denver Post*, March 1, 1989.
207 **On September 29, three people:** Fox Butterfield, "Report Finds Peril at Atom
    Plant Greater Than Energy Dept. Said," *New York Times*, October 27, 1988.
    Rocky Flats workers are accustomed: Willie Warling, interview by Doro-
    thy Ciarlo, May 27, 1998 (Maria Rogers Oral History Program, OH 0927).
    He files a report: Butterfield, "Report Finds Peril at Atom Plant." This cita-
    tion applies to the next several pages of text.

209 Shortly thereafter, a comprehensive DOE study: Adriel Bettelheim, "Flats' Buildings Top Danger List," *Denver Post*, December 6, 1994.

At the 570-acre Hanford site near Richland: "Byproducts of the Bomb: Pollution and the Weapon Factories," *New York Times*, December 7, 1988.

210 The groundwater and soil at Rocky Flats: Alan Gottlieb, "Toxicity at Flats So Bad Harm May Be Irreversible," *Denver Post*, November 25, 1988.

211 Stone has been waiting months: Jim Stone, interviews by Dorothy Ciarlo, June 10, 1999 (Maria Rogers Oral History Program, OH 0978), June 24, 1999 (OH 1979), and January 6, 2000 (OH 0980). This citation applies to the next several pages of text.

Rockwell and the DOE have always contended: Siegel, "Showdown at Rocky Flats."

212 "They blackballed me": Alicia Caldwell, "Whistle-blowers Reap, but Process Takes Time," *Denver Post*, January 11, 2005.

"I want a letter of immunity": Jon Lipsky, interview by Dorothy Ciarlo, July 23 and 24, 2005 (Maria Rogers Oral History Program, OH 1355V A-D).

213 At first, Fimberg looks at Lipsky and Smith: Siegel, "Showdown at Rocky Flats."

Lipsky hates to fly: Lipsky, interview by Ciarlo.

214 The photographs indicate: Patricia Calhoun, "Dirty Pictures," *Westword*, October 27, 2005.

Streaks of light splay out: Calhoun, "Dirty Pictures."

Based on the videotape and other evidence they've accumulated: Tamara Jones, "FBI Alleges Cover-Up at Rocky Flats," *Los Angeles Times*, June 10, 1989.

The affidavit states that the DOE and Rockwell: Tom Ruwitch, "Rockwell International: Gouging the Government," *Multinational Monitor* 11, no. 3 (March 1990).

Fimberg flies to Washington: Siegel, "Showdown at Rocky Flats."

215 His work included: Carl Johnson, "Cancer Incidence in an Area of Radioactive Fallout Downwind from the Nevada Test Site," *Journal of the American Medical Association* 251, no. 2 (January 13, 1984).

Five months after he was terminated: Johnson, "Cancer Incidence in an Area Contaminated with Radionuclides."

On December 18, 1988: Carl Johnson, "Rocky Flats: Death, Inc.," *New York Times*, December 18, 1988.

He was buried with military honors: Pamela Reynolds, "Respect in Death for Nuclear Safety, He Took a Stand," *Boston Globe*, January 11, 1989.

Chapter 5. A Raid and a Runaway Grand Jury

218 **The FBI has told Rocky Flats:** Jacque Brever, interviews by LeRoy Moore, June 4, 1999 (Maria Rogers Oral History Program, OH 1498), and by Dorothy Ciarlo, April 9, 2004 (Maria Rogers Oral History Program, OH 1210).

220 **Sanchini is reluctant to let agents:** Siegel, "Showdown at Rocky Flats," Part One.

It felt like a factory: Brever, interviews by Moore and Ciarlo.

221 **She joined the Steelworkers Union:** Brever, interviews by Moore and Ciarlo.

Jacque shut down the line: Brever, interviews by Moore and Ciarlo.

222 **They order everyone to leave:** Bruce David and Mark Cromer, "Rocky Mountain Meltdown," *Hustler* 31, no. 8 (January 2005).

They look like a bunch of chickens: Brever, interviews by Moore and Ciarlo.

"I have been victimized": Joan Lowy, "Flats Raid Spreads Shock Waves," *Rocky Mountain News*, June 7, 1989.

Romer flies to Washington: Tamara Jones, "U.S. Vows to Lift 30-Year Veil of Secrecy at Weapons Plants," *Los Angeles Times*, June 17, 1989.

A special telephone line is established: "FBI Extends Its Inspection of Rocky Flats Arms Plant," *Los Angeles Times*, June 15, 1989.

A local radio station begins: James Coates, "Government Decides Truth Is the Light on Arms Plant Pollution," *Chicago Tribune*, June 18, 1989.

223 **Now, for the first time, the controversial 1983 film:** Coates, "Government Decides Truth."

Dark Circle includes interviews: *Dark Circle*, 1983, reissued in 2007, directed by Judy Irving, Chris Beaver, and Ruth Landy.

"The plutonium that went out": Harvey Wasserman and Norman Solomon, *Killing Our Own: The Disaster of America's Experiment with Atomic Radiation* (New York: Delacorte, 1982), 165.

Billy Chisolm built his home: Tamara Jones, "Neighbors Keep a Wary Eye on Rocky Flats Plant but Resist Moving," *Los Angeles Times*, June 11, 1989.

224 **U.S. attorney general Dick Thornburgh:** Wes McKinley and Caron Balkany, *The Ambushed Grand Jury* (New York: Apex Press, 2004), 14.

It reveals that the flyover: Tamara Jones, "FBI Alleges Cover-Up at Rocky Flats: Papers Say Energy Dept. Knew of Illegal Atom Waste Disposal," *Los Angeles Times*, June 10, 1989.

Some of these entries will end up: Siegel, "Showdown at Rocky Flats."

225 **Lipsky suspects that the Criminal Division at Justice:** Siegel, "Showdown at Rocky Flats." See also McKinley and Balkany, *The Ambushed Grand Jury*.

The most seriously contaminated sites: *Rocky Mountain News*, July 23,

1989. These sites include the 881 Hillside, where dumped and buried material seeped into the groundwater; solar evaporation waste ponds that leaked chemicals and low levels of radioactivity into the soil; and the East Trenches, where flattened drums contaminated with radiation were buried in eight trenches.

225 **One of the most shocking discoveries:** "Material Unaccounted For" exhibit chart, produced for *Cook v. Rockwell International Corporation*, provided by Peter Nordberg based on information supplied by the DOE.

And despite the fact that Rocky Flats officials: Tamara Jones and Dan Morain, "Federal Probers Sound Alarms: Rocky Flats Boon Turns into Ecological Nightmare," *Los Angeles Times*, June 20, 1989.

"It was on the main road": Patricia Calhoun, "It's Toast," *Westword*, February 3, 2005.

226 **They cite an internal memo:** Jones and Morain, "Federal Probers Sound Alarms."

On the day of the raid: Jones, "U.S. Vows to Lift 30-Year Veil of Secrecy."

"Whistle-blowers," he says: Brever, interviews by Moore and Ciarlo.

227 **Rockwell decides to sue:** Tom Ruwitch, "Rockwell International: Gouging the Government." *Multinational Monitor* 11, no. 3 (March 1990).

The company also takes out full-page ads: Keith Schneider, "Weapons Plant Pressed for Accounting of Toll on Environment and Health," *New York Times*, February 15, 1990.

On September 28, 1989, the EPA: Siegel, "Showdown at Rocky Flats."

228 **"Sure," the friend replies:** Wes McKinley, interview by author, Denver, Colorado, January 19, 2004. See also Wesley (Wes) McKinley, interview by Dorothy Ciarlo, April 25, 1998 (Rocky Flats Cold War Museum and Maria Rogers Oral History Program, OH 1271V A-B).

U.S. district judge Sherman G. Finesilver takes a full hour: Siegel, "Showdown at Rocky Flats."

229 **There's so much sandy material:** Bill Kemper with Jim Stone, interview by Dorothy Ciarlo, February 16, 2005 (Maria Rogers Oral History Program, OH 1302V).

"It takes minuscule amounts": Schneider, "Weapons Plant Pressed."

John Cobb, a professor of preventive medicine: John Cobb, interview by Hannah Nordhaus, December 24, 2003 (Maria Rogers Oral History Program, OH 1302V).

When DOE inspector Joseph Krupar: Siegel, "Showdown at Rocky Flats."

230 **The case is dismissed:** "Judge Dismisses Suit Against Bomb Plant in Harassment Case," *New York Times*, September 4, 1992.

230 **"The best thing I can do":** Brever, interviews by Moore and Ciarlo.

231 **The "spray irrigation" was done:** Siegel, "Showdown at Rocky Flats."
The year after the shutdown: *Sierra Club vs. Rockwell International and the Department of Energy*, Case #89-B-1181. Mark Obmascik, "Flats Violated US Laws, Judge Rules," *Denver Post*, April 17, 1990. The court ordered that the DOE could no longer illegally burn hazardous waste in Building 771 and had to comply with environmental law. Adam Babich, attorney for the Sierra Club, stated that "EPA and the State did not have the political courage to do it [what should have been done years before]." Quoted in *The Ambushed Grand Jury*, 264. See also "Closure Plan for Mixed Residue Recovery Incinerator, Building 771, Room 149," Department of Energy, February 5, 1991, 3.

232 **Under pressure, the DOE reveals:** Janet Day, "61 Pounds of Radioactive Dust: Flats Ducts Have Enough Plutonium to Make 6 A-Bombs," *Rocky Mountain News*, March 29, 1990.
"I can guarantee if we don't move aggressively": Matthew Wald, "38-Year Plutonium Loss at Plant Equals 7 Bombs," *New York Times*, March 29, 1990.

233 **At a restaurant in Brighton:** Siegel, "Showdown at Rocky Flats: The Justice Department Had Negotiated a Rocky Flats Settlement, but the Grand Jury Could Not Keep Quiet About What Happened There," *Los Angeles Times*, August 15, 1993 (Part Two).
After twenty-one months of work: Brever, interviews by Moore and Ciarlo.
Not a single Rockwell: Michael D. Lemonick, "Sometimes It Takes a Cowboy," *Time*, January 25, 1993.
Rockwell agrees to plead guilty: Jonathan Turley, "Free the Rocky Flats 23," *Washington Post*, August 11, 1993; see also Matthew Wald, "Grand Jury Seeks Inquiry on Weapons Plant Case," *New York Times*, November 19, 1992.
Rockwell is required to pay: Patricia Calhoun, "Gag Reflex," *Westword*, July 3, 1997; see also Rudy Abramson, "Rockwell Pleads Guilty to Waste Dumping, Blasts U.S. Settlement," *Los Angeles Times*, March 27, 1992, and Bill Scanlon, "Flats' $18.5 Million Fine Stands," *Rocky Mountain News*, June 3, 1992.
Rockwell routinely received millions of dollars: Jones, "U.S. Vows to Lift 30-Year Veil of Secrecy at Weapons Plants."
Rockwell is also allowed: Turley, "Free the Rocky Flats 23."

234 **The agreement stipulates:** Patricia Calhoun, "Carved in Stone," *Westword*, April 12, 2007.

234 **The agreement also allows:** Turley, "Free the Rocky Flats 23."

They refer again to the instructions: Siegel, "Showdown at Rocky Flats," Part Two.

It's not uncommon for the Justice Department: Turley, "Free the Rocky Flats 23."

"We were studying a million pages": Bruce David and Mark Cromer, "Rocky Mountain Meltdown," *Hustler*, March 12, 2008.

235 **Three years earlier, the same judge:** Calhoun, "Gag Reflex."

But somehow, a few days later: Bryan Abas, "The Secret Story of the Rocky Flats Grand Jury," *Westword*, September 1992. Ryan Ross (aka Bryan Abas), interview by author, August 19, 2004.

Harper's **magazine publishes excerpts:** "The Rocky Flats Cover-Up, Continued," *Harper's*, December 1992.

Twelve of the jurors: Wald, "Grand Jury Seeks Inquiry on Weapons Plants Case." See also Mark Obmascik, "Flats Jurors Ask for Probe," *Denver Post*, November 19, 1992.

In January, seven of the jurors: Lemonick, "Sometimes It Takes a Cowboy."

236 **Prosecutors emphasize that they've been the first:** Siegel, "Showdown at Rocky Flats," Part Two.

In January 1993, the Wolpe committee: Turley, "Free the Rocky Flats 23." See also Howard Wolpe, Congressman, Chairman, Subcommittee on Investigations and Oversight, Committee on Science, Space and Technology, U.S. House of Representatives, *The Prosecution of Environmental Crimes at the Department of Energy's Rocky Flats Facility*, January 4, 1993.

"The most important thing": Lemonick, "Sometimes It Takes a Cowboy."

Jonathan Turley, a Washington lawyer representing the grand jurors: Turley sought a special grant of immunity so that grand jurors could tell their story directly to Congress. See also Jonathan Turley, "We Need to Unearth Environmental Felons," *Wall Street Journal*, March 11, 1993, and "Jurors Alone Can Unravel Rocky Flats Mystery," *Rocky Mountain News*, March 18, 1994.

238 **DeBoskey disputes the government's position:** Keith Schneider, "Data for Nuclear Arms Workers Cast Light on 3 Decades of Plutonium Peril," *New York Times*, November 18, 1989.

While residents wonder, the General Accounting Office: United States General Accounting Office, "Nuclear Materials: Removing Plutonium Residues from Rocky Flats Will Be Difficult and Costly," September 4, 1992.

Silverman finds himself facing: Mark Silverman, "Cleaning Up the Cold War Legacy," *Assembly*, May/June 1997.

239 *Time* magazine reports that "aging buildings": Michael D. Lemonick, "Rocky Horror Show," *Time*, November 27, 1995.

Chapter 6. Doom with a View

241 EG&G manufactures everything from: Adriel Bettelheim, "EG&G a Key Nuclear Contractor," *Denver Post*, September 23, 1989.

EG&G's four-year contract: Janet Day, "EG&G, U.S. Reach Accord on Operation of Rocky Flats," *Rocky Mountain News*, October 7, 1989.

242 Response is swift, and more than: Kristen Iversen, employee newsletter.

243 Rocky Flats officials settle: Ann Breese White, interview by author, Denver, Colorado, July 31, 2004.

He knows he's sitting: Michael D. Lemonick, "Rocky Horror Show," *Time*, November 27, 1995.

A DOE official comments: Joan Lowy, "Rocky Flats Shutdown Will Take 20 Years," *Rocky Mountain News*, February 3, 1989.

255 It was a real career: Randy Sullivan, interviews by author, February 13 and 17, 2007, February 17, 2008, and e-mails. See also Randy Sullivan, interview by Hannah Nordhaus, November 8, 2005 (Maria Rogers Oral History Program, OH 1386).

The two top managers from EG&G and the DOE walk around: Ward Marchant, "Rocky Flats' Nuclear History Leaves Legacy of Peril and Plutonium in Colorado," *Los Angeles Times*, February 26, 1995.

Debra and Diane: The names of these two employees have been changed.

256 And as in the past, Rocky Flats is involved: Doug Parker, interview by author, October 29, 2006.

259 One day Mr. K: Name has been changed.

262 When I return to my desk, Anne: Name has been changed.

264 On November 4, a memo: Memo from A. H. Burlingame, president of EG&G, "Safe Operations Contingency Plan," November 4, 1994.

The blow is softened: *Horizon* (Rocky Flats employee newsletter), November 3, 1994.

266 The narrator, Dave Marash: *ABC Nightline*, December 20, 1994. Much of this section is based on the *Nightline* transcript.

Chapter 7. Fire, Again

273 "people are scared of fires": Tom Clark, "Is Denver Safe from Rocky Flats?" *Denver Magazine*, March 1978.

The Arvada and Fairmount fire districts: Brad Martisius, "Solution to Chemical-Waste Issue Sought," *Denver Post*, December 19, 1979.

278 **Property values in surrounding neighborhoods:** Peter Nordberg, interviews by author, July 16, 2007, June 6, 2008, and e-mails.

280 **On the night of January 29, 1997:** Mykaila Nordberg, interview by author, September 15, 2010.

284 **Rocky Flats could become a sort of poster child:** Len Ackland, "The Other Cleanup at Rocky Flats: We're Burying Its Significance," *Denver Post*, August 7, 2005.

287 **He was fifty-seven, and his time at Rocky Flats:** A Russian technical team visited Rocky Flats in 1994 and 1996.

288 **"We may have done too good a job":** Mark Obmascik, "Overworked Rocky Flats Manager Quits," *Denver Post*, April 23, 1996.

   In 2000, the same year, when Silverman's cancer: Kelley Hunsberger, "Finding Closure," CH2M Hill/Kaiser-Hill, PM Network, January 2007.

289 **Dr. Harvey Nichols and others:** Harvey Nichols, interview by Hannah Nordhaus, September 14, 2005. See also Rocky Flats Coalition of Local Governments Board Meeting Minutes, April 2, 2001, http://www.broomfield.org/RFCLOG/rockyflatsminutes4_2_01.shtml.

   Paula Elofson-Gardine, a local resident: Paula Elofson-Gardine and Susan Hurst, "Stop the Nuclear Brushfires," *Earth Island Journal,* 2000. See also Paula Elofson-Gardine and Susan Elofson-Hurst, interview by Dorothy Ciarlo, February 23, 2007 (Maria Rogers Oral History Program, OH 01457V A-B).

Chapter 8. What Lies Beneath

302 **Infinity Rooms—called that because:** Mark Obmascik, "Infinity Rooms: Rocky Flats' Horror Show," *Denver Post*, February 20, 1994. These rooms include Room No. 141 in Building 771, originally a plutonium storage vault, where several pumps leaked liquid plutonium nitrate in the 1960s. In 1968 the steel door to Room No. 141 was welded shut, sealing in a ladder, a jackhammer, and hoses, due to lack of funding to decontaminate the room completely. Room No. 134-West and Room No. 141 in Building 776, the site of a major fire in 1969, together comprising 2,000 square feet, were constructed after a cleanup of the fire and have been used in recent years to cut up and package radioactive wastes from other areas of Rocky Flats. Room No. 141 can be entered only after traveling through four separate airlock chambers that seal off contamination.

304 **In 1996 a Boston University epidemiologist:** Richard W. Clapp, report submitted November 13, 1996, for plaintiff's, counsel in *Cook v. Dow Chem-*

ical and Rockwell International, United States District Court, District of Colorado.

304 **In 1989 a class-action lawsuit by residents:** Tim Bonfield, "Fernald: History Repeats Itself," *Cincinnati Enquirer*, February 11, 1996, http://www .enquirer.com/fernald/stories/021196c_fernald.html.

305 **Following Clapp's study, in 1998:** Colorado Central Cancer Registry, Colorado Department of Public Health and Environment, "Ratios of Cancer Incidence in Ten Areas Around Rocky Flats, Colorado, Compared to the Remainder of Metropolitan Denver, 1980–89, with Update for Selected Areas, 1990–95," http://www.cdphe.state.co.us/pp/cccr.ratio.pdf, published in 1998.

A radiation health specialist: LeRoy Moore, "Democracy and Public Health at Rocky Flats: The Examples of Edward A. Martell and Carl J. Johnson," in *Tortured Science: Health Studies, Ethics and Nuclear Weapons*, edited by Dianne Quigley, Amy Lowman, and Steve Wing (Amityville, NY: Baywood Publishing, 2011), 117.

The year 1999 marked the end of a decade-long study: "The Rocky Flats Historical Public Exposures Study," Colorado Department of Public Health and the Environment.

306 **With respect to water:** "Technical Topics Papers: Historical Public Exposure Studies: Water Contaminants," www.cdphe.state.co.us/rf/contamin.htm.

307 **Of the $8.7 million of federal funds:** Richard Fleming, "Glowing Reports: There's Plenty of Good News About Rocky Flats. And You're Paying for It," *Westword*, March 15, 1995.

Despite ongoing requests: "Research on Adverse Health Effects Related to Rocky Flats," Rocky Flats Historical Public Exposures Studies, Colorado Department of Public Health and the Environment.

313 **The defendants, represented by the one-thousand-plus-member law firm:** Merrill G. Davidoff, Peter Nordberg, and David F. Sorensen, "Nuclear Win Was Years in Making," *National Law Journal*, January 29, 2007.

316 **They award punitive damages:** The jury recommended that the companies pay $352 million in actual damages. With interest, the judgment ultimately totaled close to $1 billion.

Bini herself has had cancer: Hank Pankratz, "Decision a 'No-Brainer': Neighbors of the Now-Defunct Rocky Flats Nuclear Facility Had Long Worried About the Health Situation," *Denver Post*, February 16, 2006; Bini Abbott, interview by author, February 7, 2005.

"It's a tremendous verdict," she says: Miriam Hill, "Contamination-Case

Success: Phila. Law Firm Wins $554 Million Verdict, but 16-Year Battle May Not Be Over," *Philadelphia Inquirer*, February 16, 2006.

319 **The bomb test sites:** LeRoy Moore, "Rocky Flats: The Bait-and-Switch Cleanup," *Bulletin of the Atomic Scientists* (January/February 2005): 53.

They call for an independent assessment: These groups included the Rocky Flats Citizens Advisory Board, the Rocky Mountain Peace and Justice Center, the Town of Westminster, and the City of Broomfield, as well as Congressman David Skaggs.

In response, in 1998 the DOE: Seth Tuler et al., "Perspectives on Public Participation at a Department of Energy Nuclear Weapons Facility. Case Study: Setting Soil Clean-up Standards at the Rocky Flats Environmental Technology Site," Social and Environmental Research Institute, October 2003. Stakeholder involvement at Rocky Flats includes the Rocky Flats Citizens Advisory Board, the Rocky Flats Coalition of Local Governments, and the Rocky Flats Radionuclide Soil Action Levels Oversight Panel, each producing individual studies and reports. Many people believe that these citizen groups have been essential to a broader understanding of Rocky Flats and facilitating the cleanup process; others argue that public involvement and releasing information about risks associated with the plant may have exacerbated conflict between the government, contractors, and the public.

320 **A 1996 study of burrowing animals present:** Virginia Gewin, "Nuclear Site Turns Wildlife Refuge," *Frontiers in Ecology and the Environment* 5, no. 7 (September 2007): 345.

Only 7 percent of the total—roughly $473 million: Moore, "Rocky Flats: The Bait-and-Switch Cleanup," 56. During cleanup, the majority of weapons-grade plutonium from Rocky Flats was sent to the Savannah River Site in South Carolina, with some also sent to the Pantex facility in Texas. Weapons-useful uranium was sent to Oak Ridge, Tennessee. Transuranic waste went to the Waste Isolation Pilot Plant (WIPP) in New Mexico, and additional waste was sent to facilities in Nevada and Utah.

321 **In December 2004, the U.S. Fish and Wildlife Service:** Andrew Todd and Mark Sattelberg, "Actinides in Deer Tissues at the Rocky Flats Environmental Technology Site," State News Service, Contaminant Study Completed. U.S. Fish and Wildlife Service, Rocky Flats Deer Tissue Study Executive Summary, http://www.fws.gov/rockyflats/Documents/DeerTissue_ExSummary.pdf.

"Close it, fence it, pave it over": "Tread Warily, You Deer-Watchers: Turning Nuclear Sites into Wildlife Refuges Isn't That Easy," *The Economist*, February 24, 2005.

322   **Dosimeter badges, which employees wore:** Ann Imse, "Review Exposes
      Flats Data as Faulty," *Rocky Mountain News*, February 6, 2006.
      Sixteen members of the Dobrovolny family: Ann Imse, "Family Full of
      Flats Workers Deals with Death and Illness," *Rocky Mountain News*,
      April 27, 2007.
      A DOE-financed study in 1987: Thomas Graf, "Flats Widows Fighting an
      Uphill Battle: DOE, Contractors Deny Fault in Workers' Deaths," *Denver
      Post*, November 20, 1989.
      He was ordered to submit: Suzanne Ruta, "Fear and Silence in Los Alamos,"
      The Nation, January 4–11, 1993. See also Keith Schneider, "Panel Questions
      Credibility of Nuclear Health Checks," *New York Times*, February 28, 1990,
      and Gregg Wilkinson, "Seven Years in Search of Alpha," *Epidemiology* 10,
      no. 3 (May 1999). See also Gregg Wilkinson et al., "Study of Mortality
      Among Plutonium and Other Radiation Workers at a Plutonium Weapons
      Facility," *American Journal of Epidemiology* 125, no. 2 (1987).
323   **Wilkinson found that exposure:** National Institute for Occupational Safety
      and Health/Centers for Disease Control and Prevention, 2000. Wilkinson
      et al., "Study of Mortality Among Plutonium and Other Radiation Work-
      ers," and "Study of Mortality Among Female Nuclear Weapons Workers,"
      May 19, 2000.
      In 1990, testing by doctors at the National: Mark Obmascik and Thomas
      Graf, "Flats Lung Disease Discovered," *Denver Post*, January 14, 1990.
      Rocky Flats workers in general: Brittany Anas, "CU Professor Drowns in
      Mexico," *Boulder Daily Camera*, June 21, 2007, http://www.dailycamera
      .com/ci_13082169.
324   **Charlie Wolf is one of the few managers:** Charlie Wolf, interview by author,
      June 13, 2006.
      For every single pound: Obmascik, "Infinity Rooms: Rocky Flats' Horror
      Show."
325   **On the way to the event, Lipsky receives a call:** Jim Hughes, "FBI Agent
      Silenced on Rocky Flats Nuclear Site," *Denver Post*, August 26, 2004.
      Based on the compromised cleanup standards: "Last of Rocky Flats Worst
      Waste Removed," *Los Angeles Times*, April 20, 2005.
      Studies demonstrate that vegetation: W. J. Arthur III and A. W. All-
      dredge, "Importance of Plutonium Contamination on Vegetation Surfaces
      at Rocky Flats, Colorado," *Environmental and Experimental Botany* 22, no. 1
      (February 1, 1982): 33–38.
      Another study shows that: Shawn Smallwood, "Soil Bioturbation and
      Wind Affect Fate of Hazardous Materials That Were Released at the

Rocky Flats Plant, Colorado" (November 23, 1996), report submitted for plaintiff's counsel in *Cook v. Rockwell International Corporation*, United States District Court, District of Colorado, no. 90-CV-00181; see also the transcript of Smallwood's appearance in court in this case, 3912–4130. See also K. Shawn Smallwood, Michael L. Morrison, and Jan Beyea, "Animal Burrowing Attributes Affecting Hazardous Waste Management," *Environmental Management* 22, no. 6 (November 22, 1998): 831–47.

327 **"He died with nothing more than the clothes":** Laura Frank and Ann Imse, "Rocky Flats Whistle-blower Dies at 82," *Rocky Mountain News*, April 12, 2007; see also Virginia Culver, "Whistle-blower Helped Shut Flats," *Denver Post*, April 13, 2007.

330 **In 2000, however, scientists at Los Alamos:** George L. Voelz, as told to Ileana G. Buican, "Plutonium and Health: How Great Is the Risk?" *Los Alamos Science*, no. 26 (2000): 77–78.

And new studies by the DOE: U.S. Department of Energy and U.S. Department of Health and Human Services, "Agenda for HHS Public Health Activities (for Fiscal Years 2005–2010) at U.S. Department of Energy Sites," U.S. Department of Energy and U.S. Department of Health and Human Services, January 2005, www.hss.doe.gov/healthsafety/iipp/hservices/documents/agenda.pdf.

Back in 1981, Dr. Carl Johnson reported: Carl J. Johnson, "Cancer Incidence in an Area Contaminated with Radionuclides Near a Nuclear Installation," *Ambio* 10, no. 4 (1981): 176–82.

332 **Curtis Bunce's doctor recommends:** Stacy Gardalen, interview by author, December 18, 2011, and e-mails.

The indoor sample is taken from a crawl space: Samples taken on April 14, 2010, analyzed by Marco Kaltofen, PE, of Boston Chemical Data Corp. Report is available at http://archivesite.rmpjc.org/about+sampling+technical+report.

333 **Much of it is so toxic that:** Linda Rothstein, "Nothing Clean About 'Cleanup,'" *Bulletin of the Atomic Scientists* (May 1995): 34–41.

334 **In September 2004, in response to the Draft Environmental:** U.S. Fish and Wildlife Service, "Rocky Flats National Wildlife Refuge, Appendix H: Comments and Responses on the Draft Environmental Impact Statement," September 2004.

Shirley Garcia worked at Rocky Flats: Shirley Garcia, interviews by Dorothy Ciarlo, January 19, 2001 (Maria Rogers Oral History Program, OH 1023V), and November 13, 2004 (OH 1204). Interview by author, December 2005.

334 **However, current testing of wells:** Agenda for HHS Public Health Activities (for Fiscal Years 2005–2010) at U.S. Department of Energy Sites, January 2005, http://hss.energy.gov/healthsafety/iipp/hservices/documents/agenda.pdf, 93.

John Rampe, a former Energy Department: David Kelly, "Dispatch from Rocky Flats National Wildlife Refuge, Colorado: An Idyllic Scene Polluted with Controversy," *Los Angeles Times*, February 7, 2005.

340 **All documentation from the 1989 FBI raid:** The leaked grand jury report was eventually posted on the website of the Denver chapter of the Sierra Club. Judge Finesilver released a redacted grand jury report, with comments added by the Justice Department, on January 26, 1993. None of these partial and edited reports contain evidence or testimony. The full report is still sealed.

Epilogue

341 **The U.S. Fish and Wildlife Service (FWS) says it lacks:** Bruce Finley, "Property Swap Aims to Link Rocky Flats National Wildlife Refuge with Mountains," *Denver Post*, October 1, 2011.

342 **Nobody can ever safely live here, federal scientists concede:** Ralph Vartabedian, "Nuclear Scars: Toxic Legacy of the Cold War," *Los Angeles Times*, October 20, 2009.

Throughout Japan, radioactive substances: Hiroko Tabuchi, "Radioactivity in Japan Rice Raises Worries," *New York Times*, September 24, 2011. See also Evan Osnos, "Japan: The Nuclear Village," *The New Yorker*, October 10, 2011.

343 **Skeptics in Japan and abroad accused the government:** David McNeill, "Why the Fukushima Disaster Is Worse Than Chernobyl," *The Independent*, August 29, 2011. See also Evan Osnos, "The Fallout: Letter from Fukushima," *The New Yorker*, October 17, 2011.

At an antinuclear protest in Tokyo: Speech delivered by Ruiko Muto of Hairo Action Fukushima, Meiji Park, Tokyo, Japan, September 15, 2011. Translated by Emma Parker.

It was only after radiation: "Chernobyl Haunts Engineer Who Alerted World," CNN Interactive World News, April 26, 1996, http://www.cnn.com/WORLD/9604/26.chernobyl/230ppm/idex2.html.

344 **The estimated cost to clean up the "vast areas" contaminated:** "Japan: Radiation Cleanup Will Cost at Least $13 Billion, Premier Says," *New York Times*, October 21, 2011.

In the United States we currently have approximately 25,000: "Plutonium

'Triggers' for Nuclear Bombs," Alliance for Nuclear Accountability, 2009, www.ananuclear.org/Portals/0/documents/2009%20Fact%20Sheets/Pits5%20final.pdf. See also Greg Mello, "A Nuclear Facility We Don't Need," *New York Times*, November 14, 2011.

the production of plutonium pits began again: Matt Mygatt, "Los Alamos Making Plutonium Triggers," *Denver Post*, July 2, 2007.

INDEX